With very best wishes

SUITS:

Speaking of Research

SUITS:
Speaking of Research

C. GUY SUITS

Vice President and Director of Research
General Electric Company

John Wiley & Sons, Inc. NEW YORK LONDON SYDNEY

Library of Congress Catalog Card Number: 65-21432
Printed in the United States of America

Foreword

In 1935, as I was in the process of completing graduate work in physics at Princeton University and planning to join the staff of the University of Rochester, Dr. Albert W. Hull, famous for his work in X-ray diffraction and one of the associate directors of the General Electric Research Laboratory in Schenectady, visited Princeton in search of candidates for summer positions at the laboratory. The tradition of taking on a few "summer boarders" was an old one at Schenectady, going back to the founding of the laboratory at the start of the century. To anyone who had grown up during the 1920's with a teen-age enthusiasm for science, the research center at Schenectady was the counterpart of a temple on Mt. Olympus or in Valhalla. By a miracle, I managed to convince Dr. Hull that I was a worthwhile candidate. This happy event started a lifelong association with the laboratory and its staff, including a two-year period in which I was a full-time member.

During the summer of 1935 Dr. Hull turned me over to his colleague Dr. Saul Dushman, who was one of the early enthusiasts for theoretical quantum physics in our country and is perhaps best remembered today for his pioneering work in thermionic emission and the theory of high-vacuum techniques. He exuded great warmth of understanding for a new graduate who brought with him much of the new knowledge on wave mechanics; and, being steeped in the background of his laboratory, he delighted in relating innumerable incidents and adventures dating back into its early history.

Dr. Willis Rodney Whitney, then 67, who had joined the company in 1900 and, with Charles P. Steinmetz, had created the research center, was an everyday figure in the laboratory during the summer. He had retired as Director a few years earlier and was now devoting his attention to the application of physical techniques to biology and medicine and to the purification of water. He circulated everywhere with enthusiastic but modest inquiries about research under way. Probably more than any other single person, he had been responsible for the policies and traditions which had evolved in the laboratory and had made it one of the greatest scientific institutions in our country. Also present were William D. Coolidge, then the Director, and Irving Langmuir, who by that time occupied in the public mind somewhat the same position that Thomas Edison had had at a comparable age a generation earlier. Coolidge, who had done so much to develop the modern X-ray tube, was still closely engaged with a group exploring the production of the high-voltage generator, whereas Langmuir was not only in the midst of his work on molecular films, but was already ruminating on the possibility of weather modification.

Prominent among the younger generation was Dr. C. Guy Suits, then just 30 years of age, who had joined the staff in 1930, after completing education in Wisconsin and Switzerland, and was already a world-renowned authority on the properties of high-pressure arcs. It was quite obvious that the older generation of leaders had laid a solid groundwork for their own succession.

In those days, before university laboratories had the advantages of large grants of Federal money, and before the national laboratories, such as Oak Ridge and Argonne, had become as relatively commonplace as they are now, the number of laboratories comparable in standing to that at Schenectady could be counted on the fingers of one hand. What is perhaps even more to the point, the warm, informal atmosphere of cooperation which pervaded the laboratory at Schenectady at that time, and has continued ever since, was completely unique in industrial circles. It has served throughout my life as my own conception of the ideal atmosphere of a mission-oriented scientific organization. I believe our nation owes an enormous debt to Willis Whitney and his colleagues for implanting and fostering that spirit of cooperation within our country. The scores of alumni from Schenectady who as summer boarders at the laboratory experienced the spirit which has pervaded the organization there, have carried it with them throughout the country, allowing it to diffuse and become something in the nature of a national standard. In brief, the laboratory has had a profound influence on the evolution of team-structured research in our country. The fact that so many great leaders of scientific organizations in the United States, such as Karl T. Compton, Ernest

O. Lawrence, and Lee A. DuBridge, have been summer boarders in Schenectady means not only that they have stimulated the group working there but also that the laboratory has made an indelible imprint on them, an imprint which has inevitably influenced their attitudes and actions when they assumed great positions of leadership later on.

Dr. Suits has been a member of the laboratory for 35 years—a period covering more than half of his life. Moreover, he has served since 1945 as its third Director. It was he who directed its expansion in the postwar period, during which the United States finally closed the science gap which had existed across the Atlantic since colonial times. It was he who promoted the development of the magnificent Knolls laboratory complex overlooking the Mohawk River. In keeping with the traditions of his predecessors, there is no facet of the work of the laboratory to which Suits has not directed his attention. There is no scientist in the laboratory who does not know him as a friend.

In this volume, which is a compilation of a few of his addresses given over the past fifteen years, Suits has focused on a few of the topics and issues which emerge out of the activities of a great laboratory when viewed by its director. Some of the essays deal with the products of scientific research, others with matters of policy. Still others stem from philosophical reflections which, in addition to being stimulating and informative in their own right, give one a clear indication of the contemplative quality of his leadership.

As the reader will see, the technical and scientific lectures range from a review of the status of atomic power to high polymer research, from the synthesis of diamonds to semiconducting circuitry. All of these focus on areas in which his company and laboratory have blazed so many trails.

The essays dealing with matters of policy range from discussions of the relation between the engineer and the scientist to the training of physicists and engineers—matters of crucial interest to the continued advance of research-based technology.

On the philosophical side are essays dealing with the maintenance of creativity in an organizational setting and the interplay between large and small business in cases in which there are common technological interfaces.

It has been my privilege to hear the author deliver many of these addresses in his brisk and stimulating style and to read many others in reprint form. All have been stirring and informative. We are fortunate to have them bound together in a single volume. I believe that everyone interested in the evolution of applied science will be enriched by reading them.

Dr. Frederick Seitz
President, National Academy of Sciences

Washington, D. C.
April 1965

Preface

In this book there are references to my "crystal ball" and the suspicion that it may be somewhat cracked, crazed, and cloudy from overwork. The role of the scientist as a prophet is indeed precarious; nevertheless it is held by some that anyone holding the title of "director of research" is obligated to make occasional safaris into the "never-never-well-maybe-perhaps-land" of the scientific soothsayer.

After re-reading my own words of the past fifteen years, I am struck by the improved accuracy that *could* be imparted by re-writing all of them from today's more advantageous viewpoint. In truth, with appropriate editorial pruning these assorted talks and articles could be immeasurably improved for inclusion in this book, making me appear—to my friends, at least—as a new Nostradamus, uncannily perceptive about things to come.

However, it seems more prudent to publish these articles—both ancient (written fifteen years ago) and recent—in their original form. Hopefully, the result will impart an authentic if imperfect picture of some of the excitement and satisfactions of scientific progress in a particular industrial research laboratory. Rewarding experiences come from success in the discovery and understanding of new facts of nature and, equally, from success in starting new science-based technology down the road to utility in the service of mankind.

It is now evident that the scientific and inventive progress of two decades has brought into focus fantastic changes in man's material world. During this incredible period, I have been privileged to work with a great many of the scientists who have helped shape our remarkable times, and I understand and share their motivations and satisfactions. During the same period I have participated in the highest councils of a truly great industrial enterprise which pioneered industrial scientific research, and from this vantage point I have been able to view science with a management perspective. Happily, from both points of view I have gained a strong feeling of great mutual respect and appreciation, which provides the firmest possible foundation for future joint progress.

Modern scientific progress is shaped in a variety of settings and environments, to accommodate the varying interests, opportunities, and aptitudes of the researcher. If the talks included in this volume, prepared for various audiences and occasions, add to the understanding and appreciation of some of the unique attributes of the *industrial scientific research* environment, it will be a source of great satisfaction to me. My hope is that throughout this book the reader will discern a single unifying theme: the belief that scientific research in, and for, industry can produce lasting benefits for mankind.

C. Guy Suits

Schenectady, New York
May 1965

Contents

Development of Useful Power from Nuclear Fission

Reprinted from *Electrical Engineering*, American Institute of Electrical Engineers, June 19, 1950. An address presented at the National Academy of Sciences, Rochester, N. Y., October 25, 1949.

The development of nuclear fission as an important source of useful energy is one of the great scientific and technological challenges of the world today. Here is a source of energy which is potentially capable of supplying the world's needs for hundreds of years; yet the problems of developing and proving the technical feasibility and economics are beset with peculiar difficulties, for the solution of which very little precedent exists. First and foremost is the problem of secrecy and security. The fuel for nuclear power plants is also the explosive material of atomic bombs, and the consequences of this fact have a far-reaching effect on every aspect of the task. The political problems of atomic energy are, of course, paramount, but this article will be confined to the scientific phases of the problem.

Among the technical problems, the by-product radiation which accompanies fission is the most troublesome and costly. Furthermore, the technology of such plants is entirely new, so that the materials and processes and methods for design, fabrication, control, testing, and operation must be worked out from the ground up. For all of these reasons the development of atomic power is certain to be a long and expensive venture. Finally, the atomic power plant will have to compete eventually with alternative sources of heat energy which have been refined and perfected over a long period of years of scientific and technological research and development; accordingly, the atomic power plant will have to achieve initially a corresponding degree of refinement in all of the details of its operation.

Perhaps at this point we should bolster our courage by considering the case history of the development of another energy source which now seems commonplace, but which in its infancy was faced with a somewhat similar array of problems—the petroleum industry.

The first well was drilled for oil in the United States in 1859 at Titusville, Pennsylvania, but it was over fifty years before this natural resource became a significant power source here. The important use for petroleum for a period of thirty years following 1859 was as kerosene, principally in lamps; there were only four automobiles in the entire country in the year 1894. The real development of the petroleum industry awaited the internal combustion engine and was not realized until the many scientific and technical problems of this machine were overcome, if they have been overcome, progressively during a long period of years. The internal combustion engine and its progeny, the motor car, the airplane, the diesel engine; and finally the refining, cracking, and synthesis of petroleum and its scientific exploration and discovery—it was the development of the science and technology of these problems that brought the petroleum industry to the point of real significance as a power source. The development of atomic energy as a significant power source awaits the solution of a similarly imposing array of technical problems and will require a correspondingly long period of time.

Two important facts of nature condition the character of the atomic power development from the very beginning. One is the fact that the nuclear fission chain reaction is not self-sustaining on a miniature scale. There is no prospect, therefore, that the test tube and pilot plant stage of power plant development, during which practically all of the technical problems of ordinary industrial processes are solved, can be employed. I do not mean to imply that there is no need for a great deal of intensive laboratory work on nearly all details of the development; nothing could be farther from the truth. But the experimental plant which brings together all of the elements of the complete process is itself a large and costly undertaking which cannot be tested in miniature.

The second fact is the unfortunate necessity for radiation shielding for such plants and their associated facilities. The cost of the radiation shield varies as the logarithm of the power, and is, in any plant of practical size, an important item of overall cost, so that even if the fission chain reaction would operate at low power the necessity for heavy and costly shielding could not be avoided entirely. These are unfortunate facts that make the development of atomic power a very costly undertaking. The very first experiment in the building of such power plants is necessarily quite a large one which will cost a great many millions of dollars, and it must be followed by subsequent scaled-up experiments which will become progressively more costly. The first petroleum from the first oil well which was drilled in this country was sold at a price of $20 a barrel. From the profits realized from its sale, subsequent wells were drilled, more oil was sold, and from its profits the industry grew.

This is, of course, the typical case history of successful industries. They are generally profitable even on a small scale, and the reinvestment of part of the profits in the industry permits it to grow by the progressive reduction of unit costs to the consumer, frequently accompanied by improvements in the quality and utility of the product. The development of petroleum as a source of energy, as an example, had these small beginnings which, however, could be sold for a profit from which the further development of the industry took place.

There is no such analogue in the development of atomic power. It is probable that about $200,000,000 [1] will have to be spent in the process of building successive experimental atomic power plants before we will have a proper basis for judging the future economic possibilities of atomic energy. Because of the great array of unanswered technical questions there can be no assurance at present that the final result will be successful. If a profitable atomic industry ever develops, it is certainly decades in the future. Obviously this financial risk is one which cannot be supported by any private capital in the world, so it is clear from the outset that only the resources of the National Government can sustain the required effort. Not only does the great cost of the development make it necessary for the National Government to foot the bill, but the fact that fuel for atomic power plants is also the explosive material of atomic bombs makes it necessary for the Federal Government to control the security and accountability aspects of the project in great detail. All atomic energy work is thus the responsibility of the Atomic Energy Commission (AEC) and is carried out in laboratories operated by its contractors.

Many aspects of the technical problem of building reactors may be inferred from a consideration of the principal elementary process, the fission chain reaction. Figure 1 illustrates this reaction in simplified form. When a neutron strikes a fissionable nucleus, for example U^{235}, the two may be presumed to coalesce to form a compound nucleus, very much the way a small raindrop might

[1] The Atomic Energy Commission reactor program for the fiscal years 1948–49 and 1949–50 is estimated to cost $215,000,000 on the basis of authorizations approved by the Congress.

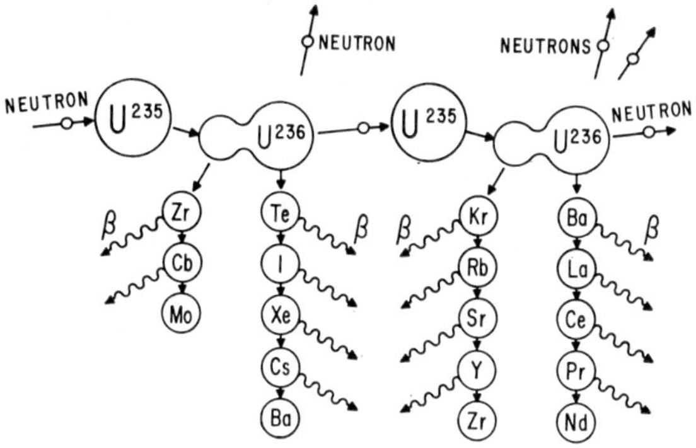

Figure 1. Simplified form of the fission chain reaction. When a neutron strikes a fissionable nucleus the two form a compound nucleus. This compound nucleus is unstable and breaks into two fragments which fly apart with enormous energy.

join a large one in falling through the air. This compound nucleus is very unstable and breaks into two fragments somewhat in the manner in which a large unstable liquid droplet might be divided by surface tension forces. The two fission fragments, of similar atomic weight, fly apart with enormous energy, to produce two lighter atoms near the middle of the periodic table.

The first feature of this reaction, of importance to power piles, is the following: of the total energy for fission of about 200 million electron-volts, 80 per cent, or approximately 160 million electron-volts, resides in the kinetic energy of the two fission fragments. Although it is true that these fission fragments carry electric charge, no one has been able to figure out how to take advantage of this fact for the direct generation of electric power, so that the only alternative at present is to permit the kinetic energy of the particles to convert to heat. This kinetic energy thus appears as heat in any fission apparatus, and hence the first observation from the elementary process: (I) heat is the principal energy production in fission.

The fission-fragment nuclei have initially the high ratio of neutrons to protons of about 1.5, appropriate to the heavy uranium nucleus but unsuited to elements of much lower atomic number. Hence, these fragment atoms are unstable and undergo radioactive decay, the end point of which is a stable nucleus. In the case illustrated in Figure 1, for example, the first fission in this segment of a chain reaction produces the unstable nuclei ${}_{40}Zr_{97}$ which decays by successive beta emissions to the stable nucleus $_{42}Mo_{97}$. The other fission fragment $_{52}Te_{137}$ decays by the same process to the stable nucleus $_{56}Ba_{137}$. The second fission in this hypothetical case results in two fragments which eventually decay to stable $_{40}Zr_{91}$ and $_{60}Nd_{145}$. The detailed fission process is not identically the same in each fission event, but it covers a considerable range of possible mass combinations which add up to the mass of the parent nucleus. The stable end product of each fission-product decay chain shows a similar variation so that the statistical result of the chain reaction is an array of radioactive and stable nuclei which covers a broad region in the middle of the periodic table. The resulting radiation is made up principally of beta emission from a neutron–proton conversion in the nucleus but is accompanied by some gamma emission, and occasionally by a neutron.

The second property of the basic fission reaction of importance in power piles is this: (II) intense radioactivity is a necessary by-product of fission. About five per cent of the fission energy does appear in this radiation. An occasional but important neutron comes from the fission-fragment decay series with a suitable delay characteristic of the series of half-lives involved in a given decay chain. One to three neutrons are produced initially with the fission of the compound nucleus. Exactly how many neutrons occur with fission has been something of a secret because it is such an important number in calculations of chain reactions.

These neutrons produced by fission are of tremendous value and are the key to the entire situation. If we waste them by leakage from the periphery of the nuclear reactor or by parasitic absorption in the materials of construction, the pile will not react at all. The basic question of whether (on the average) at least one of the one to three neutrons produced initially would be available for a chain reaction was answered for the first time in the West Stands uranium-graphite pile at the University of Chicago. This proved the calculations which showed that if the pile were sufficiently large, so that the surface-to-volume ratio (important for peripheral neutron leakage) was favorable, and if care was used to exclude neutron absorbers, on the average at least one of the one to three neutrons per fission would be available to produce the next fission and hence propagate the chain reaction. Many other successful piles have since been built so that no one is concerned now about the first neutron per fission. The progress of the atomic age thus far has been built on this first neutron. Since one to three neutrons are produced in fission, and since only one is required to propagate the chain reaction, theoretically up to two neutrons may be left over.

As will be shown, the development of useful atomic power focuses primary attention on these neutrons which are left over from each fission. For present purposes they are the most important of all, and in a very real sense the future atomic age depends on them. If we can build nuclear reactors which are sufficiently large and sufficiently well engineered to conserve the left-over neutrons, fission energy may become an important source of power some decades in the future. If not, the applications of atomic power will probably be limited to cases where economics is not a primary consideration. Hence a third observation from this elementary process points to (III) the necessity of conserving the neutrons left over from the chain reaction. Some consideration will serve to show why this is so. In the chain reaction, the fissionable nuclear fuel is used up to produce (*a*) heat energy (160 million electron-volts per fission), (*b*) fission products accompanied by much (*c*) radiation and (*d*) some stray neutrons which are lost to the structure or its shield.

Thus, to produce the heat energy, we destroy an atom of fissionable fuel and obtain as by-products fission fragments and radiation, both of which may be of future economic value, but which at present are expensive nuisances. Now let us consider the possibility illustrated by Figure 2, which shows the elementary process in the type of nuclear reactor which conserves and uses the left-over neutrons, the power breeder. Here, for each neutron which propagates the chain reaction, at least one is presumed to be available for a side reaction—the production of a new fissionable atom from fertile material.

This process, termed the breeding of nuclear fuel, or, simply, breeding, is the principal source of hope for abundant atomic power of the future. If this process can be made to work, it has many practical results. The power breeder will initially be charged with fissionable material (for example, U^{235}), which will be surrounded by fertile material (for example, U^{238}). Thus, as heat is produced from the fission of U^{235}, the exhausted fuel is continuously replaced

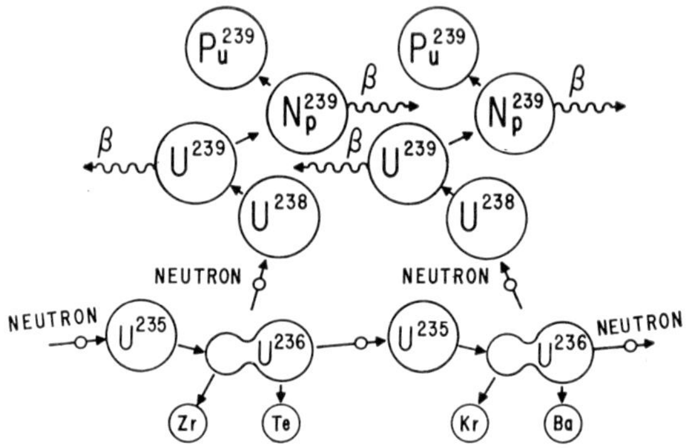

Figure 2. Fission-fragment nuclei are unstable and undergo radioactive decay. The detailed fission process is not the same in each fission event, but covers a range of possible combinations which add up to the mass of the parent nucleus.

by new fissionable fuel produced by the left-over neutrons. This atomic furnace might thus produce a pound of new fuel for each pound of fuel consumed. More properly, of course, it might convert a pound of fertile material to fissionable material for each pound of fuel consumed. Since the number of neutrons left over from the chain reaction conceivably may be greater than one, the power breeder may not only replace the fuel consumed, but may, with skillful design, produce more new fuel than is consumed in the process. The fuel investment might thus be returned with interest all during the time the reactor is producing useful heat energy.

By means of a reactor program which would reinvest the fuel dividend as it becomes available, a very large power industry might be developed. Since the initial investment is thus built up by compound interest, even a very small initial deposit will grow to a very large fuel pile in the course of sufficient time. It is necessary, of course, to have an adequate supply of fertile material to stoke the atomic furnaces over this period, and the question of supply of fertile material is pertinent. If only the isotope U^{235}, present in native uranium in the concentration of one part in 140, is available as nuclear fuel, fission energy may become a significant and possible strategically important power source, but it is not likely to compare in magnitude with common fuels. If fuel breeding can be made to work, so that 100 per cent of the available native uranium and thorium can be converted to fissionable material, the total energy supply thus created will be significant in comparison to that available in coal or oil.

Hence, the important concepts underlying present plans for atomic power development are based on the utilization of the heat energy of fission; the necessity of coping with the problems of by-product radiation; and the necessity of conserving the left-over neutrons for fuel breeding.

The first important step in the development of atomic power will be the demonstration of the technical feasibility of high-temperature breeder reactors in experimental plants. Further in the future, these pilot plants, if successful, must then be extrapolated to very large power plants, and finally the economics of the large power breeders must be developed.

The design and construction of a power reactor involve new and fascinating scientific and engineering problems at every point. In the core of the reactor the active and structural parts will be subjected to a very high temperature and to an intense flux of neutron, beta, and gamma radiation. In this environment reactor materials must not absorb neutrons appreciably and must retain strength and structural stability. In a somewhat analogous problem, materials for high-temperature steam, mercury, and gas turbines, where the parts are subjected simultaneously to very high temperatures, corrosion, and high tensile stresses, thirty years of research and development have produced an impressive array of specialized alloys for this service, most of which now can be selected from handbooks of engineering practice. The handbook of high-temperature radiation-resistant reactor materials has, of course, not been written. Much of the intensive radiation metallurgy which has been in progress in AEC laboratories since the war is providing the data which some day will fill the chapters of this book to make it as useful and practical as existing handbooks are today.

Although it is much too early to do anything but conjecture about the future economics of atomic power plants, some respects in which the first plants will be exceptionally uneconomical are quite clear. It was pointed out earlier that the fission product elements cover a broad range of the periodic table. Some of these elements have a high absorption cross section for neutrons and their accumulation in the pile will gradually poison the reaction.

This loss of neutrons is wasteful since it detracts from the number available for more useful purposes; for example, breeding. One solution of this problem which will be used in early power piles is the frequent reprocessing of the fuel for the removal of these atomic ashes.

Reprocessing the fuel for the removal of fission products is an excessively costly operation, however. In fact, any simple experiment of physics or chemistry which must be accomplished in the presence of a high level of radioactivity is expensive. Consider, for example, a simple operation such as mixing two chemical reagents to form a precipitate which must be washed, collected, and dried. For nonradioactive chemicals this operation may be performed on any scale with simple and inexpensive equipment. However, if the reagents are radioactive at a level of several curies,[2] as an example, the simple experiment becomes cumbersome, slow, and very costly in terms of the equipment, facilities, and organizational requirements.

Before the chemist may enter the area, which is equipped with heavy shielding and remote-control gear required for radiation chemical work, he acquires a film badge and two pocket electroscopes which record any radiation exposure to which he may be subjected. He must then pass through a change room where he exchanges part of his outer garments for a special smock and shoe covers. The change room, of course, must be served by a special laundry facility which is equipped to deal with radioactive clothing, including methods for the storage or disposal of contaminants. The film badges and electroscopes must be serviced similarly by trained personnel with suitable testing and recording methods. A 1-curie level of radiation requires the equivalent of about a foot of concrete shielding between the operator and his equipment. The equipment must therefore be controlled remotely by any one of a number of complicated methods.

Because of ingenious mechanical design which has been carried out in many AEC laboratories, almost any kind of an experiment can be accomplished remotely. The cost of this equipment and the heavy shielding with which it must be used is, however, very great,

[2] The curie is the unit of radioactivity and is almost exactly that of one gram of radium. More precisely, it may be defined as the amount of radioactivity when 3.7×10^{10} disintegrations occur per second.

and, worse still, the time required to set up a given experiment, perform the manipulations, and clean up later adds greatly to the cost of the work. Finally, the chemist, upon leaving the radioactively hot area, must remove his protective clothing and check his hands and feet for contaminating activity.

In the case postulated, the water (or other liquid) used to wash the precipitate will be contaminated, and the disposal of this waste raises a serious and as yet unsolved problem. An inspection of half-lives of the radioactive elements in typical fission-decay series discloses a range from fractions of a second to thousands of years. Although an initial period of storage will permit the decay of short-lived activities, any reduction in activity in the stored wastes realized in this way is obtained at the very considerable expense of the storage facilities. Some of the activity is of such longevity that no practical period of storage will serve to eliminate it as a hazard. The traditional method of dealing with wastes is to throw them away, and the trouble with radioactive wastes is that no one has yet found a safe place to throw them.

Any examination of the elements of cost of an atomic power plant will show that the radiation shielding and remote control equipment and methods imposed by the presence of radiation are inordinately large items. Our success in dealing with this problem in the future eventually will have an important bearing on the economics of atomic power.

It is true that there is some low level of waste activity which may be discharged to the air or to a river without hazard. The determination of this safe radiation level is always somewhat conjectural, and as a result present standards are probably very conservative. It is clear that zero level of radiation is not attainable, since a practical minimum is set by the natural radioactivity of the earth, cosmic radiations, and especially the natural activity of radio-potassium found in most foods and in body tissue.

At present, the solution of this problem is being deferred in favor of the more urgent matter of making a high-temperature reactor work. If power reactors do work and become economical some decades hence, the disposal of an energy content in radiation comparable to the energy production in electricity will have to be dealt with successfully. Like the atomic bomb, there is little prospect that the problem of disposing of radioactive wastes can be ignored in the hope that somehow it will go away.

The manifold difficulties in the way of the development of atomic power are more than matched by the ingenuity and enthusiasm of the scientists and engineers who are dedicating their efforts to this Herculean task. There is no dearth of ideas or plans for the solution of all of the problems which now confront us, and certainly one can see clearly, if somewhat distantly, a progressive line of improvement in power reactors which will lead far beyond any now on the drawing boards. However, it is vitally necessary to prove reactor designs, by construction and operation of suitable prototypes, so that future improvements will be based on a sound foundation of experience. The Atomic Energy Commission's reactor program, which was announced some time ago, includes four reactors, all of which will contribute important information to the design of future power reactors. Quoting Dr. L. R. Hafstad, AEC Director of Reactor Development, these four projects are the following:

1. The first has been designated as a materials testing reactor, called *MTR*. While it is itself an experimental reactor, as its name implies, it is intended also to give information on the behavior of materials in reactors so that larger and more powerful special-purpose reactors may be built ultimately.

2. The second reactor is a land-based prototype of a reactor for use in propelling naval vessels. It would be a simple, single-purpose reactor designed specifically and solely for the purpose of producing large amounts of heat under conditions which will permit conversion to propulsive power.

3. The third reactor is a single-purpose experimental reactor designed specifically to give us information about the breeding process. This reactor at the present time is the most likely to demonstrate the actual breeding of new fissionable material. It is, however, neither a high-power reactor nor designed for the purpose of demonstrating appreciable amounts of useful power though an incidental amount of power may be produced by a by-product.

4. The fourth reactor is the more ambitious project—the Knolls Atomic Power Laboratory reactor. This reactor is intended to produce a really significant amount of electric power. At the same time, it is hoped that it will be able to demonstrate at least partial success in breeding. This reactor is, therefore, a very complex device since its design represents a compromise between the demands for power production and for breeding. If successful, however, this project would represent a major step forward in the direction of the production of useful power without depleting, and perhaps even increasing, our national supply of fissionable material for any purpose.

This last reactor project has been somewhat altered in objectives since this report was issued. The application of atomic power plants to ship propulsion has become an immediate goal. Power reactor development is the largest single activity of the Knolls Atomic Power Laboratory (KAPL), operated by the General Electric Company for the Atomic Energy Commission at Schenectady, N. Y. The test site for reactors developed by KAPL is located 18 miles north of the city, on a 4,500-acre plot of marginal farm land adjacent to the Adirondack forest preserve. The construction of a power reactor is a major technical task and will require several years of intensive effort. In the development and design of such plants, great effort is being made to solve the many engineering, design, and operating details sufficiently well to permit their ready extrapolation to large power reactors.

The term "fuel breeding" will be defined further. While the one to two neutrons left over from the fission chain reaction are theoretically available for fuel breeding, whether they are practically available for this purpose depends upon the knowledge and skill of the reactor scientists and engineers. Every structural and active material that goes into reactors must be examined and tested with great care for its parasitic neutron absorption, for no matter what other virtue a reactor material may have, it must not waste neutrons.

This examination of a material, for example, iron, must extend not only to the iron itself, but to the practically unavoidable impurities of iron, some of which may be very harmful even in very small concentration. The components of a reactor structure thus contribute to the neutron economy and will in aggregate determine how many of the left-over neutrons are available in practice for fuel breeding. Some of the measurements which enter into the tabulation of neutron losses are so difficult to make with precision that no final answer to this question can be determined except from studies of an experimental breeder reactor in moderately high-power operation.

In addition to the KAPL reactor, power development reactors are being designed at the Argonne National Laboratory, operated by the University of Chicago for the AEC, and in the Atomic Power Laboratory being operated in Pittsburgh by the Westinghouse Electric Corporation for the AEC.

All of these reactor development groups will, of course, depend on the other national laboratories of the Atomic Energy Commission for assistance. Thus, the development of atomic power is a project which will draw upon the combined resources of industrial and government scientific and engineering knowledge and experience.

I shall not indulge in any predictions concerning the ultimate future of this great scientific and engineering experiment. In arriving at one's own conclusions, however, one should bear in mind that the attainment of the goal of useful power from nuclear fission will not be based solely on what we know today, but rather on what we know today plus what we are going to learn in two or three decades of scientific progress. The latter must be weighed heavily in judging the future possibilities of any development of this magnitude.

Measuring the Return from Research

Reprinted from the Proceedings of the Fourth Annual Conference on the Administration of Research, Engineering Research Institute, University of Michigan, September 11–13, 1950, Henry M. O'Bryan, presiding.

Last month I attended a series of company meetings at Association Island on Lake Ontario in which each functional activity of our company was reviewed, starting with research and proceeding through engineering, manufacturing, marketing, as well as the principal administrative activities. These meetings provide, in our company, an indispensable source of integration and balance, and, incidentally, illustrate to an excellent degree the problems inherent in measuring the return from research. The proponents of each of these activities of an industrial enterprise feel, with proper enthusiasm, that theirs is the most important component of the whole. For example, we in the laboratories feel that without our pioneering scientific work the corporation's future would be doubtful. We must admit, however, that some of the output of our laboratory is hardly ready for market in the form in which it leaves our hands and that a great deal of rationalizing and reducing to practice must be accomplished by engineers as the essential next step.

The manufacturing people, in turn, are quick to point out that no matter how well conceived and engineered, unless the highest degree of manufacturing skill is applied to the product, it will never reach a satisfactory market position. Finally, the marketing groups say, with some justification, that without skillful selling there will be little or no need for manufacturing, engineering, or research. Although it is true that most successful products draw fully upon all these industrial functions, the problem of measuring the contribution of each of them separately is not easy. Yet this problem has been studied many times in the course of the history of our laboratory, and I can summarize my talk at once by saying that in our case we have never found an objective and complete solution of the problem of measuring the return from research. My comments will, of course, apply to a research laboratory in industry and, more particularly, to the laboratory in my company with which I am most familiar, which has long-range exploratory research as its primary function. As a necessary consequence of this specialization in long-range research, we depend heavily for the practical utilization of our work upon development laboratories and engineering-design divisions, where the application of science is the primary task. The contributions of many groups of specialists are thus generally involved in the successful life history of a new idea.

I propose to discuss this question in terms of a variety of case histories of actual developments, which will serve to illustrate the complexity of the problem, without, however, leading to any very workable means of measuring the return from research expenditures in a specialized industrial environment. I hope, however, that we may obtain a better understanding of the matter from a consideration of actual examples. Obviously, the character of the industry in which the company operates, the size and kind of business in which the company engages, and, finally, the kind of laboratory operation—all have a great deal to do with the question. Hence the conclusions drawn from case histories may not suggest generalizations at all. In specific cases the return from research may bear little relation to the cost of research; the return may in some cases be very much greater than any properly allocated cost, or, in other cases on which I would not like to dwell, very much less.

Now, measuring the return from research is generally desired for reasons of good business, for, if we can accomplish this result, we can judge whether our expenditures have been made wisely or whether they should be greater or less than at present. Another equally important, related question is the prediction of the future value of the results of research, so that this may, in turn, be compared to the cost of research in progress.

Everyone would, of course, like to be able to accomplish this admirable result with precision, for it would provide the final answer to many important questions of research planning, effort, and expenditures. Because the term "research" covers a great variety of technical activities, it must be admitted that some types of research activity may be susceptible to detailed planning at every stage, including the prediction of return.

Consider, for example, the hypothetical case of the laboratory activity, principally of an applied engineering character, involved in the substitution of a pressure die casting in a product which is now being manufactured with a fabricated part. The circumstances may be such as to permit the anticipation of a considerable saving in manufacturing cost if a die casting can be employed. The cost of doing the necessary laboratory, engineering, and pilot production work may be estimated with satisfactory accuracy, as may also be the expected manufacturing cost savings. Some of the expected savings may permit price reductions and a consequent expansion in volume, which may be determined by a market survey. In other words, enough information is available or obtainable to make possible an estimate of the cost of doing the work and the return which may be anticipated from this particular example of applied research. With such facts, or reasonable facsimiles, fairly complete planning of research effort may be feasible.

Everyone is familiar with the way in which preliminary estimates, particularly cost estimates, acquire an aura of exactitude with the passage of time, frequently to the embarrassment of the estimator. Planning the commercial future of an exploratory research project in its early stages is analogous to planning the education and career of a new baby and is fraught with as many uncertainties. It would be better to wait until something is known about the child's personality, intelligence, and character traits as basic factors upon which to plan.

This is not to imply that planning of this type of research is not possible. One can plan the scope of fundamental and exploratory research and can relate the scope of the work to the scientific opportunities presumed to exist in the field of work and to the plans and aspirations of the company in relevant industrial areas. Although it is frequently true that the detailed results of exploratory research are unpredictable, some of the gross results are not subject to the same uncertainty. For example, some work in our laboratory pointing toward new methods of interrupting electric arcs is more likely to lead to results of importance in the field of switchgear than in polymer chemistry, although I know of a case in which the opposite was true. Planning for the future of such work must be done on a much more expedient basis than in the case of engineering development work. The director must expect to do a good deal of playing by ear, bearing in mind that new experimental results may, without warning, completely alter the complexion of the work for better or for worse.

At the other end of the scale of research is the purely exploratory work in the frontier areas of modern science: research at temperatures near absolute zero involving superconductivity and superflow, investigations of the phenomena of semiconduction in solids, exploration of nuclear reactions, to mention some current examples in progress in our laboratory. Since the results of exploratory research are thoroughly unpredictable, the application of planning methods which require a detailed knowledge of the time and cost of doing the work and the possible application of the results in terms of products and markets involve unknowns at every point. Planning the career of the research project in the case of truly exploratory work cannot be done in the definite terms with which one may deal in the case of applied research or engineering development. In spite of the fact that principal information on which planning and forecasting depends is generally not available in the early stages of exploratory research, engineers frequently attempt to plan for the future results of the work. I feel that forecasting in such cases—for example, estimating the cost of a new chemical product from very incomplete early research data—frequently does far more harm than good. Once numbers are obtained, one is likely to associate with them a degree of reliability that is by no means justified by the assumptions upon which they are made.

THE CASE OF THE SILICONES

I would now like to consider a relatively simple case of a research project on which a priori one ought to be able to measure the return from research. I refer to the silicones on which our laboratory has been working for quite a number of years. The initial exploratory work was instigated largely by a consideration of one of the primary weaknesses of organic materials, the carbon-to-carbon linkage. It is the failure of this bond which accounts for the long-time thermal decomposition of organic materials at a temperature of about 105°C; the carbon-hydrogen bond is much stronger. The silicon atom is similar to carbon in much of its chemistry and is not too much larger, so that silicon-substituted organic compounds were possible, and a few such materials had been synthesized. Although it was recognized by analogy with the numerous organic families that a very large number of compounds of silicon, carbon, and hydrogen must be possible, it was the potentialities of high-temperature insulating materials which were sufficiently attractive to justify an extensive investigation. It is important to point out that it was not the sale of silicone materials as such which provided the incentive for this work. It was rather the prospect of a greatly improved performance of electrical machinery which silicone materials might make possible. This project was carried through all of the stages of laboratory and pilot-plant research, and market for wire enamel is not large; it is a specialty chemical product of limited application. However, improvements in wire insulation have a far-reaching effect on the cost of electrical machinery, and it is in this area that the important results of the research appear. If the costs of the research which led to Formex are allocated against the end product, which is the wire enamel itself, the sale of the improved material readily permits a liquidation of the cost of the research. However, no accounting system we have devised measures the great importance of this development to the design, manufacturing methods, and hence the cost of electrical machinery made possible by the Formex development. Hence the return from the research in this case is apparent and satisfying, but its measurement, as that term is generally understood, is a matter of great complexity.

THE ELECTRIC BLANKET

I would now like to describe a project of our laboratory which I hesitate to call research. In this case the difficulties of judging the value of the result and developing a market were prominent. Dr. Whitney became interested in the idea of the electric blanket in the early nineteen thirties, more or less for the fun of it. I do not think he was consciously trying to develop a new electrical appliance; he rather was motivated by a curiosity to see how an electrically heated comforter would work. He had Mr. William Kearsley build a first sample, and it worked well from the beginning. A type of control was devised which automatically compensated for changes in room temperature, and a number of experimental blankets were built and used by willing laboratory collaborators. One distressing difficulty was discovered. When the blanket is properly spread on the bed, a satisfactory temperature requires a heat input of about 150 watts. If the blanket should be rolled up without being turned off, this 150 watts of heat, well insulated, will cause the center of the rolled-up blanket to become very hot and eventually to burn. Before we had a solution to this problem, a number of the early experimental electric blankets caught fire.

When this happened, Mr. Kearsley was invariably confronted the next morning, if not earlier, by the unfortunate individual who, clutching the charred remains in one hand, described his misfortune with the other. Happily, there were no serious accidents.

Mr. Kearsley began to notice that people who burned up blankets, although naturally somewhat upset by the experience, were far more upset if there were to be any appreciable delay in repairing or replacing their blanket. The obvious enthusiasm of users of electric blankets, even under the somewhat difficult circumstances I have described, convinced us in the laboratory that there was much merit in the idea, and we promptly set about to interest our engineers in the results we had obtained. The danger of the rolled-up blanket was solved by placing a number of tiny safety thermostats in the blanket, which open the electric circuit if the temperature rises excessively. The latest design of electric blankets has a far more elegant solution to this problem.

It did not take long to interest our appliance engineers in the electric blanket; but everyone recognized that the very novelty of the idea would necessitate a considerable period of public education before one could hope for any general acceptance of the idea. The development of a new electrical appliance is frequently preceded by market surveys and market testing.

But a market survey in the case of a product as new as was the electric blanket in 1936 would have been useless; any housewife confronted by the question, "If an electric blanket were put on the market at a certain price, would you buy one?"—could only have given the questioner a puzzled expression. She had never seen an electric blanket and could hardly be expected to imagine what they were like. The first regular production of electric blankets took place in 1940. Production increased every year, with a hiatus during the war, but it was not until 1947 that a profit position was achieved. Meanwhile a great deal of engineering development went into progressive refinement of this basically simple device. Ten years after the blanket was first marketed, the national market was estimated to be some $12 million.

It is easy to measure the modest research expenditures which went into this development, but for reasons similar to those found in the previously mentioned cases, the determination of a return from the development which is assignable to research is quite another matter. The contribution of many groups was essential to the success of the project, but special credit is due to the marketing people, who must be particularly skillful in the case of any basically new product such as this.

PROJECT CIRRUS

A peculiarly intricate example of the difficulty of measuring the return from research is presented by Project Cirrus, the work on weather modification, in which our laboratory is participating in an advisory and analytical capacity.

During World War II, Langmuir and Schaefer undertook a study of aircraft-wing icing, which focused attention on the problem of supercooled clouds. Ice forms on aircraft structures when they come in contact with clouds in which the liquid water particles are at temperatures substantially below the freezing point. The meteorologists were familiar with supercooled clouds, but they did not know very much about them, and the phenomena held a great fascination for Langmuir and Schaefer. So, at the end of the war, they set out to learn how the water in supercooled clouds can be made to freeze. A home-freezing cabinet was the principal laboratory equipment required for the experiments, and in it Schaefer formed supercooled clouds and studied their properties. He discovered that a particle of dry ice would quickly nucleate the entire cloud, causing it to settle out as snow in the bottom of the chamber. Later Dr. Vonnegut determined that nucleation of a supercooled cloud could be accomplished by silver iodide particles. Both methods were successfully applied to actual clouds: by "seeding" the cloud from aircraft, dispensing dry ice or silver iodide, or by silver iodide generators operated on the ground. At this point we asked the armed services (and they agreed) to undertake a contract in partial support of the work, primarily because of the legal liability aspect which is inherent in parts of the task.

This pioneering work has turned out to be a new science of experimental meteorology. Active experimenting on weather modification is known to be going on in Australia, Canada, Mexico, Hawaii, South Africa, in many South and Central American countries, and probably at many unannounced points. In the United States, in addition to the well-known reservoir-filling operation carried out by New York City, there are many local cloud-seeding projects going on in various parts of the country. It is somewhat astonishing to behold the rather casual and routine manner in which rain-making contractors carry out their operation in the southwestern part of the country. Probably far more important than local rain making is the growing evidence that continent-wide weather modification is feasible. Where this work will lead eventually is impossible to predict, but its potentialities are of the very greatest importance. It is certain that for the first time in history man has at hand a powerful means for influencing weather. These are the bare outlines of the case history of the project. Now let us try to measure the return from this research! I know of no single application of this work to the products which we manufacture. Yet it was decided in the highest councils of the company that the importance of the work to the economy as well as to the national security was such that we should continue the work. This case illustrates the necessity and obligation of considering many factors, some of which are noneconomic, in measuring the results of research of this character.

The problem of measuring the return from research varies greatly with different types of work. In some cases the problem reduces to that of separating the contribution of research from the contributions of the many skills—engineering, manufacturing, and marketing, etc.—which have by their collective efforts made possible a new development. In other, less simple cases the effect of a new material, for example, Formex, has so many ramifications in so many different areas of an industrial economy that it seems hopeless to measure it adequately. In the case of still other research, for example, Project Cirrus, the results of research extend so far beyond the boundaries of a given industrial enterprise—in this case they may well extend beyond the boundaries of the nation—that one cannot hope to measure the return from the research.

I have developed the thesis that measuring the return from research is difficult because of the manifold ways in which the beneficial results of research manifest themselves and because of the important and inter-related contributions of engineering, manufacturing, and marketing to a research result which attains a market. I would like to leave no reasonable doubt, however, on one point. Although the contributions of many groups are vital to the practical utilization of a research result, it is the scientist himself who produces the vital foundation on which the whole structure rests—the new fact of nature. The scientists and the laboratories which have the freedom to follow their curiosity in search of new facts of nature are among our most important national assets.

Power from the Atom—An Appraisal

Reprinted from *Nucleonics*, February, 1951, Vol. 8, No. 2, pp. 3–9, New York, McGraw-Hill. An address presented at a panel discussion held in conjunction with the 117th meeting of the American Association for the Advancement of Science, Cleveland, Ohio, December 28, 1950.

The development of useful power from atomic fission has been in progress in this country for about five years. As yet, no useful power of any consequence has come from the considerable effort in this direction. For this and other reasons it is clear that this is a long-term project, and the general realization of this fact represents progress.

The *next* stage of progress is the one which we are now entering. In this second stage we can achieve, as a result of the excellent scientific and technical work of many laboratories, the beginnings of a real perspective on the problem of power from the atom. It has taken many years of research on the reactor problem to provide the solid foundation upon which such a perspective may be supported.

In addition to the laboratory work on power-producing reactors, there has been a vast accompaniment of public debate, discussion, and comment on this subject. Perhaps never before has so much been said by so many with so few facts. What atomic power is, or is not—what it will do, or fail to do—every possible sociological and economic aspect of the matter—has been so thoroughly considered that little remains to be said. There is, of course, the necessity of obtaining the scientific facts upon which the whole conjectural structure rests; this will take another ten years. It is safe to say, however, that atomic power is *not* the means by which man will for the first time emancipate himself economically.

What then is atomic power; where does it fit into the picture? At present, atomic power presents an exceptionally costly and inconvenient means of obtaining energy which can be extracted much more economically from conventional fuels. If the economics of heat production were the only consideration, the great outlays for reactor research and development could not be justified. We have no shortage of fuel in this country. Whatever the elastic life span of petroleum may be, coal is certainly available for hundreds of years to come. The economics of atomic power are not attractive at present, or are they likely to be for a long time. This is expensive power, not cheap power as the public has been led to believe. The development of nuclear reactors for power production is, however, very important for quite a different set of reasons. These reasons are:

1. Atomic energy is an extremely concentrated source which presents important military advantages in a mobile power plant.

2. Atomic plants may simultaneously produce large blocks of power *and* fissionable material needed for military purposes.

3. The scientific and engineering knowledge gained from intensive effort on (1) and (2) above may in the longer term make possible the building of economical plants primarily for the production of power.

These reasons are compelling and important and are strongly accented by the troubled state of the world. They are quite different from the reasons which seemed important in 1945.

Let us now examine some of the numbers which bracket the economics of atomic power plants. Although the basic information required for a precise appraisal of these matters is not available, there are several hypothetical cases worth considering, each of which corresponds to a set of technical possibilities and hopes. As a base line to which these hypothetical cases may be referred, consider the set of figures of Table 1 which gives for a typical steam-electric plant[1] of 400×10^6 watts (or 400 MW) electrical output, (a) the estimated efficiency, load utilization, and plant cost, (b) the estimated construction cost breakdown, and (c) the operating costs. Such figures have a considerable variation between different plants at different locations, so that the term "typical" must be used with some reservations.

For example, the operating costs per kwh (exclusive of fuel) for steam electric plants show an extreme variation of some 20-fold throughout the country, while the fuel costs per kwh also show a 20-fold variation. The reasons for this involve a great variety of practical considerations, including the age of the plant, operating efficiency, load utilization, size, location relative to centers of population, and source and kind of fuel and so forth. The figures of Table 1 have been selected to represent some of the important desiderata in the present discussion. An efficient atomic plant must

[1] See for example, Federal Power Commission Report S-83, "Steam Electric Plant Construction Cost and Annual Production Expense, 1949."

TABLE 1

Typical Steam-Electric Power Generating Plant—1950

Net station electric power output	400 MW
Auxiliary power requirements	37 MW
Gross output	437 MW
Net station heat rate	9,800 BTU/kwh
Gross heat input (400,000 × 9,800)	$3{,}920 \times 10^6$ BTU/hr
Gross heat requirement (maximum boiler capacity)	1,150 MW
Estimated construction cost	$40,000,000
Unit cost (on 400 MW net available)	$100/kw

Construction Cost Estimate

Site	1.%	= $ 400,000
Coal handling, Ash handling	6.	= 2,400,000
Boiler plant	46.	= 18,400,000
Generator plant	40.	= 16,000,000
Switchgear, transformer	7.	= 2,800,000
	100 %	= $40,000,000

Estimated Electric Power Production Costs (*at 70% utilization*)

Taxes (at 2%), Interest (at 6%), Depreciation (at 3%)	1.8 mills/kwh
Fuel	2.
Operation	1.
	4.8 mills/kwh

be large—the steam-electric plant of Table 1 is large. The location is presumed to be on the eastern seaboard where the quoted costs are believed to be realistic for late 1950. A different set of assumptions could be made with corresponding changes in the figures; the numbers of Table 1 will serve our present purpose, however.

Referring now to Table 1, we may see that the gross heat requirement is 1,150 MW, which could be supplied presumably by one or more nuclear reactors. The construction cost of the entire plant is at the rate of $100/kw. Under somewhat altered circumstances this might be $140/kw or even higher. Please note that the boiler plant, which is the unit for which we would substitute a nuclear reactor, is 46 per cent of total plant cost, corresponding to \$46/kw of the net electric generating capacity, and that the cost of equipment for handling coal and ash is only 6 per cent of the total. Under the items of generating cost, note that fuel is 2 mills/kwh; operation, 1 mill/kwh; and that taxes, interest, and depreciation amount to 1.8 mills/kwh.

The depreciation is figured on the basis of 3 per cent/yr which is well supported by utility experience. The rate of depreciation of any power-producing reactor would have to be much higher; in the absence of experience I have assumed 10 per cent/yr in the discussion below.

CASE A

Now refer to Table 2 and especially the assumptions on which it is based. Here it is assumed that the cost of nuclear fuel on a heat basis equals conventional fuel. This might be realized in a variety of circumstances which would add up to the same result. The weight comparison between fissionable fuel and coal for the same heat release is $1:3 \times 10^6$. The ratio of $1:10^6$ is about the relation between a gram and a ton. Hence the cost of fissionable fuel per gram may be about three times that of coal per ton. Reliable figures are of course not available, although it was stated by Dr. Hafstad [2] of the AEC in reference to a quoted figure of \$20/gram that "the \$20 figure is almost certainly low and seems likely to increase as the supply of high grade uranium runs out." Since the cost of handling fissionable materials and especially radio-ash will be high, it is likely that the cost of producing separated fissionable materials must be brought down considerably if we are to use them in enriched power reactors.

It is altogether probable that these materials will be produced more cheaply in the future. G. R. Prout, General Electric manager of the Hanford Works, announced recently [3] that the cost of plutonium had been cut in half during the last four years. A considerable reduction in the cost of enriched fuel would help to offset processing costs which at present are very high.

[2] L. R. Hafstad, Director, Division of Reactor Development, U. S. Atomic Energy Commission—remarks before American Petroleum Institute, Los Angeles, Calif., Nov. 15, 1950.

[3] *New York Times,* December 4, 1950, p. 13.

TABLE 2

Hypothetical Nuclear Steam-Electric Plant of 400 MW Power Output in 19–

CASE A

ASSUMPTIONS:
1. Nuclear fuel cost (including handling and processing) equals conventional fuel costs on a heat basis.
2. Operation costs the same.
3. No credit for by-products.

Credit from coal fired boiler plant	–	0.8 mills/kwh
Annual from coal fired boiler plant	–	$ 2,000,000
Additional cost * nuclear boiler †	–	$11,100,000
Gross plant cost	–	$32,700,000

CASE B

ASSUMPTIONS:
1. Nuclear fuel costs negligible.
2. Operation costs the same.
3. No credit for by-products.

Credit from fuel savings	–	2 mills/kwh
Credit from boiler plant	–	0.8 mills/kwh
Additional cost * nuclear boiler †	–	$38,300,000
Gross plant cost	–	$59,900,000

CASE C ‡

ASSUMPTIONS:
1. Nuclear fuel costs negligible.
2. Operation costs the same.
3. Plutonium credited at the rate of 2 mills/kwh.

Credit for fuel savings	–	2 mills/kwh
Credit for boiler plant savings	–	0.8 mills/kwh
Credit for plutonium	–	2 mills/kwh
Additional cost * nuclear boiler and chemical plant †	–	$65,500,000
Gross plant cost	–	$87,100,000

* With interest on investment at 6%, local taxes 2%, depreciation 10%.
† To produce electric power at same net cost as a conventional plant.
‡ For a plutonium plant.

It is assumed in Case A that plant operation for the nuclear steam-electric plant is the same as for the conventional steam-electric installation. The physicist-chemist-encrusted gear which surrounds present reactors does not encourage this hope, but we have much to learn and must expect progress in this area. In the conventional plant a great volume of coal must be handled, but this is largely a mechanized operation. The nuclear plant, on the other hand, may under the best of circumstances handle a minuscule volume of fuel and one may expect that this operation will be highly mechanized in the future.

It is further assumed in this particular case that nuclear by-products are not worth reclaiming; or, if reclaimed, the cost of processing offsets their value.

Under the stated assumptions, the higher depreciation rate on the nuclear boiler means that less capital cost can be tolerated for the same net electric power cost. Accordingly, the reactor costs for the 1,150 MW unit must not exceed approximately $11 million.

CASE B

Similar hypothetical plants are discussed as Case B and Case C. In the former, the assumptions are the same, except the cost of nuclear fuel per heat unit is assumed to be negligible. This might be the case for example in a reactor in which native uranium would be "burned" until exhausted and then thrown away; or it might be realized in a nuclear economy in which the spent power reactor fuel could be sold for a by-product value approximately equal to its original cost.

In this case, the calculation takes account of the saving in fuel cost and of the omission of the conventional boiler. Credits from these items are allocated against the nuclear boiler, bearing in mind that it is depreciated at the rate of 10 per cent per year, whereas the conventional boiler is depreciated at the rate of 3 per cent per year.

Under these circumstances one might allocate a capital investment of $38 million to the nuclear boiler of 1,150 MW heat output.

CASE C

This case includes a combination of favorable circumstances which might develop in the long-term evolution of atomic power in a plutonium economy. The new assumption here is that an efficient nuclear fuel breeding cycle is established and that the "spent" nuclear fuel is sold for its plutonium content and credited against the power production at a rate of 2 mills/kwh. It is obvious that with a substantial by-product credit, a considerably greater investment in reactors can be justified. Under these assumptions, the

TABLE 3

Boiler Plant Estimated Costs
1,150 MW Heat Output

Conventional boiler	$18,400,000
Nuclear boiler (Case A, fuel and operation cost equal to conventional)	11,100,000
Nuclear boiler (Case B, fuel cost negligible)	38,300,000
Nuclear boiler (Case C, by-product credit)	65,500,000

reactor-boiler plutonium producer might cost $65 million and still permit competitive costs on power production.

REACTOR COSTS

It would be interesting now to take a look at reactor costs for the purpose of determining whether their present magnitude or future downward extrapolation is likely to fall within any of the hypothetical brackets which have been cited.

The cost figures are collected in Table 3 for convenience. Since the cost of most reactors has not been made public, and since no power-producing reactors have been built, and since the ones now being built will be much smaller than what will be finally required, a great deal of guessing must be done from this point on—few really pertinent facts are available.

I believe the only public information available [4] on a reactor for which both power level and cost are known is the experimental reactor at the Brookhaven National Laboratory. This unit, rated at 30 MW, is stated to cost approximately $25 million.

Extrapolating to high-power reactors from the baseline provided by the Brookhaven reactor involves many uncertainties. This reactor is primarily an experimental unit which produces only a token amount of power incidental to its cooling. Operation at the high temperatures required to produce power efficiently would necessitate a change in design and probably a higher cost for the same power. On the other hand, the provision for experimental facilities inevitably adds greatly to the cost of any reactor. More important than either of these considerations is the fact that these small experimental reactors properly include factors of safety with

[4] AEC Release #219, November 28, 1949; see NUCLEONICS *6*, No. 1, 94 (1950).

regard to their thermal and nuclear design, their controls, their materials of construction, and provisions for flexibility of operation and inspection which will be considered unnecessary and excessive in practical power reactors.

The mere extrapolation of a reactor from pilot plant size (10 MW–30 MW) to practical power ratings (100 MW–1,000 MW) permits substantial cost savings per unit of output. Although a reactor for higher power must generally be larger, the increase in size will be much less than the increase in power. As a broad generalization of some validity, power reactors of a given basic design tend to have a more or less constant size over a considerable range of power output. The heat transfer system must increase in size and cost as power increases, but the weight and cost of the shielding, which is a very substantial item of cost, does not change much.

It is evident from a scrutiny of the materials of construction now available and from the factors of safety which we must employ that we are paying a high price for the elementary state of our knowledge of reactor technology. A substantial factor of improvement must be expected from continuing research and development in this area.

After first considering and weighing all of these factors, one may estimate that the cost of building a power production reactor of 1,000 MW output under current limitations of technology might be in the range of $50 million to $100 million.

CONCLUSIONS

It is rather surprising to find that there is not a greater divergence between economically permitted reactor costs and the postulated costs of reactor construction. Although it should be emphasized that the underlying assumption must be borne in mind, it should also be pointed out that the economic baseline presented by Table 1 does not by any means represent the most favorable set of circumstances for reactor economics. A combination of high fuel and transportation costs coupled to the requirement for large blocks of power would considerably enhance the case for nuclear power plants, and it is probable that such cases will provide initial opportunities for exploiting this new power source.

The future solution of difficult technical problems will be accomplished partly with knowledge now at hand, and partly with new knowledge which we are certain to gain from continuing scientific research. On a long-term project it would be a great mistake to underestimate the importance of the latter to the final solution.

The Engineer and the Fundamental Sciences

Reprinted from *The Scientific Monthly*, Vol. LXXVI, No. 2, February, 1953. The Silliman Lecture at Yale University Centenary School of Engineering, October 24, 1952.

In developing my thesis, I will have much to say about technology, by which I mean that complex of science, engineering, and experience that is the foundation of industrial know-how, and which depends heavily upon engineers and scientists for its present strength and future development. Here I wish to examine the relationships between science and engineering and, particularly, to point out how highly developed and interdependent these relationships have become in the research and advanced engineering phases of a highly technical industry. To gain perspective, let us look briefly at engineering and science in the early days of American industry.

PRIMITIVE INDUSTRY AND EARLY SCIENCE

It is clear that, at the start of the Industrial Revolution, around 1770, and during the industrial development that extended into the 1850's, empiricism rather than science was the rule. During this period, neither science nor engineering was recognized as a component of primitive industry. The foundations of modern industrial technology were, however, being laid in the scientific work of this same period. In the electrical industry, modern technology is built upon the early work of Coulomb (1736–1806), Ampere (1775–1836), and Faraday (1791–1867), followed by Maxwell (1831–1879), Hertz (1857–1894), Henry (1797–1878), and many others. Their researches, largely accomplished in a university environment, provided the observational facts and interpretation that became the basis for electrical engineering practice. The mechanical industries were similarly generated in fundamental researches of Newton (1642–1727), Carnot (1796–1832), and Joule (1818–1889), among others. In chemistry, I suppose the names of Lavoisier (1743–1794), Priestley (1733–1804), Scheele (1742–1786), and Dalton (1766–1844) must be reckoned among the pioneers whose work was later to support the engineering foundations of the chemical industries. The few names I have mentioned by no means complete the list of pioneers of science upon whose work, translated into engineering terms and techniques, industry built a large structure.

The development of machines, especially Watt's steam engine, initiated the factory system and the Industrial Revolution, but the primitive industry that was thereby born drew but little of its strength from science and engineering. Watt's steam engine, Cartwright's power loom, Hargreave's spinning jenny, the cotton gin of Eli Whitney, and many similar machines employed in their development a very small portion of science and a large content of mechanical inventiveness. The later phenomenal growth of modern industry could not have taken place, however, without the foundations in early natural science which developed independently of, and concurrently with, primitive industry.

THE ENGINEER IN AMERICAN INDUSTRY

The real beginnings of engineering go quite far back in time, but the trained engineer appeared on the American industrial scene around 1850, at about the time many of our engineering schools, including the Yale School of Engineering, were founded in this country. The period between 1850 and 1900 was, among other things, the age of inventors. In addition to the trained engineer, who was making his way into industry, the early inventor played an important role in the translation of scientific facts into embodiments of practical utility. Industrial technology during this period was not yet so involved, nor basic scientific knowledge so complex, as to require intense specialization. Hence, a single talented individual could comprehend a broad area of industrial skill and related technical knowledge and could excel in conspicuous individual inventiveness. In the industry with which I am associated, it was the inventive genius of Bell, Stanley, Elihu Thomson, Edison, and others who translated the science of Faraday, Henry, Maxwell, and Hertz into products capable of manufacture and sale.

The increasing complexity of both science and its application led to the gradual development of engineering and of the engineer as the specialist who reduces the laws of natural science to the more or less exact practice of analysis, design, and construction. Of course, engineering does not concern itself solely with the application of science. Although this is a large and increasing content of engineering practice, the application of experience—and especially the extrapolation of experience—are important functions of engineering. The traditional inventor, as an individual with great natural aptitudes and a smattering of science, has almost disappeared as a logical consequence of the increasing technological complexity. Natural aptitudes of inventiveness have not, of course, disappeared. They have simply ceased to be individually conspicuous because, in modern technology, individual effort merges into a background comprising a great manifold of highly specialized technical skills.

THE BIRTH OF MODERN SCIENCE

Consider the year 1890 and some of the significant events associated with it. It is well known that the scientists of the time felt that most of the fundamental laws of matter—represented by the knowledge of electricity, magnetism, mechanics, and thermodynamics—had been observed and interpreted, and that refinement of detail rather than new discovery would comprise the future of science. The electrical industry had its beginnings at about this time, originating in inventions based on the work of early electrical scientists. With minor exceptions, however, no scientists were employed by industry—a remarkable fact, in view of the many applications of science that were being made at that time. This situation can be understood, however, in terms of the traditions of primitive industry and the philosophy of early science. Early scientists felt that the adapting of the laws of nature to the practical uses of man involved a stigma, and that science was pure only if it was sterile. The men of early industry, on the other hand, knew that their experts had been trained almost entirely by working experience; hence, the concept of academic training for industry was contrary to all practice. It cannot be said that these early ideas have been thoroughly changed even now. I hope that my readers do not share these feelings concerning the debasement of science, for I shall have occasion to describe pertinent instances of the application of new scientific discoveries to practical use.

The well-defined boundaries which science seemed to have in 1890 soon disappeared. Roentgen discovered his famous X-rays in 1895, and Becquerel discovered the radiation emitted by uranium in 1896. The electron was discovered by many investigators, but was first identified and interpreted by J. J. Thomson in 1897. In 1900 the Curies discovered radium, and Max Planck formulated the beginnings of the quantum theory. In 1903 Rutherford and Soddy correctly described and interpreted natural radioactivity.

The most significant result of this exciting period was the disappearance of the previously well-defined boundaries of science; it is a remarkable fact that, more than fifty years later, in spite of tremendous research effort in laboratories throughout the world, the boundaries to new scientific discovery cannot even now be seen. Before 1900 industry had not yet embraced science, but it was about to do so, for it was in that year and against this background that the first scientific research laboratory was established in industry. Scientists thus began to play a direct role in industrial organizations and to develop their functional relationships to engineering.

By 1900 the engineer in industry and his industrial functions were well established. To be sure, engineering was by no means as specialized then. In General Electric, for example, only four specialized skills in engineering could be distinguished: civil, mechanical, mining, and electrical engineering. The engineer in industry was likely to be engaged in a variety of technical activities involving the design, construction, testing, and operation of industrial equipment and processes. He applied what science was available to him, but large factors of experience, judgment, and courage entered into his designs. This is, of course, no less true today.

Around 1900 the engineer and the scientist could be clearly distinguished as separate professional specialists. In the first place, the engineers were to be found largely in industry, and the scientists were almost entirely in the universities. Scientists in industry were somewhat apologetic concerning their defection from professional tradition; and industrial engineers, on the other hand, were equally apologetic over any dealings with scientists, who were known to be a thoroughly impractical and visionary lot.

This distinction between the scientist and the engineer was, moreover, aided by dress and demeanor. The scientist of the times was likely to look like a scientist in the classical sense, whereas the engineer frequently looked the part and affected a manner befitting a thoroughly practical man. When I first came to industry, one could still find in our Schenectady plant a number of picturesque individuals who were reverently pointed out as "real" engineers. They were pioneers who had grown up with the electrical industry and who embodied the early traditions of the profession to a conspicuous degree. They wore their hats at work, indoors and out, occasionally spat upon the floor, and were somewhat given to profanity. They were distrustful of theories, especially those expounded by young men just out of college. New developments in science reached them by an exceptionally circuitous route, if at all, and they were almost entirely spared from direct contact with scientists in the normal course of their engineering activities. That professional complexity was to come later. They may have remembered, in contemplative moments, the period some forty or more years before when they were young engineers entering industry. They were then the impractical technicians with their theories and their mathematics, who subsequently had to establish their worth with the practical men in the infant, but rapidly growing, electrical industry.

In the early days, the technologists were the practical men, aided and abetted by inventors, and guided in analysis and design by engineers who first translated the electrical phenomena of Henry, Faraday, and Maxwell into particular embodiments of sufficient utility to permit the beginnings of a manufacturing enterprise. Engineers soon became the keepers of industrial technology, and they have for many years carried forward its development. Scientists appeared on the industrial scene fifty years after the engineers, and their work has contributed to expansion and diversification, particularly in the electrical, chemical, and communications industries, and has led to the great present-day growth of industrial laboratories.

Although the process has been going on for some fifty years, the extent to which science has now been integrated into the industrial complex has not been generally appreciated. For example, the early and successful employment of scientists and engineers in the electrical, chemical, and communications industries has not by any means established a pattern throughout industry. There are manufacturing businesses in our country, with an annual sales volume of hundreds of millions of dollars, which employ no scientists and few engineers. The extent to which science and engineering have been applied to industrial technology thus varies widely with industries. In some industries the techniques of employing engineers and scientists have hardly been explored, and in others scientists and engineers work together on complex technologies as vital to the business as its manufacturing and sales functions.

TECHNOLOGICAL COMPLEXITY AND PROFESSIONAL SPECIALIZATION

The development and increasing complexity of industrial technology have been paced by the increasing technical specialization of its practitioners. In engineering alone there is now a great variety of technical specialization. Although the engineering curriculum at many schools is extensive, I suspect that there are still some engineering specialties not included in training programs. The electrical industry originally employed only four professional categories of engineering, whereas the number is now very much greater. In the General Electric General Engineering Laboratory,[1] one can distinguish sixty-two separate fields of engineering, and the company-wide total undoubtedly would be much greater. A further index of specialization is provided by the number of national technical societies that are now in existence. In the field of engineering, for example, there are at least 135 national engineering societies in this country.

The same specialization has taken place among scientists and their national professional societies. The terms "physicist," "chemist," and "metallurgist" are now useful only as broad generalizations; they no longer denote definite fields of scientific specialization. It is not possible, at present, to be a specialist in physics. One might hope to encompass the specialized knowledge associated with a narrow field of physics, if this field is not expanding too rapidly; for example, classical acoustics, sound, heat transfer, and thermodynamics may present this possibility. But in the rapidly growing fields of modern physics this is impossible. Some years ago when the atomic nucleus was a novelty, one might have specialized in nuclear physics, but today the field is too broad and there are many specialists within this area, including reactor, radiation, and cosmic ray physicists. A similar, but less extensive, degree of specialization is probably developing currently in solid-state physics, where spectacular progress is being made in the art and science of the semiconducting elements.

It is a fact that modern developments in science and engineering are becoming more and more complex. As a first approximation to the truth, one is tempted to conclude that all the simple discoveries have already been made, but this, of course, could hardly be true.

[1] Now, the General Electric Advanced Technology Laboratories.

Each new tool of science opens up a new vista that could not be seen with the older, and generally simpler, tools. The electron microscope has unlocked a world that light microscopy could not probe. With microwaves, details of nature are seen that escape kilometer waves. Particles with energies of several hundred million volts resolve nuclear phenomena that exhibit no detail at lower energy. The storehouse of nature, once a simple structure, is now revealed as a vast edifice of ever-lengthening corridors adjoining more and more intricate inner chambers. The translation of these complexities of nature into the service of mankind is an important industrial function which employs all the subdivisions of professional specialization lying between exploratory fundamental research and production engineering.

The path from a complex new discovery to practical utility is frequently long and distressingly expensive. Bringing nylon to market is vastly more difficult than bringing cotton or wool to market. Crawford H. Greenewalt, of Du Pont, has stated that ten years and $27 million were required to accomplish this particular modern miracle. The accumulated costs, as an index of scientific and engineering effort, that have gone into the development of television (which I hesitate to call a modern miracle) have probably not been computed, but the total certainly would be staggering. One need hardly mention the cost and complexity of bringing some of the military developments, especially those in the atomic field, to practical utility. The $2 billion cost of the first atomic bomb is familiar to everyone. The continuing cost of bringing out improved yearly models accounts for an appreciable fraction of our annual tax money. Hundreds of millions of dollars have already been spent in developing atomic fission as a source of power, but certainly a billion dollars will be spent before economical atomic power plants are built. Fortunately, the military applications of uneconomical nuclear reactors are important, and their development will help to pay the bill. I mention these examples—and many more could be cited—of the great cost and complexity of important modern developments of science and engineering. There are undoubtedly modern developments of science and technology that are simple in concept and in translation to practical use, but they are not numerous, and they will be more rare in the future.

THE SCIENTIST AND THE ENGINEER IN MODERN TECHNOLOGY

Let us examine some of the consequences to science and engineering that result from this technological complexity. For example, does the overwhelming weight of planned programs of research mean that exploratory fundamental research has disappeared? I think not. The fact that vastly greater funds flow into the area of applied science does not impress me as evidence to the contrary. I think it is true that fundamental science has never before been so well supported. Although nuclear research is an exception, the fundamental research stage of the technological process is generally not an expensive one, at least not in comparison to the applied research and engineering development that may follow. Hence, in this area, a small amount of money goes a long way. In spite of the great cost of fundamental nuclear research, vast sums have been made available for this work by the Atomic Energy Commission. The enlightened policies of the Office of Naval Research and the recently established National Science Foundation are designed to encourage fundamental science, and they actually do provide generous financial support. Industry is supporting fundamental research to an ever-increasing degree, partly in its own laboratories and partly under university auspices. If truly worthwhile research proposals cannot obtain support under present circumstances, the sponsors must be remarkably inept in presenting their case.

How about the individual scientist in this age of cyclotrons, synchrotrons, supersonic wind tunnels, and atomic proving grounds: is it forever his fate to submerge his individuality into a background of programmed research projects? To the extent that he works with expensive research tools, this is likely to be the case. If a scientist spends $10 million building a particle accelerator, even discounting the probable source of the money, he must not expect to find its research use to be a contemplative solitary endeavor. The mad, three-shift confusion that surrounds some of these machines creates quite a different impression. Fortunately, not all of modern science is paced by million-dollar machines. The study of the solid state of matter is a notable example of a field of research of the greatest importance, which has not yet developed the expensive tools and inconvenient trappings of atom-smashing. Here, outstanding progress is being made by individual researchers and small scientific investigation teams.

In modern technology, at what point does science stop and engineering begin? In the early days of industry, when the two were quite separate, this question was easily answered. At present, both scientists and engineers man the numerous stations along the stream of technical progress. At the end nearest to design and manufacture, it is mostly staffed by engineers; at the other end, where the laboratories are located, the scientists predominate; in between, the professional groups intermingle, depend upon each other's technical contribution, and may be difficult to distinguish professionally.

What is the impact of modern fundamental science upon engineering training and practice? Time was—and it was an earlier time—when the engineer could assimilate, during formal engineering training, a codified set of the laws of natural science which provided the tools for a career in engineering practice. This traditional idea of stored technical knowledge sufficing for a professional career is fast becoming obsolete, not only in engineering but in many other fields. To be sure, the old rules are still true and still fundamental, but science has been adding to the rule book, so to speak, and the new editions, with appendices, come out so frequently that it is difficult to follow the game. Consequently, engineering is becoming more complex and more specialized, as new subject matter is added at an ever-increasing pace. The engineer who stays in practice finds it more difficult than ever before to remain current in his technical knowledge, particularly if he is isolated from research laboratories and development laboratories. The technical societies play an ever more important role in bringing to their membership the latest information concerning new developments. It seems to me, however, that a more direct attack on this important problem clearly is necessary. We must recognize the need of replenishing and renewing knowledge as a consequence of rapid technological progress. Perhaps college refresher courses in the latest engineering developments would make it possible for the practicing engineer to augment his fund of specialized knowledge. In any event, it is clear that occasional technical reading will no longer suffice for this purpose.

The scientists who are causing all this trouble are, in turn, in a similar plight. A good scientist, of course, keeps up with the scientific literature, at least in his field of specialization. That could formerly be accomplished by occasionally reading the technical journals in question. But the journals have multiplied in specialization and have become much thicker. Even the abstracts journals in some fields, especially chemistry, have become so voluminous that there is a need for an abstracts journal to abstract the abstracts journals. If a research scientist were really conscientious about his reading, his research project would start and end in the library, for it would take all his time to read what others are doing in the field in which he intends to work.

Is the engineer subject to more regimentation as a result of the increasing complexity of engineering practice? I think the contrary is likely to be the case. It is undoubtedly true that the efficient pursuit of large research-and-development projects—for example, the development of a nuclear reactor—requires a high degree of coordination of the many technical skills that must contribute to the final result. However, the extensive variety of technical skills employed by modern industry provides, as a corollary, a broad selection from which an individual may choose his field of specialization. Thus, a great range of individual interests may be accommodated within the confines of large coordinated project activity.

What is the saturation point for scientific and engineering manpower in industry, and when will it be reached? Consider the following figures. United States industry in the year 1900 employed, on the average, one engineer for every 250 employees. In General Electric at present there is one engineer for every twenty employees. The number of chemists in United States industry has doubled in the past fifteen years, and the number of physicists has doubled in the past eight years. Another interesting index is provided by the tabulation of the number of companies which recruit at the colleges. In 1916 only twelve companies did so. By 1929 this had risen to 750 companies; by 1942 the number was 1500; and during 1952, 5200 companies recruited at the colleges. I see no indication that this clear trend in the employment of trained people for the development of industrial technology will not continue long into the future. *Technology is based on science, and science seems to be boundless.* Continuing research will undoubtedly lead to new developments in science that will lead to new applications to engineering. This technological process appears to be an inexhaustible natural resource.

SOME NEW TECHNOLOGIES

The relationships I have been discussing between fundamental science and engineering will perhaps be clearer if I cite some examples of important and rapidly developing technologies. Many cases come to mind, but a particularly good example is provided by recent research in the field of solid-state physics, where spectacular progress has been made in understanding some of the detailed electrical and mechanical behavior of matter, and where equally spectacular and important progress is being made in the application of the knowledge to practice. The semiconducting elements, of which silicon and germanium are important examples, have provided the best foil for the attack on the electrical properties of solids. Here theory preceded practice (the converse is generally true), and it was only with the discovery of means of drastically purifying the semiconducting materials that theoretically predicted properties were realized. Chemically pure germanium may be hopelessly contaminated, since one part of boron, as a typical impurity, in 100 million parts of germanium will double its electrical conductivity. Skillful experimenters finally discovered means of reducing the impurities in germanium to one part per billion.

Ingots of high-purity germanium have been made in our laboratory that are sufficiently pure to satisfy even the theoretical physicists, who never really expected that anyone would be able to make germanium as pure as they had specified. After we went to all this expensive trouble, the theoreticians said that they did not really want pure germanium after all, but that they wanted to put in their own impurities; and they were particularly partial to indium and antimony. If you put an occasional antimony atom in the germanium crystal lattice, where it fits nicely, the valence electron is surplus and it becomes a conducting electron. Or you may put an occasional indium atom in the germanium lattice. In this case, the *lack of an electron* leaves a kind of "hole" which moves like a positive charge, much as a bubble (which is an absence of something) moves through a liquid.

Thus, if you put some antimony-germanium in contact with some indium-germanium the result is a rectifying junction. This in itself is not remarkable because there are other rectifying junctions—for example, copper-copper oxide. But the germanium junction is remarkable because: (*a*) people can understand why it rectifies; and (*b*) it is an almost perfect rectifier, passing a very high current in one direction and almost none in the reverse direction. This has been a wonderful thing for the theoretical physicists, because it worked out just as they had said it would; but it has been a wonderful thing in many other respects, too, for the phenomenon of semiconductors is adding a new dimension to industrial technology. For example, rectifiers made on this principle are very efficient for use in high-frequency electronics circuits, where they are known as germanium diodes. Or, they may be made for real industrial power purposes, where their high efficiency makes possible rectifiers of very small size capable of carrying hundreds of amperes of current. The Bell Telephone Laboratories discovered, a few years ago, how to put the equivalent of the control electrode of a vacuum tube into a semiconductor, thus making power amplification possible in a device called a transistor. Although these transistors were originally thought of as low-power devices suitable only for electronics circuits, we have made a transistor recently for high-power applications. In the span of a few years, there have come from this hitherto neglected field of semiconductors a flood of new technical capabilities which have already developed great importance, and which will in the future exert a profound influence upon electrical technology. The particular capabilities I have mentioned, embodied in practical devices for controlling electric power, are certain to have important implications in industry. But it would be a mistake not to recognize that the real significance in this development resides in the fundamental understanding that has been achieved in the basically important field of the solid state of matter.

In the hands of metallurgists, the study of the solid state is yielding new knowledge that is destined to have a correspondingly important bearing on mechanical technology. The strength of materials, especially of metals and alloys, is a primary limitation on the capabilities of the vast galaxy of machines that pace modern civilization. The speed of a motor, the power of an engine, the efficiency of a turbine, the endurance of an airplane, and the vital properties of a host of mechanical things all depend on, and are limited by, the properties of the metallic materials of construction. Without exception, these metals and alloys are crystalline solids, and the problem of improving and extending their properties reduces, in its most basic form, to the problem of understanding the mechanics of crystals. Until recently, attempts in this direction have been frustrating. There is no doubt that a crystal is a regular three-dimensional array of atoms, and, from a knowledge of the atoms and their arrangement, it should be possible to deduce simple mechanical properties. The density of a crystal is, of course, correctly calculated from the known atomic weight of the constituent atoms and their arrangement, but that is about all that the theory of crystals does for the mechanical properties of solids. If one attempts to calculate the tensile strength of a crystal, the calculated result differs widely from the measured tensile strength. The theoretical strength of an aluminum crystal is 600,000 psi. Pure aluminum in practice, however, attains only approximately one-hundredth of this calculated value, and the best aluminum alloys are only about one-tenth as strong as the calculated strength of a perfect crystal of pure aluminum. There are similar difficulties in interpreting other practical properties of crystals in terms of perfect crystal structure. Measured rates of diffusion, rates of solution, rates of crystallization, and rates of phase change, for example, all very important in metallurgy, defy interpretation on the classical basis of crystalline regularity. The key to these problems of great theoretical and practical importance lies in the defects of crystals. Everyone has known for a long time that crystals are generally not perfect, but it now appears questionable whether any example of a perfect crystal has ever been seen, at any time, by anyone. Hence it is not the geometric perfection of crystals, but their invariable imperfections, which determine the practical properties of crystalline solids such as metals and alloys. Concentrated study of the detailed nature of these defects is now revealing for the first time the fundamental nature of the properties of the materials upon which our mechanical civilization rests. An excellent example of the role of the crystal defect is shown by the growth of crystals from a saturated solution. A perfect crystal would grow very slowly indeed, because the perfect plane crystal faces are atomically smooth, and hence resist the attachment of an atom from the solution. Practical crystals are, however, full of defects and structure discontinuities, in the neighborhood of which almost all the phenomena of crystal growth occur. These growth phenomena have been observed under the microscope, and have even been photographed in motion pictures. Such motion pictures illustrate in a graphic way the crystal growth associated with a stepped, or screw, dislocation. This particular dislocation appears to be a very important one in many practical cases. The geometry of the stepped-wedge dislocation is illustrated schematically in Figure 1. Consider a perfect cubic crystal in the shape of a perfect cube. Make a cut into one face parallel to the cube edge and halfway through the crystal. Then displace the layers of the crystal planes on one side of the cut relative to the layers on the other side (Fig. 1). As compared to the atomically smooth faces of the crystal, the wedge dislocation is a highly preferred area for growth phenomena to occur. Once this was realized, scientists all over the world discovered spiral growth patterns arising from wedge dislocations in a great variety of crystalline materials. Crystals of silicon carbide that had been in the Mineralogy Department of the British Museum for many years were found recently to exhibit beautiful spiral

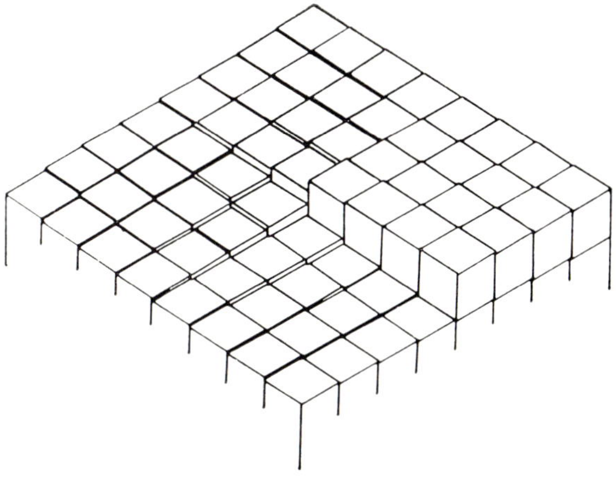

Figure 1. Schematic illustration of stepped-wedge, or "screw," dislocation.

The origins of metallurgy lie in antiquity. One of the early handbooks of metals was published in the year 1556 by Agricola, who described how to smelt the ores of tin, lead, iron, copper, antimony, silver, gold, and mercury, and how to make crude alloys from them. Agricola's smelter for cinnabar, an ore of mercury, is shown in Figure 1. He leaves little to the imagination. *B* is a door, while *C* is a little window. The text explains the function of the trees within the smelter. As the ore is roasted the mercury vapor rises and collects on the leaves as a silvery dew. When the chamber cools the leaves are shaken and the droplets fall and are collected. Apparently the efficiency of this condenser is not exceptional, for

Figure 1. Mercury smelter from Agricola's *De Re Metallica.* A. Enclosed chamber. B. Door. C. Little windows. D. Mouths through the walls. E. Furnace in the enclosed chamber. F. Pots.

Agricola notes that one may observe the very sweet odor of mercury on the leeward side of the smelter. In this event, he advises, go to the windward side, for people who smell too much mercury lose their teeth. His precaution was well taken, for it is now known that one part per million of mercury vapor in the atmosphere is quite poisonous. In spite of occupational hazards then, and now, the results of successful metal working have been so important that people have been willing to go to great pains to achieve them, for metals and alloys pace material human progress. The properties of metallic materials impart the ultimate limit to the power of a steam turbine, the pressure of the boiler, the electrical output of the generator, the efficiency of the transformer, the thrust of the jet and rocket engines, the speed of the airplane, and to a host of machines upon whose endurance, power, and capabilities modern mechanical civilization relies.

The working of metals had become an important and highly developed art before the advent of modern physical science. It has only been in relatively recent times that some of the important phenomena of practical metallurgy have been explained in terms of fundamental concepts based on atoms and molecules. The primary progress in this direction has come from an application of new knowledge of the physics of the solid state of matter. It is too early to say that efforts in developing the basic understanding of fracture, deformation, diffusion, grain growth, and similar phenomena have been reflected in great advances in metal and alloy technology, but there is a justifiable expectation that this will be so, and substantial progress can be noted. The great antiquity of the metal-working art and the complexity of the modern alloy technology have not made easier the erection of a scientific basis for metallurgy, for a great deal of back-tracking and re-examination of old established principles has been necessary. The task of the metallurgical scientist has been somewhat analogous to that of the contractor who has had to put a new foundation under a very large and very ancient mansion. It would have been easier, and more logical, to erect a sound foundation first, and build the house upon it subsequently, if that choice had been available. The anachronism of the development of an art preceding its science is by no means peculiar to metallurgy, but this field provides an exceptional illustration which is in strong contrast to some of the newer sciences. For example, nuclear science preceded, and still leads by a wide margin, the practical development of nuclear technology.

Because of the historical circumstances surrounding metallurgy, including the fact that its practice had attained a great importance before any real science of metals was known, some careful scrutiny has been required to discern just where practice left off, and science began. Some concepts that have been understood intuitively have, upon closer examination, been shown to involve basic fundamentals that were quite unclear, and not at all intuitive. As an example of a simple but very important property of metals, which until lately has not been explainable in terms of basic concepts, consider tensile strength. Under sufficient tension a wire first stretches elastically like a spring, then it stretches plastically, and finally it breaks. From the pull, expressed in pounds, and from the cross section of the wire, one can express the tensile strength in pounds per square inch. For lead, at room temperature, this number is approximately 3000 pounds per square inch; for aluminum, it may be 10,000 pounds per square inch; for copper, 60,000; for iron, around 100,000; for the strongest alloys, it is about 500,000 pounds per square inch. Now, we shall look at this important property in more detail.

From the fact that all metals and alloys are crystalline solids one might seek an understanding of the strength of metals in terms of the properties of crystals. The crystals, which may be large or small, fit together and produce an interlocking structure, much like building blocks in a masonry wall. The example may be simplified if one considers single crystals rather than polycrystalline material for, by going to some trouble, it is possible to grow large single grains of a metal. This single-crystal metal, which of course cannot fail inter-granularly, is not stronger, but is much weaker, than the polycrystalline material. As a matter of fact, if the experiment is done with great care and with very pure metal, the single-crystal strength is found to be almost zero.

These puzzling results are not easier to understand from a theoretical crystal structure point of view. The tensile strength of this pure single crystal, which has practically no strength, may be calculated by several methods, all of which lead to very high values. The calculation takes into account the known arrangement and spacing of atoms in the crystal lattice, and the known forces between them, and it is not immediately apparent why the calculation should not lead to a correct result. Single-crystal aluminum is shown to have a strength of a few hundred pounds, but if the experiment is done carefully it may be much lower. Polycrystalline, but unalloyed, pure aluminum may have a strength of 7,000 pounds per square inch, while the best aluminum alloys show a strength of 75,000 pounds per square inch. However, the calculated strength of pure single-crystal metal is over a half million pounds per square inch. This failure to account for crystal strength is not confined to the case of single crystals of aluminum; no other single crystals of metal or other substances appear to be substantially better in this respect.

The difference between calculated and measured values is particularly disturbing, because concepts of crystalline regularity, as related to structure, phase change, solidification, grain growth, deformation, and a host of important phenomena are fundamental to metallurgical science. In fact, the growth of the modern science of metallurgy can be traced to the origin of the crystal concept, yet, in its application to strength, it fails by a wide margin to account for this basic property. Of course, the failure of the crystal concept to explain strength does not, in any way, detract from its success in explaining quantitatively many other important properties of metals, such as specific heat, thermal expansion, compressibility and energy of formation, all of which depend upon *average* properties of a crystal lattice. However, strength is an exceptionally important property of metal crystals; so much so that serious attention has been directed to a study of this question in recent years. A clear outline of a quantitative picture of these phenomena may now be seen.

The case cited, and others which could be mentioned, have emphasized the quantitative difficulties of the crystal concept in its application to some mechanical problems in the study of metals and alloys, and have focused attention in recent years on the defects in crystals. Pioneering work on this problem has been done by F. C. Frank and N. F. Mott at the University of Bristol, by A. Cottrell at Birmingham, at the University of Illinois by F. Seitz, at Stuttgart by Kochendorfer, at the Bell Telephone Laboratories by Thornton Reed, and at the General Electric Research Laboratory by J. H. Hollomon, J. C. Fisher, and their co-workers.

The crystal defects which explain these mysteries of crystal strength and crystal growth are generally not gross defects of structure, visible to the naked eye, but are more subtle departures from atomic regularity, on an atomic scale. What had been somewhat overlooked these many years is that the crystal lattice is only a model, and practical crystals must contain many defects, one of the simplest of which is an impurity atom. An impurity of one part per million, which is seldom attained with metal crystals, would still permit 10^{16} impurity defects per cubic centimeter. The contribution of one impurity atom per million to specific heat, or to other crystal properties which depend upon an average effect of lattice elements, obviously will not be of much consequence. However, crystal strength is an example of a property that is extremely sensitive to defects. A valuable tool in visualizing such defects has been a special kind of bubble model of crystals, first employed by Sir Lawrence Bragg. In this ingenious experiment, bubbles of uniform size represent atoms in a crystal lattice; their arrangement and behavior simulate to a remarkable degree the perfection and imperfections of actual crystals.

A simple defect may be imparted to an otherwise regular lattice by an impurity atom or by a vacant lattice position; more complex defects involve more or less extensive mismatch between rows and planes of the crystal structure. All such defects and dislocations may produce local lattice strains and distortions of substantial numerical magnitude, which have not been taken into account in calculations based on the strength of perfect crystals, and hence are believed to be accountable for the practical departures from perfect crystal properties. Some calculations have been carried out recently in our laboratory by Dr. John C. Fisher to determine in detail the quantitative effect of some typical lattice defects on tensile strength. For the case of alpha brass they show that the magnitudes involved are sufficiently large to account fully for the experimentally observed tensile strength of the actual crystalline material. This work has, for the first time, made possible a quantitative understanding of this most important property of crystals. It would be difficult to overestimate the value of this work, for it seems certain that great progress may be expected from a full development of current studies of crystal defects.

The study of the growth of crystals has provided a particularly good opportunity to observe the role of defects and dislocations. When crystals are grown from solution it is observed that the growth rate may vary widely from one crystal to another of the same chemical composition, and this difference can be explained by the relative abundance and character of the dislocations present. A particularly important defect in this case is the lattice displacement known as an edge dislocation. This defect arises from the displacement of many layers of atoms relative to their neighbors, on a line perpendicular to a plane of the lattice. Consider the function of this edge dislocation in a crystal experiencing growth from a supersaturated solution. Atoms from the solution migrate to the crystal where they are held by surface forces, but these are particularly weak on an atomically smooth plane of a perfect crystal. In the neighborhood of the edge dislocation, by virtue of the geometry, these forces are very much greater, with the result that practically all of the crystal-growth phenomena take place at these points. The abundance or scarcity of such defects can thus account for a vastly different rate of crystal growth in different crystals of identical chemical composition, the most perfect crystals experiencing the slowest growth.

The relationship between an edge dislocation, and its geometric result—a spiral growth pattern—was first recognized theoretically, by F. C. Frank at the University of Bristol. The first observation of such a spiral, after Frank's prediction, was made by L. J. Griffin on a crystal of beryl from the British Museum. A. J. Forty of Bristol first observed the spiral growth process with cadmium iodide.[1] The crystal growth around the edge dislocation leads to a spiral growth pattern, and when this was first realized, many natural and artificial crystalline materials were scrutinized for growth spirals. Such patterns have been found in many substances in addition to cadmium iodide—for example, natural graphite, silicon carbide, gold, silver, magnesium, paraffin and beryl—from which it must be concluded

[1] Some pictures of this interesting experiment, taken in our laboratory, were shown at the meeting, together with motion pictures of the growth of cadmium iodide crystals by means of growth spirals and color photographs of growth spirals on the surface of silicon carbide.

that the edge dislocation is a practically important type of defect that is involved in many crystal phenomena. Some particularly striking growth spirals are found on ordinary commercial silicon carbide crystals of the kind employed for industrial grinding and abrasive purposes. Although pure silicon carbide is a clear transparent substance, the material produced in the industrial electric furnace has a bluish-black sheen due to surface interference colors, by virtue of which the growth spirals are revealed in a striking manner.

Frontiers in Applied Science

Centennial Forum "Tomorrow and Science," Omaha, Nebraska, March 10, 1954.

A century is a long time when measured in terms of the life span of a human being. It becomes a much shorter unit when recording the history of a nation. And, when viewed as a stage in the progress of mankind, a century shrinks to such proportions that only the most significant highlights remain to give the picture perspective.

The changes in the American way of life that have taken place over the past hundred years are vital and far-reaching. That this hundred years has been a significant period in the history of our nation no one here will question. But who can assess the importance of this one century in the total course of civilization? Although the American way of life has undergone profound changes, the way of life of vast numbers of people throughout the world remains essentially unchanged. To what, then, can we point with confidence as the contributions of this period of a hundred years? What will the critical light of historical perspective eventually reveal as the highlights of this century?

It seems to me that one distinguishing characteristic of the past hundred years is the *technological awakening*—the birth of an era of technological progress that has not yet encompassed all the peoples of the earth, but that will surely do so as its benefits multiply. I think that it is a fact beyond contradiction that the past century has witnessed a greater extension of the control of man's environment by man than any comparable period in history. Scientific research itself, which is the source of our knowledge of the laws of nature, predated the development of technology by more than a hundred years. But technology as the industrial application of science, engineering as a function of the industrial process, and all the specialized practitioners in the fields of technology, engineering, and industrial science—these things came into place and developed their logical role in the century just past. For reasons that will be brought out, I predict that the scientific and technological conquests of the next hundred years will far outdistance those of the last hundred years.

I should like to take a look ahead—to make a scouting, so to speak, of the future prospects in the application of science to human welfare. But, first let us take a backward glance just to fix our landmarks.

If my grandfather had made the journey from Schenectady to Omaha—say to participate in the laying of the cornerstone of the territorial capital in 1855—he would have been one of the first to make the trip by rail. For just one hundred years ago, in 1854, the Chicago and Rock Island Railroad was the first to reach the Missouri River. I made the trip by air in relatively few hours. Had my grandfather decided to go on to the West Coast, as I plan to do, he would have had to travel the rest of the way by covered wagon, and would have been many weeks en route. If some prophet in the wagon train had predicted that a mere hundred years later people would make the same trip in six hours by flying a mile above the earth's surface in the belly of a huge metal bird propelled by windmills, my grandfather would have muttered something about corn likker, which, I am told, was occasionally available in the club car of the wagon train. If the venturesome traveler leaving Omaha to cross the vast western plains by covered wagon could have had his secret wish, it probably would have been for an arrow-proof vest and a tomahawk-resistant hat—for those were pioneer days, and Omaha was a frontier town.

In this country, the frontiers imposed by geography are gone, and, in a similar way all over the world, geographical frontiers have been disappearing. I suppose some odds and ends of unexplored geography still exist somewhere on the globe. However, the modern explorer must live in the constant fear that he will discover, in addition to native artifacts, positive evidences of civilization, such as a faded copy of *Punch*, or a rusted tin can.

The frontiers of science and technology are being pushed back, too, but there is one remarkable difference here. *No one can see the limit to the frontiers of science.* Each new discovery leads to new mysteries, the solution of which leads to new discoveries, and so on, *ad infinitum.* For example, at one time, the indivisible particle of nature was thought to be the molecule. Upon study, it was revealed to be a complex structure built from atoms as the ultimate building blocks of nature. In time, these indivisible entities were revealed as complicated miniature solar systems—systems of electrons surrounding a central nucleus. For many years the nucleus was an impregnable unit—so compactly put together that no fragments could be chipped off. The multimillion-volt atom smashers ended that stalemate, and now the nucleus itself is showing a complex structure. The list of "fundamental particles" has grown from the original molecule to a present total of some twenty or twenty-five identifiable elementary particles, and no one knows for sure whether *any* one of them is really an indivisible entity, or a "fundamental" particle. And so, not only in the structure of matter, but also in nearly every direction in which the scientist and researcher looks, the frontiers of science appear to be, not boundaries, but long corridors of many doors, each of which opens upon a long corridor of many other doors.

Perhaps some examples will make this clear. Let us first take a look at the field of metallurgy. The use of native metals in the construction of weapons, utensils, and ornaments dates from prehistoric times. Indeed, man's use of metal is one of the mileposts of antiquity in the long history of civilization. Thus, metallurgy is one of the oldest of the industrial arts. A hundred years ago, the metal practitioner was the blacksmith who shod the horses and formed the iron tires for the overland wagon trains, such as were being outfitted in Omaha in the early years of her history. A critical factor in the success of those primitive caravans was the forging of axles of sufficient durability to withstand the long, rough journey. With the growth of the vast industrial development whose origin is symbolized by the invention of the steam engine have come great developments in metals and alloys; they literally pace our mechanical progress in a mechanical age.

In this connection, it is interesting to compare the typical locomotive that might have been seen in 1854 with the modern locomotives that ride the rails today. The steam locomotive of 1854 weighed 28 tons, developed 300 horsepower—not much more than a modern motorcar—and had a maximum speed that frightened no one. A diesel unit of today develops 1600 horsepower and weighs 120 tons, and such units are, of course, used in multiple, at speeds up to 100 miles per hour. The gas turbine locomotives now operating on the Union Pacific between Omaha and Salt Lake City are rated at 4500 horsepower, and weigh 260 tons. The power, endurance, and efficiency of these modern rail giants, as well as the performance of the machines that propel our aircraft, the huge turbines that generate our electric power, and the host of machines that use, transform, and generate power for industrial America all depend heavily upon the metallurgist and his metals and alloys.

And yet, in spite of astounding progress in adapting the metallic elements of the earth's crust to the physical needs of man, and the impressive science and technology that has been built up around this important subject, this whole field is now on the threshold of great new progress based on new fundamental knowledge of metals. Old doors have been unlocked, and exciting new corridors lead in every direction. They will yield new materials and new processes that will affect every technological area in which metals play a part.

I want to tell you about one of the very exciting developments in metallurgy that has just taken place. It comes under the heading of "perfect crystals," and it is not yet important in a *practical* way. We do not know how to put perfect crystals to use in any product for home or industry at present, yet they seem to be of great significance and are certain to lead eventually to new practical progress in metals and alloys.

These metals and alloys, as you may know, consist entirely of crystals. The chromium that trims your motor car, the steel in machines and structures, the copper in electric motors—all metals and alloys are crystalline aggregates. Hence, most of the problems of metals revert to the problems of the component crystals, and one of the most important properties of metals and their crystals is *strength*. It has, for years, been a great mystery as to why the theoretical strength of metal crystals, which is readily calculated to be millions of pounds per square inch, is, in the case of actual crystals, found to be at least a hundred times less. That problem has been solved, and we now know that the weakness was caused by crystal defects of a special kind. The crystals that we thought were so perfect, because of their obvious crystalline regularity, actually had atomic irregularities on an atomic scale, and these defects caused all the trouble. We have learned to make some really perfect crystals for the first time, and they have astounding strength. We have just made in our laboratory some perfect crystals of pure iron in the form of tiny needlelike whiskers that are stronger than any previously known metal or alloy. Finely divided iron, or tiny wires of ordinary iron, rust almost immediately upon exposure to air. But these tiny, perfect crystal wires of iron do not rust at all. The same atomic perfection that gives them strength also prevents oxidation.

No one knows today how to put these perfect crystal wires to practical use. We certainly cannot use them, at present, to support a suspension bridge across the Missouri River. But they were only discovered yesterday; and, in time, applied science and technology will find a practical use for this form of metal, for perfect metallic crystals provide a new and exciting dimension to metallurgy.

From the great strides being made on the frontiers of this science, one can predict that many new metals will come into practical use and that innumerable new alloys will be available to our engineers. The present gap between ceramics and metals, both crystalline aggregates, will be filled with new materials different from any available today. More important still, the basic structure of metals may change, because we have available today a great wealth of new knowledge of what makes metals strong.

Let me mention another corridor of science upon which many doors have been opened recently. This is in the field of semiconductors. The elements of nature which have become so important as semiconductors in the last few years—germanium and silicon—have been on laboratory shelves for years as curiosities. Germanium is not an abundant element, but it has been known for a long time as a shiny, brittle, and useless substance. Silicon looks much like germanium, but it has important uses in some alloys, especially in magnetic steels, and we use it also in a new class of chemical compounds known as silicones. Since it is the most abundant solid element in the earth's crust, it is too bad we cannot think of more uses for silicon. When you look at the vast supply of rocks between here and the Pacific coast, and wonder why we cannot find more and better uses for rocks as an abundant, raw material, you are really thinking of silicon, for most of the earth's rocks are silicon, tightly bound to oxygen.

The new use of germanium and silicon as semiconductors will not require much tonnage of these elements, but some very important things can be done with tiny bits of these materials. The vast electronics industries are built around the vacuum tube as the basic component, and in semiconductor devices we have, for the first time, a nonvacuum device which performs many of the duties of vacuum tubes. Some of these electronics chores it does better than a vacuum tube, some not quite as well. But transistors and diodes do not require a vacuum; they are very tiny and they are exceptionally efficient. As a matter of fact, the primary function of the semiconductor is to replace the vacuum. It sounds a bit strange, but the semiconductor really acts like a sort of solid vacuum that can be cut up into chunks. This fact imparts an entirely new look to many electronics functions and to the electronics devices that perform them.

What semiconductor devices do is destined to be very important technically and industrially in military equipment, in communications gear, in radio and television, in computers, and in a host of industrial and consumer devices. And all of this has been built on neglected elements of nature, which we have had in laboratories for years. It came about because skillful scientists turned the right keys and opened new doors on one of the many corridors of science. When you add to worthless silicon or germanium a bit of industrial technology based on fundamental scientific research, a new value is created. From this new value come new products to satisfy human needs, which lead in turn to new industries and new employment. Were there time, I would like to relate many more case histories about the workings of the technological process, because in almost every direction exciting new developments are coming into being.

One of the most fundamental objectives of applied science and technology is the production of power itself, because all industrial activity has a common denominator in power. Power is produced today from the free potential energy of water resources, or from the combustion of fuels in heat engines. Although no competitor to these power sources is in business today, some candidates appear on the horizon, moving in our direction. One of the most promising is power from atomic fission, generated through a reactor and a heat engine. This atomic power is making great strides, and will undoubtedly assume an important role in the next two decades.

On the more distant horizon are some theoretically interesting alternatives for power production. One failing of all heat engines is a basic limitation to efficiency. Of the energy put into any heat engine—turbine, diesel, gas engine, or what have you—only about one-third will come out as useful power. Some electrochemical means of generating power are free from this basic heat-cycle limitation. For example, the electrochemist's dream for power generation is the fuel cell, into which combustion gases are fed and from which electric power flows, as from a sort of a fuel battery. Theoretically, this device may be nearly 100 per cent efficient, but there are very serious technological problems in the way of its practical use today. It may be an important power producer in the future.

The photovoltaic cell, the semiconductor device that powers your photoelectric exposure meter, is in some ways an attractive power generator, since its source of power is free sunlight. In our laboratory, we have had some tiny electric motors running for about twenty years with free solar power generated from these semiconductor cells. Photovoltaic cells cannot be ruled out as a future generator of power, for we are going to learn a great deal about the semiconductors from which they are made in the years immediately ahead.

Of course, photosynthesis is an important mechanism for the utilization of free solar energy, and great progress is being made in the understanding of this basic process by which growing plants convert the energy of sunlight into food.

Beyond the spectacular progress that is being made in specific fields of science and technology—in metals and alloys, in semiconductors, in electronics with its manifold applications to so many human needs, in chemistry, biology, and medicine—beyond all of these is the most striking fact of all. That fact is that applied science in the form of the *technological process* is an inexhaustible source of new wealth. Somewhere in the study of mathematics, one meets for the first time the symbol infinity, as a mathematical concept of something which is immeasurably great—which is large without limit. The storehouse of nature seems to have that infinite dimension.

There appear to be no limits to the discovery of new laws of nature. Through applied science, this leads to new materials, methods, and processes which, developed as new industrial technology, lead to new machines, process capabilities, and new limits to efficiency, endurance, and output. Finally, at the end of this technological process are the new products which supply human needs and create new industry and employment. This technological process is certainly one of the highlights of the hundred years just past.

The next hundred years will be certain to see striking progress in supplying material human needs from the exploration of the limitless frontiers of science and technology.

The Science of Commonplace Things

A reprint from *Electrical Engineering,* American Institute of Electrical Engineers, September 1955. Essentially full text of a banquet address at the AIEE Summer General Meeting, Swampscott, Mass., June 27–July 1, 1955.

"Acres of diamonds" was one of the most famous popular lectures ever presented. Russell H. Conwell, pastor of the Baptist Temple in Philadelphia, Pennsylvania, and president of Temple College, wrote it about the turn of the century and subsequently presented it more than 6,000 times. Proceeds from this staggering example of oratorical endurance are said to have made it possible for thousands of young men to attend and graduate from Temple, men who would have been unable to afford college training except for "Acres of Diamonds."

This famous lecture took its title from a fable that Dr. Conwell told at its beginning. It is the story of an ancient Persian who lived near the shore of the River Indus. The Persian, it is written, "was contented because he was wealthy, and wealthy because he was contented." However, the story continues to say that,

> One day this old Persian farmer was visited by one of those ancient Wise Men of the East, who sat down by the Persian's fireside and told how this world was made. The Wise Man said the world was once a great bank of fog, and that the Almighty thrust His finger into this bank of fog, and began slowly to move His finger around, and then increased the speed of His finger until He whirled this bank of fog into a solid ball of fire; and as it went rolling through the universe, burning its way through other banks of fog, it condensed the moisture, until it fell as rain upon the heated surface of the world, and cooled the outer crust; then the internal fires, bursting the cooling crust, threw up the mountains, and the hills, and the valleys of this wonderful world of ours. "And," said the old Wise Man, "if this internal mass burst forth and cooled very quickly, it became granite; if it cooled more slowly, it became copper; if it cooled less quickly, silver; less quickly still, gold; and after gold, diamonds were made. And," said the Wise Man, "a diamond is a congealed drop of sunlight."

The basic physics of this particular process for making diamonds is somewhat obscure at several points, but we must not be too critical of the Wise Man because he was thinking in broad, cosmic terms, unencumbered by trivial details, and everyone knows that clever technicians in the laboratory can take care of the details, once the broad picture has been painted. And, as a matter of fact, until quite recently *stirring fog* was as good as any other process for making diamonds; it did not work either.

The story goes on to describe how the Persian farmer, who had been up to this point quite happy, decided he wanted to be not merely wealthy but immensely rich, so he sold his farm, and went away in search of diamonds. He wandered about the world, lost the money he had, and ended up throwing himself into the sea, in despair.

Meanwhile, the man who bought his farm one day noticed a curious flash of light from the white sands of the stream running behind the house. The piece of rock he took from the stream was, his neighbors told him, a diamond. He went back to the stream and found more and more of these beautiful rocks, and he became immensely wealthy. The story concludes:

> Had the Persian farmer remained at home, and dug in his own potato patch, instead of wretchedness, starvation, poverty, and death in a strange land, he would have found Acres of Diamonds.

This is a fable and of course has a moral; in fact, it has several morals. The one that fits this theme is that in research, as well as by the farm on the Indus, a careful look at the commonplace and familiar may reveal some exciting gems of scientific knowledge. Many everyday phenomena can be made to yield a glittering array of new facts under the scrutiny of modern scientific research.

PROMISE OF SILICON

Less than a decade ago, the development of the first transistor by the Bell Telephone Laboratories stimulated the growing interest of scientists in semiconductors. The first transistors depended on a relatively rare element, germanium, and the detailed study of this substance opened the door to many practical applications of semiconductor devices. Scientists immediately began a systematic study of other elements that might show semiconductor properties. In the General Electric Research Laboratory we decided to put a major effort into the study of silicon.

Silicon is literally in everybody's front yard. It is the most abundant solid element in the earth's crust. The vast supply of rocks on this planet is mostly silicon tightly bound to oxygen.

The new use of silicon in semiconductors is converting this almost worthless raw material to a substance as valuable as the semirare element, germanium. Thanks to recent attainments in the preparation of extremely pure silicon, it is now apparent that this element may at some future date even outclass germanium as a useful semiconductor.

Silicon has been made in our laboratory with an impurity content of only one part per billion. This is achieved through the "zone melting," or recrystallization, method. Current practices of zone melting consist of repeated recrystallizations of silicon as the ingot is slowly drawn through a gas-filled quartz tube. Induction coils are wrapped around the outside of the tube, causing the slowly moving ingot to melt in narrow zones. In this way impurities are swept to the end of the bar, leaving behind silicon which is of highest quality.

Evidence of the new level of perfection in silicon crystals is a quantity called "lifetime." This is not how long the crystal itself will last, but the length of time that an excess of electrons, purposely injected, takes to disappear. Until fairly recently, the "lifetime" of the best silicon was a ten-thousandth of a second. However, in the new crystals, "lifetime" is in excess of a thousandth of a second. This is not long by the clock, but to conduction electrons, it is practically forever. To semiconductor scientists this is one of the most exciting measurements of recent years.

This same common element, silicon, is the basis of an important new segment of the chemical industry which is rapidly coming of age—the silicone chemical industry. Probably no new material in history has been applied by engineers to so many diversified products in such a short time as these new chemicals called silicones.

Basic research on the silicones continues at an accelerated rate as new ways are found to use these wonderful rubbers, resins, and oils. Resistance to high and low temperatures is an important property of silicones, a property which was predicted by scientists who pioneered in this field of chemistry. However, in addition, many remarkable properties were discovered which were completely unpredictable and unexpected. For example, a drop of silicone added to lubricating oil prevents foaming; a drop of silicone on a bread pan or tire mold prevents sticking; while a few drops on fabric, leather, glass, or ceramics impart remarkable water repellency. Silicone chemicals will continue to provide a succession of surprises as new members of this numerous and versatile family come to light in the laboratory.

We have found that under conditions of intense corona, the durability of silicone rubber is unique. For example, it will last for thousands of hours where any other natural or synthetic rubber will fail in minutes. The mechanical strength of silicone rubber grows each year as new structural secrets are discovered; until now it is in the range of the best natural and synthetic rubbers. This has been thrilling progress because the first silicone rubber we saw in the laboratory was about as strong as wet newspaper.

We have long wanted a silicone rubber with a handle, a chemical handle that would allow us to join it to other elastomer molecules. We now have a silicone with a vinyl group that serves just this purpose, and it can be copolymerized with other synthetic and natural rubbers in any proportions.

FRACTURE PROBLEM IN METALS

For commonplace phenomena in everyday experience, it would be difficult to find a better example than the property of all materials of breaking under stress. Fracture in a variety of forms imposes an ultimate limitation on the performance of a vast number of structural and mechanical bulwarks of modern civilization—engines, motors, generators, turbines, trains, airplanes, bridges, buildings, and machines. There are few subjects more important and basic than the study of the strength of materials, which comes to focus in the study of the phenomena of fracture.

Economics exerts a relentless pressure on the designer of mechanical things to approach ever nearer to the limit of the strength of his materials of construction, or even to exceed this limit, as in the case of World War II vessels.

Of the 1220 ships of one class built during 1941–1944, a substantial number of them broke in two, and more than 400 developed serious cracks.

It has been customary to distinguish several different types of fracture in metals on the basis of the appearance or the conditions that caused it. We talk of cleavage, or brittle fracture, shear fracture, high-temperature rupture, and fatigue fracture. It now appears that all of these "types" have one feature in common, namely, that all are preceded by some plastic deformation resulting from slip, in which the parts of a crystal slide over each other on definite crystallographic planes. It seems likely that the slip starts from an area of crystal imperfection—the first nucleus of fracture.

Actually, until recently little has been known of the fundamental reasons why metals break. Now scientists are studying the basic phenomena in great detail and the outlines of the picture are becoming increasingly clear. Dr. J. R. Low of our laboratory recently traced the complete history of the development of fracture from its origin at an atomic level to the final failure of the metal specimen, and this study is very revealing. The final break was preceded by a tiny but microscopically visible crack which appeared at only 10 per cent of the ultimate strength of the material. So the metal which, when perfect, fails ultimately at 50,000 pounds per square inch, say, had generated visible seeds of failure at only 5000 pounds per square inch stress.

If we follow the tiny crack back to its atomic origins, we find a crystal dislocation, or irregularity. A group of these is shown in Figure 1 in the microscope photograph of a region of a crystal which subsequently developed a crack. The tiny pits were formed by etching, and serve to locate irregularities of atomic dimensions in the metal crystal. The irregularities in the arrangement of a few atoms are revealed by the etch pattern; they are the ultimate cause of failure of a mechanical structure. Hence these atomic irregulari-

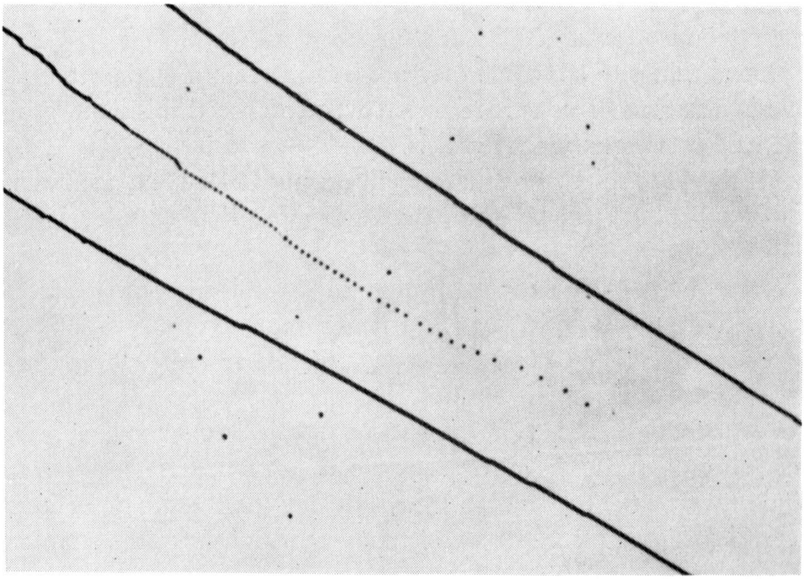

Figure 1. Microscope photograph showing crystal dislocation.

ties, or dislocations, have become a key of great importance, and are being studied intensively.

If we could make crystals without atomic irregularities, perfect crystals, they would be about 100 times stronger than imperfect ones. We have succeeded in doing that during the last two years, and now have perfect crystals of zinc, copper, molybdenum, tungsten, iron, and many other substances. These perfect crystals of pure metal will support up to 1,000,000 pounds per square inch in tension, and are stronger than the strongest alloys, but thus far we can only make them in the form of tiny filaments. We hope they will point the way to new levels of strength in practical structural materials.

SOURCES OF POWER

Sunlight, uncongealed, is another familiar thing in our life, and, for years, scientists have sought a practical way of obtaining electric energy from sunlight. This is an especially intriguing target, for, like water power, the reward is free energy—"free" in the sense that there is no fuel cost. Within the past year, interesting, but still small-scale, solar power sources of substantially improved efficiency have been demonstrated. The home and industrial use of electric power from sunlight is still a long way off, but it is an intriguing area for basic research into something close at hand.

There is, of course, another power source much in the news these days; but the subject of this article is how scientists are digging into common, familiar things, and for all its tremendous importance, atomic energy still is produced by elements that cannot yet be classified as common.

Related to the studies of semiconductors and solar power are a whole group of important new scientific projects in the field of light itself. A recent dramatic development is the direct light amplifier—a phosphor surface that actually emits more light energy than strikes it, and does this proportionately so that a picture image projected on the surface is brightened many times without loss of tone gradation or sharpness.

Direct amplification of light has been a long-sought scientific goal, but the basic idea of what it means to amplify light is sometimes difficult to explain. The easiest demonstration involves projecting a regular black-and-white lantern slide on the special phosphor film, using an ordinary lantern slide projector equipped with an ultraviolet lamp. The light amplification cell itself consists of a thin film of a special phosphor between two conducting layers, the front layer being transparent. Everyone is familiar with turning on a projector, but in this experiment, in addition, one "turns on the screen," and thereby obtains a great increase in brightness. Very little current is required, and small dry batteries are more than adequate.

Before the screen is turned on, the image is very dim. The picture is brightened not by increasing the intensity of the projector lamp or by opening up the projector lens; rather, it is done by applying a voltage to the screen itself. The brightness increases gradually as the voltage is turned up.

Many who see this for the first time ask, "What is it good for?" This is, of course, a very important question, but scientists, by experience, know that predicting the future application of new basic knowledge is highly speculative. Light amplification in a single phosphor film is a fundamental scientific achievement; in this regard it is very similar to Lee De Forest's discovery of the 3-element vacuum tube amplifier in 1906, which made it possible for the first time to amplify electronic currents. At that time De Forest was not only asked, "What is it good for?" but also, seven years after the discovery, he and his associates were brought to trial for using the mails to defraud. They were accused of selling stock "in a company whose only assets," according to the Government prosecutor, "were De Forest's patents, chiefly directed to a strange device like an incandescent lamp, which he called an 'audion,' and which device has proved to be worthless." In summing up his case, the prosecutor stated that De Forest had said ". . . in many newspapers and over his signature that it would be possible to transmit the human voice across the Atlantic before many years. Based on the absurd and deliberately misleading statements of De Forest, the misguided public, Your Honor, has been persuaded to purchase stock in his company."

There eventually may be many practical applications of the light amplifier, including the so-called "picture-on-the-wall" television set, but it is no more possible to predict these with certainty now than it would have been to foresee the extent and ramifications of modern radio and television when the first audion tube was invented fifty years ago.

MAN'S ATTEMPTS AT DIAMOND MAKING

Allegorically speaking, the items which have been discussed—pure silicon, vinyl silicone, perfect crystals, solar power, light amplifiers—are all diamonds of the type the Old Wise Man from the East said we might find by digging in the right places.

Some fundamental digging in the right places also has produced some real diamonds in our laboratory. Diamond is a crystalline, transparent form of carbon, and carbon is a very common substance, so that in this case, too, research has added glamour and value to a very ordinary material.

The story of man's attempts to manufacture diamonds is one to capture the imagination. It involves famous chemists as well as charlatans, claims of success and admissions of failure, simple gadgets and striking experimental techniques; all in more than a century of confusing but fascinating history.

Many of us were taught in early science courses that diamond could be made from carbon under high heat and tremendous pressure, and many reference sources credit Moissan with accomplishing this feat in 1894. But a well-known mineralogist, writing for an encyclopedia, has said, "For many years it was thought man had made small diamonds in the laboratory. When, during World War II, industrial diamonds became of strategic importance, these claims were thoroughly investigated in the hope that the natural supply from the South African mines could be augmented. All claims advanced prior to 1950 that diamonds had been made synthetically by "artificial" methods have been demonstrated to be false, and many government and industrial research laboratories both here and abroad are still working on the problem. If it should become possible to produce in quantity a comparatively cheap form of synthetic diamond, tremendous strides could be made in the machining and working of metals and glass."

The early attempts to make diamonds are interesting enough in themselves to deserve attention. Henri Moissan dissolved sugar charcoal in molten iron and quenched the solution in cold water in order to crystallize the carbon under the great internal pressure supposedly generated by contraction as the mass cooled from the outside. Upon dissolving the metal from the solidified melt, he found traces of a hard transparent material in the residue. The famous French scientist assumed he had made diamonds. We now know that the pressures required could not be attained by this method.

Following Moissan, Sir Charles Parsons spent over twenty-five years of extremely diversified experiments, including the firing of a high-powered rifle into a tiny hole at the range of six inches. In 1920 he believed he had confirmed and extended Moissan's experiment.

In 1924 Mellor's "Comprehensive Treatise on Inorganic and Theoretical Chemistry" said that laboratory production of diamonds was well authenticated. Nevertheless, the story was not universally accepted. And Parsons himself at about that time began to entertain doubts about his own earlier identifications of what he thought were diamonds. He undertook a scrupulous re-investigation and submitted his new evidence to the editor of *Nature*. He told how he had been misled into regarding as diamond various transparent, singly refracting minerals which were very resistant to chemical reagents and would not burn. These so-called spinels have undoubtedly been the basis for many claims of success which later proved unfounded. Spinel crystals are, in fact, the "fool's gold" of diamond-making history. It was Parsons' final conviction that neither he nor anyone else had ever succeeded in making diamond in the laboratory.

Another Britisher figures prominently in diamond-making history. J. B. Hannay reported in 1880 that he had made diamonds by heating a mixture of hydrocarbons, bone oil, and lithium at red heat in sealed wrought-iron tubes. Parsons said in 1920 he could not repeat Hannay's work. By 1950, Professor N. V. Sidgwick of Oxford, in his "Chemical Elements and Their Compounds," reviewed the pro's and con's of Hannay's claims and concluded that the artificial production of diamonds had never been shown to have been a success. He added that recent thermodynamic calculations which yielded the temperature and pressure requirements showed them to be so extreme that it was very improbable that any of the efforts to produce diamonds thus far could have succeeded.

These recent studies of thermodynamics provided our research team with the springboard they needed to undertake the diamond-making project some four years ago. The carbon-phase diagram of Figure 2 will aid in understanding the problems involved, since the diamond phase of carbon is thermodynamically stable only at very high pressures. Here pressure is plotted vertically, temperature horizontally. Parts of this diagram have been established very accurately by experimental measurement, while some parts of the diagram rely entirely upon thermodynamic calculations. There is very little direct experimental basis for the boundary line between the graphite and diamond-stable regions, particularly in the high-temperature end of the curve.

The upper limits of pressure and temperature that scientists were known to have reached in controlled experiments are shown by

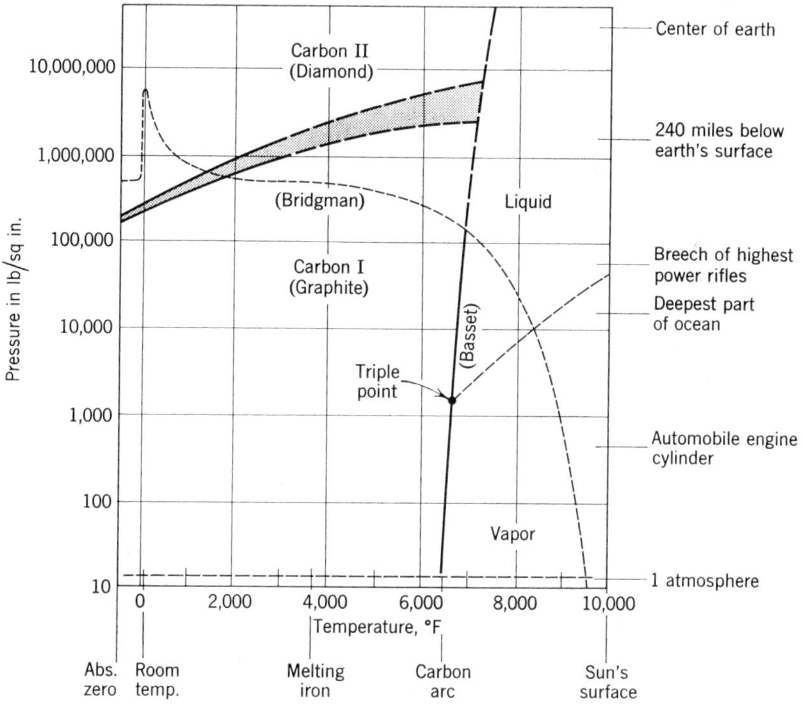

Figure 2. Phase diagram for carbon.

the dotted line running across the diagram. Professor Bridgman of Harvard University is the outstanding investigator of superpressure phenomena, and the region he has studied is shown, while Basset of Paris has been the most active worker in the very-high-temperature–moderate-pressure region.

Here a limitation of the phase diagram must be pointed out. The diagram shows pressure and temperature, but not time, and in phase transitions, time is a very important quantity; unfortunately, stability calculations do not yield the time required for the phase transition to take place. As the diagram shows, graphite is the stable form of carbon at room temperatures and pressure. All natural diamond thus exists as an unstable crystal form and at ordinary pressure it should really revert to graphite. The diamond lattice is pretty firmly frozen in place. It will thaw out, however, when heated to about 1700°F, and it then converts to graphite. So the phase diagram delineates the regions of stability for the various forms of carbon; it does not guarantee that the attainment of the temperature and pressure corresponding to the diamond phase will cause graphite to convert to diamond. The temperature may be too low to shake the carbon atoms loose and permit them to regroup in the diamond lattice; or too much time may be required for this to take place.

From a thermodynamic standpoint, it appeared in 1950 that there was a fair possibility of making diamonds in the region above 1,000,000 pounds per square inch and several thousand degrees F, provided these conditions could be reached, and held long enough. The job was to develop apparatus capable of operating at these levels, not just momentarily but for extended periods of time. With the attainment of a new region of high temperature and superpressure as an objective, our team of scientists began an all-out attack on the problem.

A large hydraulic press was used in this research but this size tends to be misleading. The size of the press, or the force it is capable of exerting, has very little to do with the maximum pressure that can be generated. The pressure vessel is the limiting factor.

Professor Bridgman had done considerable work on using multiple shrink bands similar to those employed in making large gun barrels. By making use of some of the known pressure methods of Bridgman, and by developing some new ways of distributing stress and giving support to critical parts, our research group succeeded in making a pressure vessel that can be operated up to at least 1,500,000 pounds per square inch. This pressure has been increased to 2,700,000 pounds per square inch. Some completely new and ingenious techniques permitted them simultaneously to achieve temperatures up to 5000°F. Finally, they found that they could maintain these tremendous combined temperatures and pressures—equivalent to those existing some 240 miles beneath the earth's surface—for prolonged periods of time.

Our first Man-Made diamond is shown in Figure 3, with a diamond-tipped phonograph needle present to give an idea of its size. It is about 1/16 inch long. Most of the thousands we have now produced are even smaller, but all are well within the size range needed for the bulk of industrial uses in cutting, grinding, and polishing. Some are clear white; others have the wide variety of colors found in the diamonds dug from the earth. They are, in fact, identical to natural diamonds. They are not "imitation" diamonds or "diamond-like"; they are purely and simply diamonds. All of

Figure 3. (Left) First Man-Made diamond, with diamond-tipped phonograph needle to indicate its size.

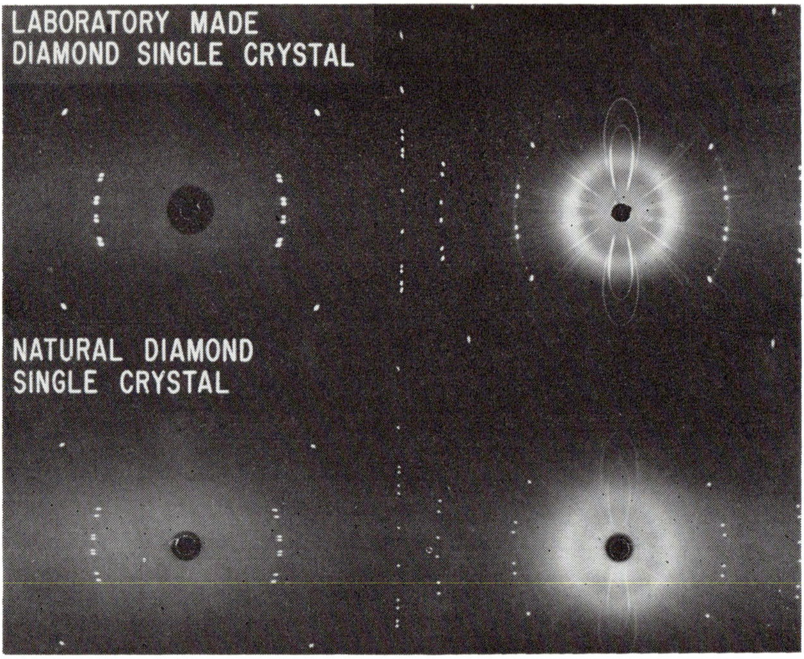

Figure 4. (Right) X-ray diffraction pattern for diamond.

the variations in color, crystallinity, and clarity found in natural diamonds are observed in the laboratory product.

The X-ray diffraction pattern provides a simple "fingerprint" for diamond, as shown in Figure 4. The internal arrangement of the carbon atoms in the diamond crystal, not the external appearance of the crystal, produces the characteristic X-ray diffraction pattern. At the top is the pattern for a Man-Made diamond and beneath is the pattern for a natural diamond. The process by which these diamonds have been made has by now been repeated hundreds of times with successful results.

The fascination and beauty of diamond gem stones sometimes obscure the real importance of this unique substance. Diamond is, by a wide margin, the hardest substance known. Therein lies a tremendous potential for manufacturing industries. Cutting and grinding operations are basic to a large segment of industrial manufacture, and the speed and efficiency of these operations are a multiplier which exerts great leverage on the American economy.

This becomes increasingly true as industrial technology produces ever stronger and harder materials in support of the inexorable competitive demand for higher temperature, higher speed, higher stress, and higher mechanical performance.

In addition, diamond is a strategic material of great importance. We presently depend entirely upon the supply from Africa, which produces 98 per cent of all of the world's diamonds. From this source there is not enough naturally occurring diamond to permit widespread use in industry, and this scarcity supports the present price of $7,000 per pound for the lowest industrial grades. For industrial cutting stones, the price runs up to $1 million a pound.

Although photographs of Man-Made diamonds have considerable beauty of their own, to the scientist the most beautiful picture to come out of our diamond research project is the simple graph of Figure 5.

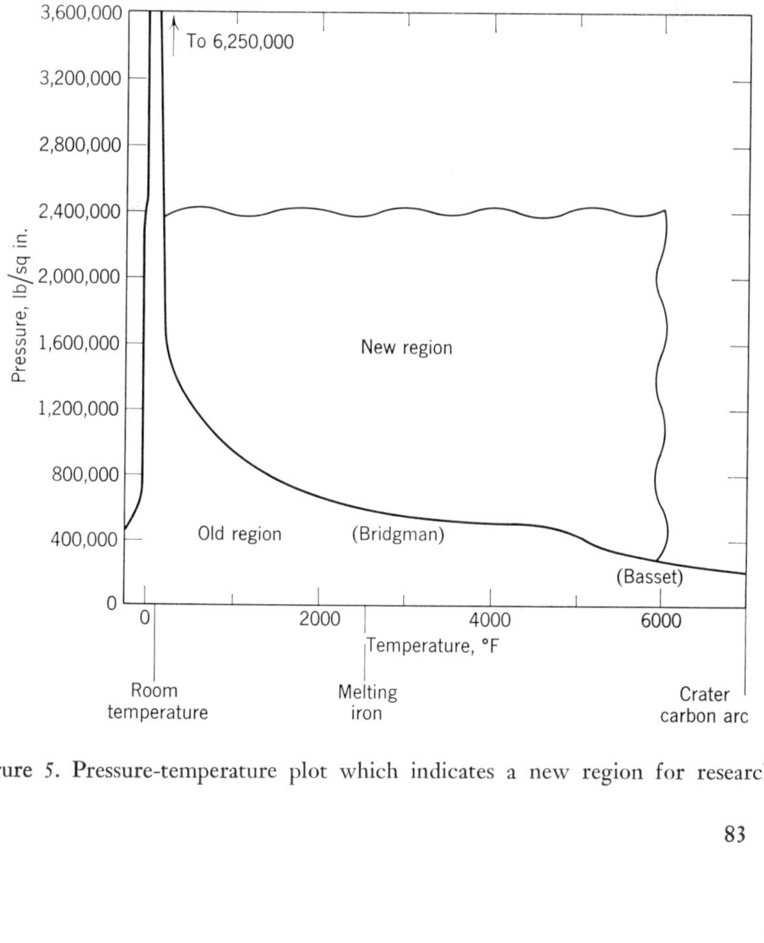

Figure 5. Pressure-temperature plot which indicates a new region for research.

It will be obvious that substances other than carbon can be subjected to these never-before-attained pressures and that pressure-temperature-time cycles never before available now lie before us. We have available a large new and exciting scientific frontier. Acres of diamonds could hardly have a more thrilling interpretation. From these new acres we may be able to extract scientific gems of even greater value than diamonds.

Recognition of the Scientist

An address given at the 1956 spring meeting of the Industrial Research Institute at White Sulphur Springs, West Virginia.

The most important element in the motivation of successful scientists is undoubtedly curiosity. The overwhelming desire to learn and understand the secrets of natural phenomena provides a drive that cannot be matched by other stimuli; indeed, history is replete with examples of brilliant scientific achievement in which natural curiosity was the primary motivation, if not the sole tangible motivation, for inspired creative work. However, the great importance of curiosity in the achievement of scientific progress is likely to obscure other motivations that are basic to human progress and of practical significance to the individual. Important in this context is recognition for the individual scientist, the so-called individual contributor, who is engaged primarily in technical work and who wishes to continue in such work in preference to possible alternative paths of progress in management and administration. Although this factor is but one of the many elements that together add up to a favorable environment and climate for creative work, it is an important one well worth careful study, discussion, and resolution. I propose to examine this matter in its professional, organizational, and local-community aspect.

For the professional man, scientific or otherwise, recognition and identification by his professional associates and by the professional organizations in his field are generally high on the list of attainments to which he aspires. Although my point of focus is the scientist and some of his environmental considerations, because they are somewhat familiar to me, similar remarks apply with little change to the engineer and to professional specialists in a host of fields. For the scientist, recognition by the scientific community is undoubtedly at the head of his list of professional aspirations. This is true, not only for reasons involving the satisfaction of identification with his professional field, but also because, as I will try to show, the exposure of one's scientific work to the appraisal and evaluation of other scientists serves an important scientific need for the specialist.

It is a fact that the appraisal and evaluation of scientific progress—for example, a new discovery—is a matter of some difficulty. Frequently many people of ability and experience contemplate a new development in science for a long time without comprehending its real significance. This is illustrated by many examples, but consider for a moment the time interval between the discovery of nuclear fission and the development of the atomic bomb, which was eight years. However, only three years of the eight-year period were concerned with the actual process of building the atomic hardware. A good deal of the balance was required to comprehend the full consequences and portent of this significant discovery; and I suppose it could be said with some conviction that substantial uncertainties still exist on that score.

Many other illustrations could be cited. For example, we now know how to make perfect crystals in the laboratory, but I doubt whether anyone can say for certain what the real significance and future importance of that discovery will be. At present the great strength of perfect crystals seems to be important—we have iron crystals with a tensile strength of nearly two million lb/in^2, which is far greater than any known alloy and immensely greater than any previously known form of iron. In addition, however, perfect crystals are really a new form of matter. Because of their lattice perfection, they evaporate or dissolve with great difficulty, and are quite inert chemically. They may be bent, and upon heating they will resume their original shape. To foretell which of these unique properties will be of future importance is quite a speculation at present.

As another example, I think that the discovery by Langmuir and Schaefer of cloud seeding, and thus artificial rainfall and the newly developing field of experimental meteorology, is one of the outstanding scientific advances of the last half century. The President's Advisory Committee on Weather Control has issued an authoritative and very favorable report which confirms the scientific reality of this important, if controversial work. I think, however, that the many fascinating results of this work could, and probably will, lead into quite unpredictable directions in the future. Can anyone say that we really understand what this discovery portends?

I started out to say that recognition in the world of science is an important aspiration of the scientist, but that beyond personal satisfactions, such recognition has a deeper functional significance, which I hope will now be plausible.

It is difficult for a scientist, because of his proximity to the work, to properly judge a new scientific result in relation to the work of others, to progress in the whole field, and to important progress in related fields or to fields that are competing in scientific interest with the author's field. The professional, scientific, and technical societies comprise in aggregate a supreme court of informed scientific opinion. The verdict of this court concerning the importance and significance of scientific work generally transcends the opinions of the author or the sponsors of the work. To be sure, the verdict may not be explicit or prompt, but it is generally final and inescapable and of substantial significance to the career of the author and to the reputation of the sponsor.

I would like to point out that the participation of a scientist as a contributor to the work of the technical societies also develops some important advantages for the sponsor. The sponsor of the research, whether it be an industrial, university, or government laboratory, shares in the recognition accorded the individual; and this is an important corollary benefit to both the scientist and the sponsor. From the standpoint of attracting capable people to the sponsor's organization, it is doubtful whether any public relations activity equals the effectiveness of the inherent public relations benefits that result from a policy of not only permitting, but also encouraging, the prompt publication of scientific work.

For the reasons given, the scientific and technical community plays a functional role in the progress of science and scientists. Beyond this, however, the sponsor of research should have in mind that, with practically negligible exceptions, modern technical progress is built on prior work. It is the gradual advance by progressive steps, each built upon a foundation of previous progress, that is the almost invariable pattern of scientific and technological gains. To be sure, the glamour of the individual brilliant discovery tends to obscure this fact of research life. Patents, of course, are always granted for "improvements," since prior art is always disclosed by an examination of the patent files. The obligation of the research sponsor to contribute to the common pool of knowledge upon which future progress will be built can be justified on purely selfish grounds; for he thereby encourages the mutual stimulation which is an important part of professional activity. Communication is always difficult if the technical gap is wide; hence the technical man needs to communicate with those working in the same field, wherever they may be located. The give-and-take associated with the activities of the scientific community is seldom a one-way street, but is rather a mutual opportunity for learning.

The thesis has been developed here that curiosity is a basic source of motivation for scientific research; indeed it may be assumed to be an important attribute of all successful investigators. Beyond this basic qualification, recognition in the professional and scientific sphere plays an important role in motivation. It is pointed out that the significance of professional recognition extends beyond questions of prestige and acceptance and has deeper roots based on the appraisal, evaluation, and development of scientific perspective which are inherent in this process and are of great value to the scientist. We believe that under the heading of recognition for the scientist the opportunity to participate in the activities of the scientific community must therefore be very high on the list.

Although I would like to look in detail at the many problems of the research environment related to recognition, there is hardly time to do so in a comprehensive way here. Some of these problems relate to the intangible factors of environment and atmosphere which are determined by the traditions and the philosophy of the organization. Others are concerned with the ever-present tangible symbols of status and achievement. In the latter category, the question of titles, particularly for individual contributor members of a research organization, lies squarely in the middle of this picture, and it would be safe to say that this subject merits discussion.

The practice with respect to personnel titles varies widely in successful American laboratories. Some laboratories in industry, for example, have no titles whatever for individual contributing members regardless of age, experience, or professional status. Other organizations, equally successful in their fields, employ a complete array of graded titles with matching job descriptions or—to use a somewhat more dignified term—position guides. Apparently there are many ways to skin a cat, and the real question is what way is proper and appropriate for a given research organization.

The question of titles—and here, as in the foregoing, I have in mind titles for individual contributors and not managerial titles—cannot be resolved out of context and must be decided in relation to other elements of recognition available in the research organization. An internally conferred title—for example, "senior research associate"—is a recognition by the organization of professional status and attainment. It is my own present belief, subject to possible future change, that the need and usefulness of internally conferred titles are the greater in research organizations where restrictions exist with respect to scientific and technical publication and ready access to members of the profession outside of the organization. The researcher may be working in a classified field where publication is impossible or infrequent, or he may be working in an industrial research laboratory where the company confidential nature of the work precludes prompt publication, or he may be working in a highly specialized field of technology where there has been no tradition or practice of publication whatever. Internally conferred titles may serve a very important need in such cases, and if they are functional and widely administered, and if the reputation of the sponsor is good, they will achieve a desired acceptance outside of the organization.

In considering the desirability of internal titles and their acceptance, the policies and practices of the company outside of the laboratory (if in industry) must be carefully considered and evaluated. It may be necessary or desirable to confer an internal rank on research personnel which will be meaningful in relation to company practice.

Internally conferred titles will not, in my belief, serve in lieu of, or partly in lieu of, salary. In fact, in spite of much thoughtful effort, no good substitute has been or is likely to be discovered for salary and other direct compensation. I have forgotten which wise man put it this way, but he said, "Whether you're rich or whether you're poor, it's always nice to have plenty of money." Since the adequate treatment of this important fundamental subject would justify a separate comprehensive discussion, there is hardly time to do more than state a few conclusions. One conclusion is that a sound compensation plan, together with an equitable and practical method for its administration, should be the starting point in any consideration of recognition for technical people. Furthermore, those affected by the plan should know what the plan is and should understand its theory and approve of its practice. Basic deficiencies here are not likely to be compensated for by titles, or by reserved parking space, a private office, engraved stationery, or a coffee break twice a day. In other words, this subject *should* be considered out of context, and with little relation to prestige privileges.

But to return to a serious consideration of internal titles, it seems to me that the local ground rules with respect to publication policy, relations of the laboratory to the company operation and company practices and needs, all enter into the structure. In the laboratory with which I am associated we have thus far de-emphasized titles. At present nearly all members of the research staff are called research associates, and we are likely to change to a simpler, more functional designation in the future.

There are some other aspects of organizational recognition that I hesitate to mention because in our laboratory we have had little working experience with them. Many research organizations make substantial monetary awards in cash, or a medium of exchange, for outstanding work. In some cases, this may be for work that leads directly to new products or product improvements, and in other cases for outstanding patent results in important areas of technology. The feasibility of making such awards must be greater as the technical work partakes more of the character of an engineering development carried out in close proximity to product and marketing plans. I am sure that such awards would be difficult to administer in the laboratory with which I am associated. This is because of the long-range nature of much of the work and the fact that its importance lies in its broadly supporting character rather than in product development.

With respect to patent awards, either associated with the filing of patent applications or their subsequent issue, I feel that moderation is appropriate. A separate monetary award for patents presumably has as objectives: (a) the encouragement of patent filing and patent consciousness, and (b) the recognition of productive inventive activity. The encouragement of patent activity, in our experience, can be accomplished by straightforward management effort; at least in industrial laboratories it is an integral part of the planning and execution of creative technical work. If the purpose of the award is recognition, there is implied an ability to evaluate the present or future worth of patents granted. To make that evaluation equitably and fairly with respect to the multiple contributors often essential in a complex technology presents problems which we feel limit the effectiveness of such awards.

Little has been said about the importance of recognition of the contributing specialist at a community level, and I think there can be no doubt that this area of opportunity is frequently neglected. It is a fact that the knowledge of the accomplishments of one's neighbors frequently reaches us by very circuitous routes. This is especially true if the achievements are those of scientific and technical specialists. The work of our scientific or technical neighbors is generally presented at technical meetings in distant cities, and always requires the most skillful interpretation and translation to become newsworthy. The problem is still greater if, for any of the reasons mentioned, the individual is not in a position to publish the results of his technical work, or if these results appear principally in the patent files. Similarly, some extremely valuable members of research organizations contribute principally through consultation and counsel. Although their advice is much sought and the results readily apparent to technical associates, there may be little or no "paper product" that could serve as the basis for interpretive publicity concerning their work.

In any event, recognition by the community, friends, and neighbors should be a part of the objective of any overall program in the employee relations area. I think it is clear that internally-conferred titles may play a role here. In addition, however, interpretive news reports, company advertisements which feature personal achievement, suitable public announcements on the occasion of promotions, and recognition of personnel milestones such as twenty-five years of service and the like will serve to develop these objectives in the local community.

Prior to writing down my thoughts on the subject of recognition, I consulted with many of my associates concerning their feelings and convictions, and I am grateful for many thought-provoking suggestions, not all of which I have been able to include here. However, a number of my colleagues presented the idea, which was new to me, that we ought to consider the problem of recognition by one's immediate family. This suggestion has in it some provocative possibilities, and I realize that some of the things we have been doing already provide progress in this direction. For example, once a year we invite to the laboratory the families of scientists who have come with us during the preceding year. Members of the research staff and myself tell about the work of the laboratory and its relation to the company. In this way we help our people to get acquainted with each other. These meetings are appreciated not only by the family but also by the scientist, who finds that his status and prestige on the home front experience a substantial, although temporary, enhancement. If this transient aura that comes from such meetings, and from news reports of father's work, or his recent promotion at the laboratory, could be sustained at home it would have enticing possibilities; but I have no such speculative hopes. The world of science, with all of its glamour—electrons and holes, crystal dislocations and defects, superpressure carbon phase changes, flow phenomena at Mach 12—this is a world that touches upon everyday reality at only a few points. But when one has to touch down on reality, it might as well be earthy and solid, if need be at the kitchen faucet, "which is still leaking, dear."

The local ground rules at home, and especially the appraisal factors for achievement, are quite different from those in use in the laboratory. In the world of science, we consider that the ability to understand, and especially to explain, natural phenomena is high on the list of attainments. However, this cuts little ice at home. If the latest chrome and stainless addition to the household machinery develops an irritating buzzing sound, the complete technical description of the source of this odd phenomenon will elicit very little interest. To gain character and prestige in this league, the kudos go to the man who has a way with a screwdriver; better yet, who can eliminate a buzzing noise with a few well-placed whacks on the cabinet. It is of course desirable for him to carefully conceal his amazement at a successful result.

The theory of electricity receives but lip service on the home front, but if you are good at the practice and can fix a reading lamp that ought to have survived a cocktail party, but did not, you are in the *upper echelon* and can take your meals with the local executives.

Similarly, junior has some appraisal factors of his own, and is little impressed by doctoral hoods and engraved scrolls, no matter how beautifully embellished. However, if you are knowledgeable about model aviation fuels or can without hesitation name any foreign sports car on sight, you may qualify in his league and thereby gain recognition.

Whether any well-designed laboratory management program can successfully cope with the intricacies of the home recognition problem is a question for the future. I regard it as unpromising.

Can Scientists Be Prophets?

New Year's Eve, December 31, 1999, should be a big occasion. Almost all of you who missed the last turn of the century have thought—privately, at least—about your chances of being on hand to start writing a letter by putting "January 1, 2000" at the top of the page. Probably you have already computed how old you will be.

Although some of you might have to be a bit optimistic about longevity, you probably still find yourselves guessing what the world will be like when you prepare for that special New Year's Eve. Will you watch—on wall-size three-dimensional color television—crowds cavorting across Times Square enjoying the balmy winter weather specially adjusted for the occasion? Will the program include remote pickups from places where years do not have 365 days? And what will you actually think about the Twentieth century as you sing *Auld Lang Syne* to a whole 100 years?

UNIVERSAL GUESSING GAME

Less than a century ago, Jules Verne had a virtual monopoly on this business of guessing what science might do in the future. Today he would find the competition tougher. Everybody seems to be a prophet, amateur or professional, and this crystal-ball gazing is not limited to children's television shows or the more fantastic pulp magazines. Some of the most respected and dignified journals have also succumbed. Wherever you look these days, there is an article, or an artist's conception, or a fanciful TV stage setting telling us what life will be like tomorrow.

Certainly there is nothing wrong with "looking ahead" and, although an easy cliche, looking ahead represents one of the most important activities in modern civilization. For example, unless American industry looks ahead today, there may not be anything to see tomorrow. But, as with almost everything, this passion for prophecy can be overdone. The danger: constant repetition sometimes makes prophecy *seem* to become fact.

To the man on the street, the practicing scientist may appear to be behind the times; and when he does something new and significant, surprise stems not from what he has done but from the length of time it took him. "I read about somebody doing that

10 years ago" is a comment the scientist hears more and more. Actually the *somebody* was probably the figment of a science-fiction writer's fertile imagination. *Today's fact may seem less dramatic than yesterday's fiction.*

WHAT IS IT GOOD FOR?

Recently a group of General Electric Research Laboratory scientists discovered an entirely new physical phenomenon: amplification of light within a simple phosphor film. As a new phenomenon, it holds many future promises—some difficult to predict precisely. However, you cannot announce something new in science without expecting to be asked that excellent question, "What is it good for?"

To some extent, predicting what the solid-state light amplifier will "be good for" is not any easier than it would have been for, say, Lee De Forest to foresee in 1906 the ramifications of the electronics industry made possible by his vacuum-tube amplifier.

Actually, seven years after inventing the audion tube, De Forest was taken into Federal Court on a charge of using the mails to defraud because "this strange device like an incandescent lamp . . . has proved to be worthless." In summing up the case, the prosecutor stated, "De Forest has said in many newspapers and over his signature that it would be possible to transmit the human voice across the Atlantic before many years. Based on the absurd and deliberately misleading statements of De Forest, the misguided public, Your Honor, has been persuaded to purchase stock in his company."

In announcing the new light-amplifying phenomenon, accusations of fraud were not the problems; because of the widespread science-fiction fad, we had the opposite kind of trouble. Some people thought the announcement meant more than it did; the simple facts of a significant scientific achievement were not enough for everybody. Some insisted on reading into it the development of a device the science-soothsayers had looked forward to for so long that their readers feel certain it is long overdue—the "picture-on-the-wall" television set. This is thought of as a thin screen hanging on the wall like a picture frame, with a small box full of transistors, and with the appropriate control knobs, resting on a nearby table.

Picture-on-the-wall television will surely be with us someday, and the new phenomenon of light amplification in a thin phosphor film may be an important clue to its development. To help answer the what-is-it-good-for question, we said as much in our announcement. But because a thin, flat screen was used as a model to demonstrate the light amplifier and because all good scientists are presumed to be looking for the flat television screen, a surprising number of people thought we had something we did not have—despite our efforts to explain in precise terms exactly what had been accomplished.

However, many do manage to keep the present in perspective admidst the profusion of prophets. Some of this group believe that when crystal gazing is to be done, the scientist should be the one to do it. And scientists have achieved a certain reputation for clairvoyance, but actually their best soothsaying results from the distressing fact that the transition time between basic discoveries in the laboratory and everyday use often takes five or ten years. In other words, they are not guessing about the unknown, they are prophesying that the known will be put to use.

THE SCIENTIST'S LIMITATIONS

Of course, it is a speculative matter for scientists to claim that they know a particular new fact will be found or that a particular new invention will be made based on knowledge not yet discovered. However, you can ask the scientist what new things may be available *to the public* in five, or even ten, years from now and expect a reasonably firm answer. The science that will provide these new things already exists in the laboratory. If you extend the period to ten or twenty years, you can expect a reasonably good *educated guess.* The science that will provide these things is at least in the planning stage. For guesses beyond this, you had better turn to a crystal ball.

The scientist does not make a particularly good long-range soothsayer: partly because of his familiarity with certain physical limitations—the speed of light, for example; or the need for a positive magnetic pole where there is a negative one; or the basic restrictions that reside in considerations of energy, such as its quantity, availability, and convertability. Although still cautious about it, the scientist probably will say that "some things can be done"—even the fantasies of the writer of a futuristic television program. At the same time, because he can visualize the laboratory activity required to attain a particular scientific goal, he may be appalled at the cost of converting papier-mache into solid reality.

THE DECADE AHEAD

If the scientist can look confidently at the next ten years or so, what does he see?

The things that are just ahead range from the indispensable gadgetry of modern living to the fundamental energy generation and conversion processes of modern industry. For one thing, in the next decade we will be completely saturated with electronics. And although it seems remarkable, the electron—discovered back in 1897—is still a lusty child, far from maturity.

Tomorrow's electronics not only will be in the office, the factory, and the farm, but also will perform dozens of functions in the home, including cooking. In this next decade, factories will undergo important progress in automatic manufacturing, with electronic technology as the vital key to automation. By the end of this period, automation will have become an absolute necessity, because there will not be enough people to manufacture by present methods all the things we will need. This is arithmetic, not clairvoyance. A simple industrial forecast will show that we must produce at least 50 per cent more goods and services in 1966, and we will only have about 10 per cent more human effort with which to do it. The inevitable conclusion: automation.

Semiconductors represent the rich relatives in the electronics family—rich in promise and potential. Semiconductors in the form of transistors and similar devices with electronic capabilities will be found everywhere during the next ten years. In 1966, you will wonder how we ever got along without them.

From the long list of things that will certainly come into being during the next decade as the result of scientific research, let us consider some of them. Chemicals are manufactured today principally by the controlled application of heat, pressure, and catalysts; production by means of these processes forms the basis of the vast chemical industry. We now have a new means of promoting chemical reactions—radiation—that will become increasingly important during the next decade. Light is radiation. We have known for years about chemical reactions that could be promoted with light; but atomic radiation, especially electron radiation in the form of electron beams, presents a new chemical action. The Research Laboratory has been working with *electron chemistry* for more than thirty years, and a few materials produced by electron irradiation—irradiated polyethylene is an outstanding example—have recently come to market. But in the next decade when electron chemistry comes of age, it will make possible a host of new materials and products that can be produced in no other way.

The most publicized use of electron chemistry has been the sterilization of food by irradiation, without heat. Many optimistic reports state this technique will eliminate refrigeration in the future. Progress has thus far been very slow. But a recent report by the U. S. Army Quartermaster Corps states that more than forty different foods, including several kinds of meat, when sterilized by this process were found *after cooking* to be acceptable by Army standards. Much remains to be learned before the process achieves widespread civilian acceptance. Looking two or three decades ahead, you may see the general use of radiation for food sterilization become possible, but no one can predict when the required new discovery will be made.

The specific examples just described offer a general area of great scientific promise—but one requiring a sharper focus. To illustrate, let us consider a few places where scientific progress is in the making.

Historically, understanding and knowledge of matter were first developed in diffuse form—for example, the kinetic theory of gases and the gas laws mark early milestones in scientific progress. Later more complex forms such as liquids yielded to study. Presently we are making great strides in understanding the still more complex interactions of matter in the solid state.

Further progress in this field certainly will hold important practical consequences. Energy conversion represents a promising area for solids. A few of the presently possible conversions: electricity can be changed to light, light to electricity, atomic radiation to electricity, heat to electricity, and electricity to cold.

Another area of solid-state progress with exciting possibilities is in the mechanical properties of crystalline solids. The role of defects in crystals and the remarkable properties of "perfect" crystals provide a hint of the great new knowledge to come from this research, inevitably affecting the world of mechanical things.

PLANNING DISCOVERIES

Now let us take a look at the second decade ahead. Because we cannot simply look around the laboratory and see the accomplished results upon which it will be based, this extension—from 1967 through 1976—is much more difficult than the one just discussed. Particularly, one should hesitate to guess about the gadgets of this future period, although scientists can generally tell you what scientific discoveries the gadgets of tomorrow will depend on.

For the most part, the basic scientific discoveries for this period have not yet been made. Just what they will be and when they will be is a somewhat speculative matter. There are some plausible reasons for thinking that in the future scientists may be able to do a better job of predicting scientific developments.

Increasingly, scientific discoveries occur in areas of human need; for today's scientist is not completely at the mercy of a capricious nature. The tools of modern scientific research have become powerful indeed; and under the influence and pressure of a frontal research attack, an unwilling nature may frequently be made to yield a desired result. Generally speaking, you will find some validity in the statement that to an increasing extent in the future scientists will determine what discoveries need to be made and then plan to make them.

The plans will, of course, not always be successful. Mother Nature has the final word and may cast a veto. But the question of the final outcome of a research investigation is becoming more and more subordinate to the practical questions of probable cost, time required, and the availability of required scientific skills to accomplish the research. This will probably be true to a progressively greater extent in the future. Scientists plan to make discoveries in areas of great opportunity and need, and the powerful tools and skills of modern research will implement the planning.

But do not get the impression that basic research and pure exploration on new frontiers of science will disappear. Presently and in the foreseeable future, these frontiers seem to be limitless; they will continue to provide an outstanding challenge to the mind of man. In applied research where you seek a specific objective, the variety and effectiveness of the scientific techniques that can be brought into focus on a given problem will increasingly bend nature to our will.

Projecting the future—at least the future that will be based on science and technology—looks relatively easy for the next decade. The second decade confronts us with the prediction of discoveries which have not yet been made, making only general outlines discernible. However, those general outlines will in substantial degree conform to human needs and wants, and scientific research will provide powerful tools for shaping this future.

Looking beyond 1976, the crystal ball really becomes clouded. Even a guess at what *research projects* will be twenty years from now would be presumptuous. The scientist just cannot and should not try to say what the world will be like in 2000 A.D.

PROGRESS—A SURE THING

But the answer to "Will there be as much scientific progress during the second half of the Twentieth century as there was during the first?" is, "Yes, definitely. There should be *more* scientific progress between now and the year 2000 than there has been since 1900."

This is not inconsistent. Often a general question about the future can be answered but not a specific one. Ask if a shutout will be pitched in the National League on July 23 of this year, and the answer will have to be, "Nobody knows." But ask if there will be at least a dozen shutouts pitched in the National League during the coming season, and you will find plenty of baseball experts ready to say, "Yes, of course." Statistically, this seems to be a sure thing.

Similarly, great new scientific progress during the next three or four decades also seems to be a sure thing. Over the years, technological results have been produced pretty much in proportion to scientific efforts. You do not have to be acutely perceptive to assume that this correlation between research and progress will continue. Exactly *what* all this effort will produce cannot be foreseen, but *something* will be produced.

To justify great expectations for the future, you need only compare the scientific effort of the present with that of the year 1900. That year marks the beginning of the golden age of technological progress. But who would have guessed it from the amount of scientific research then being conducted?

If two men working on a flying machine in a bicycle shop constituted the promise of a multibillion-dollar industry fifty years later, what can you anticipate from the aviation industry's present research budget of nearly half a billion dollars annually? If "proton" and "electron" did not appear in most dictionaries published in 1900, how shall you compare progress in atomic energy during the past half century with what we can expect during a fifty-year period starting with years when the annual expenditure on the atom alone reaches nearly $2 billion?

In no area do you find the difference or the future promise greater than in *industrial* research. Today, thousands of businesses —large and small—realize that they must invest in technological progress to meet competition. Many industries, some more than others, realize that new fundamental scientific knowledge represents as important a raw material as the tangible stuff from which products are made. But in 1900 industrial research as we know it now was nonexistent. A handful of firms had chemists trying to improve products, although the word *chemist* was more closely associated with the drugstore—a place where, surprisingly enough, you purchased mostly drugs. *Physicist* was not only a word unknown to the general public—it was practically unpronounceable.

Let us try placing the history of research in some kind of proper perspective. One famous idea for organized research dates back 300 years: Sir Francis Bacon, in *The New Atlantis,* described a "college of research" called Salomon's House. Its purpose was to ascertain "the knowledge of causes and secret motions of things and the enlarging of the bounds of the human empire to the effecting of all things possible." However, many years passed before Bacon's prophecy began to be fulfilled. During the subsequent Age of Invention, man—who had been making inventions of one kind or another from the time he first lighted a fire—finally began to become scientific by collecting some fundamental knowledge about how nature worked and then applying logical processes of reasoning to these discoveries.

As early as 1875, the application of science to industry began to receive recognition—but little support—from industrial leaders. The discoveries of science made their way into industrial practice in a variety of ways. The inventor, alert to the latent potentialities of any new scientific discovery, combined the work of academic scientists with his own inventive talents. Thomas Edison, for instance, was a careful student of scientific literature in a wide variety of fields. But the use of science by industry was not organized. Industrial research really began about 1900, but in a very small way. Until 1915, the *Readers' Guide to Periodical Literature* did not list industrial research as a subject.

EARLY INDUSTRIAL RESEARCH

The earliest growth of industrial research occurred in the industries that actually had been born in the laboratory or that were directly dependent on new knowledge for their growth. Even as late as 1920, the electrical, chemical, and rubber industries employed two-thirds of all the research workers recorded in the first survey of the National Research Council. This first survey, taken in 1920, showed that about three hundred laboratories were engaged in industrial research.

By 1940 the figures increased to more than 2,000. The total number of people employed by the laboratories jumped from a few thousand to more than 70,000. The annual expenditure increased sharply to approximately $400 million. Even the depression failed to stop the march of research progress. Bankers, industrial executives, even stockholders took keen interest in the research activities of their companies and became convinced that "research is the best form of industrial insurance." If historians feel that the word *phenomenal* best describes the growth of industrial research during this period, what word satisfactorily describes what has happened since 1940?

Fortune said recently, "Industry has been on a research *binge*." Nearly 3,000 companies now maintain industrial laboratories and employ over 250,000 people, including well over 100,000 scientists and engineers. Today you can measure the expenditures in billions; last year industry spent $2.7 billion for research and development. Admittedly, this figure covers everything from the purest fundamental research to some fairly routine engineering. Also this phenomenal growth in total effort exists more in the areas of applied research and development than in the area of basic research—a matter of proper concern to many analysts of the present situation.

TODAY'S OUTLOOK

But even assuming that slightly less than 10 per cent of the nation's total research budget of $4 billion—including government and academic research as well as industrial research—goes for basic research aimed at finding new fundamental knowledge, this figure still represents an almost unbelievable increase over the corresponding figures of 1900. Certainly the comparison provides a reliable background for optimistic future forecasts. Although you would probably agree that too small a segment of the total research and development budget finds its way to basic research, the growth of the entire circle has been so great that even a narrow slice of the pie contains considerable nourishment.

In view of the established correlation between research and progress, who will say that science cannot produce more in the next half century than it did in the last?

No specific predictions as to what life will be like on New Year's Eve, December 31, 1999 can be made. But the prospects are good. Scientific progress—if it continues to depend on the best in men's minds and to survive in spite of the worst—will provide things not found in today's wildest dreams.

I cannot guess how it will be, but I would certainly like to be here.

Handfuls of Progress

An address at the 13th annual meeting of the Southern Research Institute, Birmingham, Alabama, November 26, 1957.

It must be apparent by now that no one is allowed to speak on the subject of science these days without discussing Sputnik, and I bow to that necessity. We must face many necessities, and the Sputniks pose some urgent problems in vital need of thoughtful solutions. I am sure that free people will come to the right decisions on these problems, as soon as we understand the problems. In my view, there have been at least five Sputniks, not two, and each has carried a large warning flag and an urgent message.

Sputnik I was the detection in September of 1949, later confirmed by the Soviets, of their first atomic bomb. This significant event arrived much earlier than we anticipated. We explained away our surprise partly by attributing their success to spies. Sputnik II was the announcement of the first Soviet hydrogen bomb in August of 1953.

I do not recall that we successfully explained away that significant event, except in terms of inevitability. Sputnik III arrived gradually, not explosively. It was the growing realization of the truth and basic importance of the tremendous programs of Soviet technical education and scientific and technological research. The most recent Sputniks are merely IV and V. They may be the most important ones of all because their spectacular appearance has finally shocked public consciousness into a feeling of impending doom. I can only say that I regard them as flying blessings with a silver lining. Because of them, we are now engaged in an examination and appraisal of our whole national posture on science and technology. To do so thoughtfully requires a study of not one, but all of the aspects of our dilemma which, in aggregate, adds up to the problem of national survival in a scientific age.

We must start our inquiry far back in the educational system, and carry it through the whole range of activities up to and including the final utilization of the results of new scientific progress. At almost every point in this chain of events, there are outstanding opportunities for improving the performance and efficiency of our effort. We do not, for example, except by accident, detect technical aptitudes in our young students. How many potential Langmuirs and Einsteins have been lost to science no one will ever know. If we did detect these aptitudes we would not know what to do about them, for our technical subjects are taught with so much apology and so little inspiration that only the hardiest enthusiasts among our students survive the ordeal.

Could it be that our preoccupation for pedagogical requirements has left no room in the curriculum for specialization in science and technology on the part of our teachers? At odd moments, I find myself thinking so, when I brood over the fact that in our staff of 400 outstanding research scientists at our Research Laboratory, there are only a very few who could presently obtain regular employment as high school teachers.

Following the chain of events, which we must have the courage to examine critically, we come to the controversial issue of basic versus applied research, and the question of whether our Department of Defense has any legitimate business in basic research. Under policies which have been undergoing change, we were formerly behaving very naively on this issue. We acted as though the only conceivable interest of our Defense Department in science was after it was a proven and demonstrated fact that the new science was capable of producing a new weapon. This idea was obsolete before World War I. We must now start new weapons in basic research programs, not in plumbing shops. Inquiring science will provide the primary source of the new foundations upon which can be built new technology that will serve the national defense. And our competitor will not give us time to be patient with people who arc still surprised and skeptical when this happens. Our Defense Department *does* have a legitimate role in this kind of research because the success of its future mission will depend increasingly upon its awareness of and participation in scientific progress. I wonder if people living during the early days of the industrial revolution knew what was happening then, and understood its real economic and social significance? We have a similar problem in perspective. *We are now in the midst of a technological revolution*, and we must seek to understand its broad implications.

In the technical industries, we have learned that new products, new business, and new industries are to an increasing extent born in the laboratory rather than in the fields or forests or mines. Not to participate appropriately in this source of industrial progress leads to a shrinking market and economic oblivion. For closely analogous reasons, not to participate appropriately in scientific progress on a national basis may be a fatal mistake. The Sputniks are helping to bring the realization of this fact to all of us.

The utilization of the results of successful scientific work is the final event in technological progress, and it is here that engineering and engineers play a major role. Progressive engineering education recognizes the eminence of technological change and its scientific origins, and reflects this trend in the college curriculum. I personally do not regret to find that the old standby of the freshman engineer, Surveying I, is being dropped in favor of subjects more urgently in need of understanding in the atomic, electronic, and space-flight age.

Because Soviet Russia is training three times as many scientists and engineers as are we, it would be easy to come to the conclusion that we should immediately treble our technical education programs. If this should take place suddenly, and American industry should have available two or three times as many engineering graduates, it would create havoc and demoralize the profession. In the solution of this problem, as in others created by the Sputniks, we have to start farther back in our free enterprise system. We undoubtedly need a more adequate supply of technically-trained people, but we are *not* out of balance by a factor of three. The Soviets started from almost nothing and have a great deal of catching up to do.

Our problem is basically one of achieving balance between the many factors which in aggregate determine an effective program of exploring and utilizing science for the national defense. Before we *train* three times as many technical people, we must therefore give some thought to how we would *use* three times as many technical people if we had them.

First and foremost is the urgent necessity of determining what the gross magnitude of our technical work should be—that is, scoping the job. This is not made easy by figures which are conveniently available today.

Reliable figures from the National Science Foundation show, for example, that the United States is spending about $5 billion per year on "Research and Development." This is a huge sum, and it would be easy to jump to the conclusion that in comparison to our total defense expenditures it must provide an adequate allocation for research. These figures are misleading, however, because we have lumped together "research" *and* "development." There is no clue in the "$5 billion for R & D" as to what portion is for research, and particularly what portion is for the basic research that provides the real foundation for our technological progress. It is certain, however, from other available information, that the basic research portion of this total is extremely small. In any comprehensive program based on technology, be it industrial or government sponsored, an effective *balance* should be sought between the basic research, the applied research, the engineering, and the final production activity.

Some of the trouble lies in definitions. The simplest distinction between research and development is that research seeks *new* scientific knowledge, and development uses *existing* scientific knowledge. Although there is no common definition of basic research —or, as I would prefer, *learning work*—there is a strong and growing feeling in civilian and government scientific circles that we are dangerously deficient and hence out of balance in this area. The Hoover Commission report of 1955 emphasized this. Numerous documents issued by high-level advisory committees of the Defense Department have raised other warning flags.

It may be argued that the Russian satellites are the result of a tremendous *development* effort, do not necessarily incorporate any new fundamental knowledge that we do not possess, and that their appearance has nothing to do with the basic research issue. This statement of the case includes some truth, but is completely wrong in its implication. To put our satellite in an orbit in all haste does not require that we do any basic research; this can be accomplished with existing scientific and technical information. However, far more important than competing with Sputniks already overhead is to learn the lesson taught by all five Sputniks. *The lesson is that we must mend our scientific and technological fences, all the way from the schoolroom to the missile launching pad, and we must do this so effectively that a future Sputnik X, carrying an atomic warhead, will never be launched.*

. . . .

When I had gotten to this point, in preparing notes for my talk, I made a list of scientific developments that I wished to describe briefly as examples of recent significant progress. I then noted that in each case, with possibly one exception, we had convincing evidence that the work was well in advance of corresponding work in Soviet science. This is testimony on a point that has been overlooked in our Sputnik excitement: in many important fields of science the Soviets are *not* ahead, but way behind.

For convenience, I have come with my hands full of ten tiny things that are physically small but big in research. I hasten to point out that some of these things are small because we wanted them that way—but others (some Man-Made diamonds, for example) are small in spite of our efforts to make them larger. I can simplify my presentation by pointing out that one of these hands—the right hand—is a handful of today. The left hand is a handful of tomorrow. In my right hand are five objects that were but recently new in science, are still new in technology, but are rapidly coming into practical use.

First, as you might expect, is a transistor. Most people are familiar with the use of transistors in hearing aids and tiny portable radios. The transistor has become the symbol of the many new developments in the field of semiconductor physics, but I would like to discuss the transistor first as an outstanding example of how modern science can add great value to the most common raw materials, and, second, as an outstanding example of how modern science not only can create entirely new business enterprise for the sponsor of the original research, but also can create a vast new industry with broad industrial participation.

In the facing illustration is a small ingot that is more valuable than an equivalent amount of gold. This cylinder is not silver, but silicon, the most abundant solid element on the earth. There is more silicon in a worthless handful of sand than there is in this small ingot, but this ingot is worth $250 because it is almost entirely silicon, prepared in a scientific laboratory so that it is 99.99999999 (ten 9's all together) per cent pure.

This small cylinder of silicon is valuable because 5,000 transistors can be made from it. This ingot *also* is a symbol—a symbol of modern scientific alchemy that is producing a "technological revolution" in America. In the case of semiconductors, which were nonexistent in 1950, the new industry based on these devices will gross $140 million this year and is projected to reach the billion-a-

year figure by 1967. But actually the industrial impact of this new industry is not adequately measured even by one billion dollars, for semiconductor devices, as new and versatile electronic components, have stimulated the growth of many older businesses and have created entirely new ones. Portable radios and communication systems, improved hearing aids, compact computers, automation equipment, improved industrial power conversion equipment, and numerous important applications to weapon technology are among the products which have been born or have been rejuvenated by virtue of semiconductor components.

It is clear that the new market created by semiconductors and related products thus benefits the sponsor and undoubtedly justifies his research and development expenditures in this field. Note, however, that a *broad* participation in the new market is a corollary benefit which accrues to many participants and is a particularly attractive opportunity for the small company, many of which have been born or have matured in this particular new market.

When one talks about transistors, it is fashionable to say that they are going to replace "bulky vacuum tubes." But I hasten to assure you that the vacuum tube is far from a dead duck—and the reason is a very healthy situation most easily described as "scientific competition." One of our newest experimental tubes actually is even smaller than the latest transistors—but it is a good deal more than just small. Because the elements of this tube are so close together, because it is constructed with very precise internal geometry, and because of some very special materials of construction, it has exceptional capabilities at microwave frequencies that the transistor cannot yet approach. It is the latest in a long family of so-called "disk-seal" tubes designed originally during World War II to extend the range of early-warning radar.

This tube also is something of a symbol—a symbol of *serendipity* in science. The word serendipity *is* in the dictionaries—at least, in the big dictionaries. It means finding something when you are looking for something else.

These tiny tubes involve the use of titanium metal, new ceramics, and new methods of sealing the metal and ceramics together. Because of this special construction, they can operate at temperatures several hundred degrees higher than ordinary glass or metal tubes. Similar construction can be used for other electronic parts, such as resistors, capacitors, coils, and so on. Guided missiles and supersonic planes now on the drawing boards will travel so fast that their insides may actually get red-hot from air friction. They must have electronic devices inside them, and the development of high-temperature electronic circuits is considered a major breakthrough of what the aviation industry now likes to call the "thermal thicket." There was some serendipity here—we were not looking for high-temperature electronics when this work was started.

I have only mentioned two of the five items in my right hand—this handful of today—so I will describe the rest of them quickly. Here is a magnet—the thing that is unusual about it is that it has been created out of invisible iron "dust" and is potentially many times stronger than the best existing magnets. The unique properties of this magnet are achieved by precisely controlling the size and shape of individual iron particles so small that it takes a *billion billion* to make a pound. Theoretically, magnets like this may be made ten times stronger than the best available magnets. However, the best we have made to date are about the same as, or a little stronger than, Alnico. These magnets will result in electric instruments that are smaller, lighter, more accurate, and more rugged.

They also will help make better photographic exposure meters and better aircraft instruments. Unlike ordinary magnets—which are very tough to shape and machine—these fine-particle magnets can be drilled, tapped, soldered, and molded precisely into any shape.

Also in my "handful of today" is an inch-square piece of thin glass—a memory cell—upon which it is possible to store up nearly a million bits of information. Smaller electronic computers with bigger memories are among the anticipated applications. At the heart of a new information-storage device is a thin sheet of glass in which small holes have been etched and then filled with metal. In actual use, information is written onto one side of the honeycomb by an electron-beam scanning method similar to that used in television. A "reading gun" picks up the information from the opposite side. It provides a new capability for computers and all of the complex gear for tomorrow's automation.

Early in 1955, we made a formal announcement in Schenectady that a group of our scientists had discovered a reproducible process for making diamonds in the laboratory, and that our exhaustive tests had shown that these diamonds were identical with those found in nature. As you may recall, this announcement created a certain amount of excitement not only in the scientific world but also in the world of business. There also were some people who thought Man-Made diamonds were just the result of a laboratory stunt, and not something with a practical value for industry.

Also, our Metallurgical Products Department made an announcement in Detroit which made it clear that Man-Made diamonds have actually come to the industrial market place. The New York *Times* said:

> General Electric is . . . in commercial production of industrial diamonds. By the end of 1958 . . . GE expects to be able to produce 3,500,000 carats a year of industrial diamonds, which are vitally important in grinding and sharpening carbide tools. This quantity would be about half of the total now being imported every year from the Belgian Congo and South Africa . . . the achievement is . . . a major step toward freeing the nation's industry from the uneasy situation of being forced to depend upon a single source of supply . . . The company displayed 60,000 carats of Man-Made diamonds . . . and reported that more than 100,000 carats had been produced.

The glamour of diamond gem stones tends to obscure in the public mind the importance of diamond's single most important characteristic—its hardness. Diamond is fantastically harder than anything else ever found in nature or made by man. (I will have more to say on this later.) On the familiar mineral hardness scale, the hardness difference between diamond and sapphire is greater than the hardness difference between sapphire and chalk. Let me put it another way. Diamond is so useful to industry that it can support a present-day price of about eight thousand dollars per pound. The next best abrasive for grinding wheels, silicon carbide, costs about 25 cents per pound. From a defense standpoint, this discovery means that our nation has a new source of an essential strategic material—a material for which America has previously been wholly dependent on imports. I think it is significant to note that this contribution to our national security has been wholly financed by private industry operating in a free society.

. . . .

Now let us see what the handful of the future contains. These are things that exist in the laboratory but have not yet been applied to actual products.

First, a green, hard cluster of silicon carbide—exactly the same material I have just discussed as the abrasive that is the "next-best-thing" to diamond. There is nothing new about this silicon carbide abrasive. Although it is not found in nature, man learned how to make it before the turn of the century. If silicon carbide crystals are grown in tiny laboratory "furnaces," under controlled conditions, the material is suitable for use as a *semiconductor* at high temperatures.

At our laboratory a silicon carbide rectifier recently was operated at a temperature of 1200°F. I have already discussed the military importance of high-temperature electronic equipment. But you may have noticed that I talked principally about vacuum tubes and other vacuum devices. It is clear from this development that semiconductor devices may also become available for high-temperature applications.

Now let us talk for a minute about why long-hair scientists study short whiskers. Scientists who have been trying to understand the basic reasons back of the strength of common materials—

metals and alloys—have made some spectacular progress. They have developed a detailed picture of what happens, on an atomic scale, when metals break. And to give substance to the theory, they have made tiny filaments of pure metals which show unbelievable strength in actual test. Although this great strength cannot yet be made available in bulk form for structural parts, we are very excited about this look into the future of super metals. The fairy tale of super metals, or perfect crystals, will take too long to tell in detail, but in too few words it has been found that actual metals, which are of course crystalline in character, have always had defects in their crystals on an atomic scale. That clue led to a method of growing "perfect crystals."

Conventional iron alloys have strengths ranging from 30,000 to 200,000 pounds per square inch. Iron whiskers have been made in the laboratory with a strength of nearly 2,000,000 pounds per square inch, at least twenty times stronger than most structural steel and very much stronger than even the toughest piano wire. These so-called "perfect" materials have thus far been made in sizes up to the diameter of a human hair—and about crew-cut length. "Perfect" metals have not yet had any "practical" application in actual products, but they are proving extremely useful in the laboratory. They are an aid in studying magnetism, conductivity, low-temperature effects, corrosion, and a variety of other scientific problems.

Metallurgists these days also are learning a great deal about the old mystery of magnetism. One result is our "four-square" silicon-iron, a new kind of magnetic sheet material designed to increase the efficiency of transformers, motors, and generators. This is a doubly-oriented magnetic sheet material, which means it is quickly and easily magnetized—and de-magnetized—in four directions. In a transformer, the core's resistance to these changes causes a loss of energy, reduced efficiency through the generation of heat, and actual mechanical vibrations in the core which result in annoying "transformer noise." The new material improves all of these properties significantly.

Next in my handful of the future is a small piece of a polymer. In our discussion of perfect metal whiskers, it was suggested that metals get their strength from the regular arrangement of their atoms—the absence of defects. But a polymer is very much different from the highly-ordered metal crystal. Typical polymer molecules are long chains of carbon atoms, strung out like beads on a string. Numerous chemical adornments—such items as hydrogen and oxygen—are fastened on to this string of beads to form "side chains" and these side chains give much of the chemical and physical character to the polymer. But while the polymer molecules themselves are strong, the aggregate mess is not, since the individual molecules run in random directions like a plate full of spaghetti.

One way to increase the strength is to tie the molecules together—the chemist calls it cross-linking—and one way to do this is by bombarding certain plastic materials with high-energy electrons. The first product actually on the market based on an electron irradiation technique is a polyethylene insulating tape that is much stronger at higher temperatures than ordinary polyethylene.

Electron irradiation is also being used in many other ways. A popular prophecy connected with this work is that electron sterilization of foods will make it possible for us to keep meat on the shelves—unrefrigerated—for weeks or months. Frankly, there is a lot of work to be done before this electron sterilization will be attractive for food. Meat and bread and fruit and other foods can be sterilized, but it is not much fun eating this food, because the process changes the taste and odor with generally disagreeable results. But there is a lot of work being done on this application—much of it under the sponsorship of the Army Quartermaster Corps.

When I was talking about Man-Made diamonds—a small item that is important *today*—I mentioned that the same superpressure research might produce some astounding new materials for tomorrow. And here is such a material—the last item in my handful of tomorrow. One of the scientists at the Research Laboratory who had worked on the diamond project got to thinking about a material called boron nitride, which is similar to graphite—soft and slippery and with the same hexagonal crystal structure.

Dr. Robert H. Wentorf, Jr., of our laboratory, asked himself: could boron nitride be squeezed and heated and put into the cubic crystal form? The thirty-year-old scientist has pointed out a couple of deterrents that confronted him. "First," he says, "there was none in nature to prove that such a thing existed; and second, there were good reasons for believing that it could not be made. In this respect the problem was more difficult than that of making diamond, for we knew all along that diamond could be made because some already existed in nature—all we had to do was find a way. Trying to make cubic boron nitride, however, was like trying to catch a new, special kind of fish in a strange lake. You weren't sure the lake had any such a fish, or whether *any* lake has such a fish, and the other fishermen were dubious. But our faith was finally rewarded and we unexpectedly found ways to make this cubic crystal of boron nitride."

Dr. Wentorf called the new material borazon—*bor* for boron, *azo* from a French word for nitrogen. In spite of its different appearance, borazon seemed similar to diamond in several other ways—the X-ray diffraction patterns showed that both have cubic structure with the atoms spaced about the same distance apart. The density is about the same, too. One striking—and important—difference was noted. Ordinary diamond burns in air at about 1600°F, but borazon easily resists temperatures of more than 3000 degrees.

This heat-resisting ability is only the first half of the story that tells the possible importance of borazon. The second part pertains to a matter I discussed earlier—the hardness of diamond. For centuries men have been saying that only diamond scratches a diamond. Early this year, that time-honored belief became forever obsolete—when Dr. Wentorf demonstrated publicly that he could easily scratch a diamond with a piece of borazon.

We believe borazon's resistance to oxidation will make possible superior methods of mounting stones in industrial tools and may allow bits and wheels to be operated at higher speeds, performing their cutting and polishing jobs more quickly and effectively. Borazon is—so far—a scientific achievement and not yet a commercial product. It's another example of some of the many wonders we can expect from superpressure research—and it is the last item in my handful of tomorrow.

I have talked about two handfuls and have had my hands full keeping track of them. But there is one more item of interest that I want to discuss concerning a major advance in man's long search for ways to generate electricity from heat without using machines that have moving parts. It is a new kind of electronic device—the thermionic converter—which changes heat energy directly into electrical energy. Applications of the thermionic converter are *definitely for the future*, but to demonstrate the principle in a device with appreciable efficiency is a scientific achievement today.

Experimental thermionic converters developed by Dr. Volney C. Wilson at our Research Laboratory take advantage of the simple fact that electrons can be "boiled out" of a hot metal surface and used to produce an electric current directly. Laboratory devices have converted more than 8 per cent of the applied heat energy into useful electrical power. This is too far from the 40 per cent efficiency of modern steam turbine-generators for us to suggest that this converter will have early application in power plants, but the efficiency obtained in our experimental tubes is very encouraging and the theoretical limit appears to be much higher.

The inventor of the thermionic converter, Dr. Wilson, is a forty-seven-year-old member of our Metallurgy and Ceramics Research staff. He distinguished himself early in his career as director of instrumentation and control for Dr. Enrico Fermi during construction of the first atomic reactor in Chicago during 1941 and 1942. Subsequently he has made important contributions to the study of ferrites and to the use of the neutron diffraction spectrometer as a research tool.

Most previous methods of converting heat directly into electricity—without rotating machinery—have been based on the thermocouple, a device long used by scientists for measurement and control functions. The thermocouple, which is simply a junction of two different kinds of metal, produces a very small electric current when heated, but its efficiency is well below 1 per cent.

We are investigating several approaches to the problem of converting heat directly into electricity. There are opportunities for using solid-state semiconductor devices—thermocouples employing semiconductors—as well as the thermionic method incorporated in the thermionic converter.

I cannot overemphasize the fact that the thermionic converter is an experimental laboratory device only and is not ready for production. But, it is a significant research step toward a goal scientists have been seeking ever since electricity was discovered. We must increase the thermionic converter's efficiency and adaptability to a variety of conditions if it is to become a major new electric power producer.

. . . .

What next in science? There are, of course, many who manage to keep the present in perspective these days amidst the profusion of prophets. Some of this group believe that when crystal gazing is to be done, scientists should be able to do it. Scientists have achieved a certain reputation for clairvoyance, but actually their best predicting is based on the distressing fact that the transition time between basic discoveries *now* in the laboratory, and everyday use, is often five or ten years—or more. This most certainly will be true in the case of many of the items I have talked about in my "handful of tomorrow." So, I am willing to prophesy that stronger metals and alloys, better semiconductors, harder abrasives, stronger polymers, and many other important developments—*because they are in the laboratory today*—will come into practical use tomorrow. But, in addition, it is equally safe to predict that nature will declare many extra dividends in the form of unexpected and essentially unpredictable scientific discoveries that are certain to come from research tomorrow.

Polymers, Crystals, and Plasmas

International Symposium of Science, Industry, and Education, Oklahoma City, Oklahoma, June 17, 1957.

I became engaged in a conversation about my favorite subject—research in science—with a young friend who was particularly impressed with some new information he had just picked up at school concerning the existence of "more than a hundred" chemical elements.

If I were going to become a scientist, which elements would be the best ones to study? I don't suppose I'd have time for all of them.

Since my young friend is only ten years old, he will have a good deal of time to further his research, but as an immediate answer to his question I mentioned five elements and suggested that these five might be especially interesting to study.

After a brief session with his junior encyclopedia, my friend returned, and his disappointment was apparent. My stock as a scientific advisor had obviously dropped several points.

I looked up a list of all the elements, and, gee, some of them were just discovered last year and the year before. But it says all of the ones *you* told me about were "known to the ancients."

In an effort to salvage my reputation, I explained that it was true that the elements I had mentioned had been discovered long ago. "But," I said, "they are also very common, very cheap, and very plentiful and because of that they are the ones to which we can add the most value by research. Almost anything we can do to them makes them more valuable."

By way of emphasis, I told him that there was a billion dollars worth of silicon right in his back yard, and that some of our scientists had actually made diamonds out of peanut butter, and by then his eyes were fairly popping. Just plain old iron, all by itself, I assured him, has turned out to be the strongest metal of all, and even the strongest metals may not be as useful for tomorrow's world as some materials made out of new combinations of those wonderfully common elements: oxygen, hydrogen, and carbon. And one member of this simple trio, hydrogen, may contain the ultimate answer to our need for safe and inexpensive energy.

We are at a place in the cycle of scientific progress where it is highly profitable to apply our newest knowledge to the study of the oldest elements—some so old that they were "known to the ancients."

I intend to focus on three areas of science where exciting progress is being made, areas where the key elements are the simple ones I have mentioned. Sometimes they appear in combination and sometimes very much alone. These are not the only areas of great scientific progress by any means, and I would not belittle the importance of studying elements with fancier names than oxygen, carbon, and hydrogen—that is an important and exciting frontier, too.

But these fields which I will discuss are particularly interesting and important to industry—and especially to the electrical industry—and they appear so complex that there is every chance that there will still be opportunities in them even when my ten-year-old friend is weighing the approximately 500 job offers he can expect after finishing his graduate studies in science in 1970.

POLYMERS

First, let us consider polymers, because they touch upon human experience so many times each day, and have so much to do with our future. I trust the term "polymer" does not bother you as much as it did a business associate of mine who is better known for his managerial skill than for his technical prowess. Recently, he called me to say, "Why didn't you tell me that this conference on polymers you had at the laboratory last week was all about plastics? I would have come."

Perhaps the toothbrush is the first place, each day, where we contact a polymer, although of course we may have slept on one the night before. The toothbrush handle, and probably the bristles, too, are made from polymers, as are all plastics. There was a time, about two or maybe three decades ago, when critical users thought that polymers were not much good for anything but toothbrush handles. To be sure, these early polymers had some deficiencies, particularly in strength. Most anything you could make from them would fall apart in your hands, or under foot, and if you wanted to make something really strong you had to use a metal or an alloy. This difference in strength between polymers, as synthetic chemical solids, and metals, as refined natural solids, includes a basic difference in structure.

If you take any metallic material, polish it, and etch the surface, you can readily see the characteristic crystalline structure, generally made up of many small crystals packed together in random array. Consider one crystal in this aggregate; its atoms are arranged in regular layers and rows, much like the eggs in an egg crate.

If, on the other hand, we could look at a polymer in the same atomic detail it would look vastly different from the highly-ordered metal crystal. Typical polymer molecules are long chains of carbon atoms, strung out like beads on a string. Numerous chemical adornments—invariably including hydrogen and frequently our old friend oxygen—are attached to this string of beads to form side chains that impart much of the chemical and physical character to the polymer. An aggregate of such polymer molecules, if we could see it under a super microscope, might look something like the arrangement of fibers in a sheet of cotton batting—a tangle of long-chain molecules, running in every direction. This classical structural characteristic of polymers is principally accountable for the poor strength of these materials. The molecules themselves are strong. But the aggregate is not, since the individual molecules run in random directions with very little co-ordination between them. We may tie the molecules together, with chemical cross-links, and improve the strength. Another and more successful way would be to make them more crystalline, by lining up the long molecules, or portions of them. Then they would gain strength from their structural regularity, somewhat analogous to the manner in which a woven fabric gains strength in comparison to our cotton batting.

. . . .

With this word of introduction, the new look in polymers is very simply explained; they are becoming more crystalline—that is, more highly ordered—and hence stronger. The newest and strongest polymers are a remarkable improvement over the older ones.

The new progress—partly in Ziegler process polyethylene, partly in Montecatini polypropylene, partly in numerous developments in this country in new silicones and polycarbonates—is in aggregate a very significant step forward. There is no doubt that polymers have come a long way from the old toothbrush handles.

These new high-strength polymers have led the chemists to make some exciting extrapolations into the future, where they see progressively higher strength and better high-temperature performance. This certainly should result in much wider use for polymers in years to come.

No doubt polymers will replace *metals* in many applications. In fact, they are doing it now. Most of them are not yet as strong as most metals in common use, but many metals in use are stronger than they need to be to do the job at hand.

To tell the truth, in our research laboratory, we have a friendly fight on right now between the chemists and metallurgists. The chemists say that in another ten or fifteen years or so our company will be using more polymers than metals in our products. That obviously depends to quite an extent on whether it is measured in pounds, cubic feet, or dollars, but the fact that there is room for an argument is in itself significant and indicates clearly that polymers are getting into the running as structural materials.

Need I add that the building blocks of these new polymer materials—carbon, oxygen, hydrogen, and silicon—are the ones my young friend said were originally discovered a long time ago.

CRYSTALS

Just as polymers are getting better by becoming more crystalline, the metals are improving too through the development of more perfect structure in their crystals. The striking development of so-called "perfect" crystals of the common metals is now well known. We have measured in our laboratory the tensile strength of a perfect crystal of pure iron and found it to be nearly two million pounds per square inch—vastly stronger than any metal or alloy in common use. This important progress is sure to have a practical impact on technology, but just exactly where is not yet certain. Here is another case of value added by science and technology—it is the same old iron "known to the ancients" merely rearranged atomically.

Another important class of crystalline solids, the semiconductors, has become famous for perfection in purity. The rapidly growing new semiconductor industry is presently based on the two elements germanium and silicon, refined by special processes to a total impurity of only one part per billion. When a new development like this comes about, people interested in minerals start looking for a good ore supply for the new industry, and people interested in speculating start looking for stock in the company that finds the new mineral. In the case of germanium, it was soon found that the principal source was the flues of zinc refineries, where it collects as a minor impurity and is available as a by-product. More recently, primary germanium ores from Africa have become available.

When methods were worked out for refining *silicon* to semiconductor grade, a few years ago, some prospecting for ore took place in speculative quarters, but not for long, for the basic supply of silicon ore is excellent; in fact, we have far too much of it. Practically all of the rocks and sand and dirt on earth, with minor exceptions, are full of silicon, which is the most abundant solid on our earth. Thus nearly everyone who owns real estate has extensive mineral rights to silicon, which of course is not worth much until it is refined to semiconductor-grade material, when its value goes up to as much as one thousand dollars a pound. This value has been added by research and processing and is a striking illustration of making something of great value out of something of almost no value, by the application of modern technology.

Now let us look into that billion dollars worth of silicon that my small friend and you have in your back yards. First we will have to dig the silicon ore. The most convenient way might be an open pit or surface mine which will of course ruin the lawn and flower beds and will be quite unsightly. The neighbors might object to having a mess like that on your street, and they might worry about the kids falling in the excavation, and there is probably a local ordinance against mining in your neighborhood, and a lot of other things. But let us overlook all of these practical matters, and see what the silicon might be worth, potentially.

If the lot is 100 feet wide and extends back 100 feet from the street, and we excavate to a depth of 100 feet, we would have approximately one million cubic feet of low-grade silicon ore, commonly known as dirt, sand, and stone. If this were high-grade ore, it might yield, after processing, about 25 pounds of silicon per cubic foot, but to be conservative we will plan on only one pound, to allow for some waste and inefficiency. However, that still yields one million pounds of silicon which is *potentially* worth *one billion dollars*. But wait! Do not spend the money—there will be a small processing charge; in fact the processing cost will be very nearly one billion dollars. This theoretical mining operation, right at home, points up the tremendous ability of modern technology to add value and utility, to supply human needs and wants, and to create wealth from common sources; in short, to perform modern alchemy.

Superpressure research—which a while back resulted in a process for making diamonds, and, more recently, that still more glamorous and valuable substance, borazon—is another prime example of value added by science and technology. It is no secret that diamond, whether made by nature *or* in the laboratory, was transformed from carbon, which in comparison to diamond has almost no value. Even a 49-cent jar of peanut butter has a healthy content of carbon. Since scientists, like everybody else, cannot resist having some fun with their work now and then, one of ours tried to make diamonds out of peanut butter. He succeeded, but I hasten to add that it is not quite as simple as it sounds. Also, peanut butter, in comparison with some other carbonaceous compounds, just is not a very good material to start with—at least not good enough to justify further investigation into the relative merits of creamy style versus crunchy style.

PLASMAS: STELLAR TEMPERATURES, AND FUSION

As far back as anyone can remember in the arts and sciences, the attainment of high temperatures has been important. High and ever higher temperature continues to be a primary objective of modern scientific research in many fields, with significant technological progress dependent upon the results. Scientific news these days seems dominated by items about high-temperature alloys, high-temperature vacuum tubes and electronic circuits, high-temperature semiconductors, high-temperature strength in polymers, and—if you will excuse the expression—many other similar hot topics. Nowhere is this goal of high temperature more enticing and exciting than in the study of high-temperature gas plasmas. And nowhere are we talking about such high temperatures, since here techniques are emerging for heating gases to millions of degrees—temperatures that are known to exist in the stars but which have until recently not been available in the laboratory.

The primary objective of these techniques is to bring under control the process of the hydrogen bomb so that it can be peacefully employed for the production of energy. This process—called a thermonuclear reaction—works by fusing together light atoms to form heavier atoms, with a resultant release of energy. The fusion process is not only the basic process of the hydrogen bomb, but also is the accepted explanation of the prodigious energy production of the sun. The first public discussion of the possibility that the thermonuclear process might be controlled and used for energy production was made by an Indian scientist, Dr. H. J. Bhabha, at the Geneva Conferences on the Peaceful Uses of Atomic Energy in 1955.

Subsequently, in April of 1956, the Russian scientist Kurchatov, speaking at the British Atomic Energy Establishment at Harwell, described experiments with stellar temperatures with controlled fusion as an objective. In July 1956, some work by the United States sponsored by the AEC was revealed in a paper by Dr. Richard Post of the University of California Radiation Laboratory.

From all of these sources there is now developing the outline, or at least a shadow, of future scientific discoveries potentially of great significance. Because many important scientific events in the past—for example, the nuclear chain reaction—have cast a prophetic shadow in advance of their arrival, there is a strong temptation to jump to the conclusion that the controlled thermonuclear reaction is practically here, or just around the corner. If that is the case, the announcement has not come to my attention. In any event, there is a little too much optimism in some quarters about fusion power, and one would be well advised to relax a bit and thoughtfully consider the nature of the discovery we contemplate will be made, and its possible consequences. And—most important—we should consider how long it will take for this proposed discovery to be felt at home. In this particular scientific challenge, the goal is so immense that there is a correspondingly great need to clearly delineate the difference between hope and accomplishment, for thus far research on a controlled fusion reaction has been much more productive of hope than of power. I am thoroughly convinced, however, that in the longer term these hopes cannot be denied and that eventually a reluctant nature must be made to yield a controlled thermonuclear reaction for energy generation. But vast harm can be done by a myopic view of fusion, and in the appraisal of this not-impossible technological event there is a clear need for perspective and balance.

For perspective, let us review the chronology of the fission, or atom-splitting, process, on which all present atomic power plans are being based. The process was discovered in 1938 and subsequently underwent intensive development in the atomic bomb project. The first appreciable consideration of the possible application of fission to power production dates from 1942, when Fermi's group actually accomplished a chain reaction. Extensive efforts and corresponding expenditures have been made since then to bring this older process of the atomic age to the start of practical use. The *start* of practical use will be approximately 1960. I am overlooking several earlier small-scale pilot power plants and also the important fact that the large-scale plants which will then be in production will at best show only marginal economy. Thus it will have taken approximately eighteen years to bring fission, as a technically feasible new energy process, from Fermi's laboratory to a position where power from Commonwealth Edison's Dresden station may begin to help light that historic room under the University of Chicago stadium.

Now, let us look at the timetable for the hydrogen process. Since a technically feasible (not necessarily economically feasible) *fusion* process has not been announced, it is safe to assume that we are not yet at the starting point corresponding to 1942 in the fission process. Furthermore, it is a fact of technological life that a very long time is required to develop technically complex, large-scale power machinery, and the development of fusion, *after a successful process is in hand*, will be subject to this reality.

My own view is that: (1) five additional years of research will be required to make possible a realistic appraisal of the fusion process, (2) in ten years we may be at the point of technical feasibility, and (3) pilot-plant production of fusion power will not begin for twenty years. Competitive economic power production lies beyond that. In this picture there is a clear role for power production by fission in the near future, and power by fusion for the longer term. Fission power is technically feasible today, will rapidly become competitive with older energy sources, and is supported by a fifteen-year investment in specialized technology as a firm foundation for growth. The final phasing-in of fusion power, if it eventually becomes feasible, will be greatly facilitated by the accumulated atomic power experience of the utility industry, which by that time will add up to at least two, and perhaps three, decades.

It is now appropriate to turn from the possible *chronology* of fusion to the *science* of fusion as it looks in the laboratory. To the researcher in the physical sciences, it would be hard to find a more fascinating and challenging problem than fusion. To bring stellar temperatures into the laboratory for study is a dream that, until recently, few astrophysicists would have had the courage to express. Yet the work already reported gives considerable substance to this dream. Perhaps the most important single concept so far is that of magnetic containment. At first sight it would appear to be impossible to contain a gas at a temperature of many millions of degrees, since all known materials melt and vaporize at very much lower temperatures. But stellar gas is a very special gas indeed, so much so that it has been referred to as a "fourth form of matter," to be considered along with conventional gases, liquids, and solids.

At room temperatures, common gases are electrically insulating—that is, they prevent the passage of appreciable electric current. As the temperature of a gas is raised, it becomes progressively a better electrical conductor because the gas atoms ionize and become electrical in character. This in itself is not new because an ionized gas as a conductor of electricity has long been an article of commerce. Ionized gas fills fluorescent lamps and generates the ultraviolet radiation which causes their phosphors to give off light. Ionized gas is the basis for a wide variety of electron tubes which are important in radio, TV, communication, industrial and military equipment. Finally, ionized gas plays a fundamental role in the interruption of circuits carrying electric current. In the large electric circuit breakers employed in power systems the ionized gas, or arc, is generally controlled by a magnetic field, and this fact is the clue to stellar temperatures. A gas at a temperature of a million degrees is an electrical gas. It is fully ionized—that is, each gas atom is an electrical charge carrier. This fully ionized or electrical gas is a good conductor of electricity (at ten million degrees its conductivity would be comparable to copper). It also can be controlled with a magnetic field. In this fact lies the hope of bringing stellar temperatures into the laboratory, and eventually of achieving a successful fusion reaction. The idea is to make a "magnetic container" that will hold a bit of cosmic plasma so that it never touches the walls; just how to do this still stumps the experts, and how to fill the container is another sticky problem to be worked out. And to make it more difficult, it is not a million degrees, but 100 million degrees that we must have eventually. But we have a foot in the door, and in response to what will eventually become an overwhelming need, nature must, if at all possible, be made to yield a successful answer to this effort.

In spite of the reservations I have mentioned about the length of time it may take to realize the goal of practical power from fusion, at our research laboratory we are convinced that the ultimate importance of safe and inexpensive power from hydrogen makes it essential for us to study the problem right now. We have established at the General Electric Research Laboratory in Schenectady a substantial research program to study the fusion process. It would be presumptuous for me to compare the magnitude of our effort with that of the AEC's famous Project Sherwood, but we *do* expect to focus the traditional skills of the electrical industry on this primarily electrical problem. We believe it is important for both industry and government to be engaged in studying matters of such great importance to our future. Our work on fusion has been going on for over a year and is expanding. We do not yet have any results to announce publicly and—frankly—have no expectations of building fusion pilot plants in the near future.

As Dr. Henry Hurwitz, who is heading up our fusion program, has put it, "The processes for generating and containing a thermonuclear plasma employ magnetic fields, electric discharge phenomena, power circuit design, and current-handling methods. Fusion technology, as now conceived, thus employs the most basic technical skills of the electrical industry and it is inevitable that the industry can contribute heavily to this important development."

My ten-year-old friend may not have to look very far down the list of elements on the periodic table to find something to concentrate on. The very first on the list, and supposedly the "simplest," is hydrogen. Right now I would be willing to bet that—if he wants to become an expert on hydrogen—when my young friend is looking over those 500 job offers in 1970, he will find that one of them is from us.

Science—The Fabulous Pitcher

It is the great beauty of our science that advancement in it, whether in a degree great or small, instead of exhausting the subject of research, opens the doors to further and more abundant knowledge, overflowing with beauty and utility. MICHAEL FARADAY

Reprinted from *The General Electric Review*, General Electric Company, November 1957.

There is an ancient Greek fable which tells of a miraculous pitcher of wine that never emptied, but renewed itself of its own accord as fast as the wine was poured. This presents an enticing concept–one quite contrary to experience, which teaches that good things tend to run out or to run down. A good well may supply water for a long time, perhaps indefinitely, but no practical wine pitcher lives up to the one in the fable. Even the best mines finally give out and have to be abandoned.

I was reminded of this fable of the wine pitcher not long ago when a friend of mine expressed misgiving about the capacity of science to expand indefinitely. "Is it realistic to expect," he asked, "that the astounding avalanche of scientific discovery which the past half-century has witnessed can continue indefinitely into the future? And is it likely that the technological progress of the next half-century will equal the revolutionary technical achievements of the last 50 years?"

These are thoughtful questions. Possibly there are others who have wondered whether the magic pitcher of science is in danger of running dry. How does the fabulous vessel renew itself? Certainly not by any supernatural power as related in the fable, but by the infinite multiplication of human endeavor. Perhaps nature's storehouse of knowledge is truly an infinite source, and the real limit on what we draw from it is the human mind–its resourcefulness, its ingenuity, and, most important, its comprehension. In any event, science is a powerful attribute of modern life, and it seems that the more that its substance and its methods are employed to the benefit of man, the greater is the reservoir of "abundant knowledge, overflowing with beauty and utility."

FROM IDEAS TO ASSETS

Technological progress is based upon the translation of scientific ideas and concepts into practical assets. It is interesting to note that Michael Faraday–himself a researcher in pure science–emphasized *utility*. The practical benefits of science are more than a characteristic feature of American life today; they exert a profound influence on the national economy, on our society, and on our national defense. In the electrical industry our technology is based partly on classical researches of the last century and partly on a growing host of new physical phenomena that have emerged from modern researchers and have more recently traversed the long path through applied research and engineering.

Through Fundamental Research

The starting point is *fundamental research*, perhaps better described as *learning work;* it is a primary wellspring for creative ideas and innovations that eventually enlarge our tangible assets through subsequent applying work. No better example of the eventual utility of new knowledge and understanding which originates in learning work could be found than the researches of Michael Faraday himself.

Our present electrical technology derives directly from the ideas generated by Faraday and his contemporaries, Maxwell and Henry. The basic ideas of communication by means of electromagnetic waves are directly traceable to the fundamental researches of Hertz. The extensive medical and industrial uses of X-rays stem from the original idea of Roentgen in 1895 and the essential contributions of Coolidge in 1913. J. J. Thomson identified the electron in 1897; and the important work of many later contributors, including Irving Langmuir, brought it "to market."

From a relatively few basic original ideas may come an infinite variety of practical assets; at the time of the original idea the ultimate practical impact of the concept may be hard to predict.

Engineering

Adapting technological assets to human needs is the role of engineering. This is a vital phase in technological progress, and one that demands the close interrelationship of science and engineering. Although the original creative ideas in electricity and magnetism came from fundamental research, we owe our vast electrical resources to the application and continued improvement of engineering skills in the generation, distribution, and utilization of electric power. The great diversity of chemical products and processes that enrich our technology depend upon the happy wedding of chemical research and chemical engineering. Evidence of the important role of engineering in present-day technology is so impressive and so widely recognized that there is no need to review it here.

My objective in this paper is to emphasize the importance of the *combined impact* of scientific research and engineering upon the American way of life, both now and in the years ahead.

Throughout my professional life it has been my privilege to be associated with an industrial research laboratory whose primary responsibility is the search for new ideas. There is no more exciting quest than the pursuit of a new fact of nature, and one of the greatest sources of satisfaction for anyone closely associated with industrial research is the transition of ideas into assets.

During the fifty-seven years of its existence, the General Electric Research Laboratory, in cooperation with engineers and scientists throughout the company, has participated in a wide variety of technological enterprises of great public utility. In many of these the ultimate value of the original idea could not have been foretold with any degree of certainty. Looking back, however, it is not difficult to trace the development of an idea from its conception in research to its useful application through the manifold contributions of engineering.

Take, for example, the fundamental researches of Irving Langmuir. Like Faraday, Langmuir was a scientist whose work in pure science—although motivated primarily by insatiable curiosity—was directly responsible for widespread practical benefits. And like Faraday, Langmuir was always alert to the utility of scientific knowledge. Speaking of his own work, Langmuir once said: "Whatever has come as industrial applications has come incidentally to experiments followed for their interest alone."

That practical benefits do surely derive from pure research is well demonstrated by Langmuir's own career. From his early researches on the conduction of heat in gases and related phenomena, we got the gas-filled lamp, which increased the efficiency of the incandescent lamp by a very large factor. Estimated savings to the American public resulting from the development of the gas-filled lamp amount to $1 billion annually—a spectacular example of the translation of ideas into assets.

Langmuir's basic researches on thermionic emission and electron space charge are now classic. They laid the foundation for manifold developments in electron tubes and, incidentally, won for him the highly coveted Faraday medal of the Institute of Electrical Engineers of Great Britain, an appropriate citation for a true disciple of Faraday. The industrial art of hydrogen welding grew out of Langmuir's studies of the dissociation of hydrogen, and the condensation pump was the result of his investigations in high vacuum.

This brief reference to Langmuir's work reminds me of a question that I am frequently asked: "Your elder scientists are firmly established in the hall of fame, but where are your Langmuirs and Coolidges of today?" To this challenging inquiry I reply unequivocally: "They are right here in the Research Laboratory." They are young. They are becoming recognized in the scientific fraternity, but their ideas have yet to be translated into generally recognizable assets. Forty years ago Langmuir's genius was recognized only by his immediate scientific fellowship. If, in 1915, my distinguished predecessor, Dr. Willis R. Whitney, had been asked: "Where are your Newtons and Maxwells?" would anyone have appreciated his probable reply, "We have Langmuir, Coolidge, Hull, and Dushman"?

If you looked into Langmuir's laboratory about 1912, you would find his quarters somewhat more primitive than the modern laboratory today. But in these simple facilities some very important researches were accomplished.

What about the work he was doing? As typical examples, I have listed here the titles of several of Langmuir's early papers in technical journals:

Chemically Active Modification of Hydrogen.
Chemical Reactions at Very Low Pressures.
Effect of Space Charge and Residual Gases on Thermionic Currents in High Vacuum.
Vapor Pressure of Metallic Tungsten.
Dissociation of Hydrogen into Atoms.

How many of them would have been recognized outside scientific circles as portending great practical benefits? For that matter, would anyone thumbing through the pages of Faraday's notebook in 1825 have been able to predict that the ideas contained therein would change the way of life the world over? Look at a modern steam turbine-generator and then at a page from Faraday's notebook, and you cannot help being impressed with the impossibility of predicting the eventual utility of fundamental scientific research.

The measure of a scientist is the originality and fruitfulness of his ideas. The great names of today are those with an accumulation of achievements through productive ideas. The great names of tomorrow will emerge from among the young scientists who are today pursuing their abstruse ideas in the seclusion of their laboratories. I cannot give you their names today, but they will be found in the scientific headlines tomorrow.

Time

Sometimes a long period of time elapses between the conception of the basic idea and its development into a practical asset. A case in point is the magnetron, the principle of which was discovered by A. W. Hull in 1925. The idea of controlling the flow of current in a vacuum tube by means of an external magnetic field was a new and promising one, but it was nearly twenty years before a device based on Hull's original idea, further developed and extended by others, was actually put to use.

The opportunity came during World War II, when the pulsed magnetron emerged as the best generator for the microwave radar, and the continuous-wave magnetron was developed as the most powerful means of jamming enemy radar. But it required the necessities of war and many important contributions by other scientists and engineers to complete the transition from idea to asset.

It is now likely that a modern version of the magnetron, still further developed and extended, will play a basic role in electronic cooking. None of these applications of the magnetron was dreamed of by the most imaginative scientists and engineers of 1925.

And Team Effort

Important among new chemical products of the past decade are the *silicones*, polymers which have provided us with a large new family of synthetic rubbers, oils, resins, and varnishes for a great variety of applications. The crucial idea that gave impetus to a long and interlocking chain of developments which eventually brought silicone products to the marketplace was the discovery of a substitute for the *Grignard* reaction for producing methyl and phenyl chlorosilanes, the basic materials from which silicones are made.

The fundamental discovery was made by Dr. E. Rochow; and,

as in much of present-day work, the important subsequent steps were the result of teamwork involving many workers. Incidentally, the team effort required in so much of modern research and the interplay of diverse scientific endeavors are factors that make it difficult to individually identify today's young Langmuirs, Coolidges, and Hulls.

Magnetic materials are of great importance to the electrical industry, and our laboratory has been engaged in research in this area almost from its inception. For many years the performance of electric equipment was limited by the inherent magnetic properties of ordinary iron. In the 1930's a new idea—introduced and developed by our metallurgists under the leadership of William E. Ruder—culminated in a way to improve upon the natural magnetic behavior of iron by rendering it especially susceptible to magnetization along a given axis. The resulting "oriented" magnetic sheet provided an additional degree of freedom to the designers of electric equipment and increased the efficiency of the equipment significantly.

Another significant step along this line is a new magnetic sheet that is "doubly oriented"; that is, it has two lines of ready magnetization at right angles to each other. This material, which will make possible many improvements in design, was the result of the cooperative effort of young theoretical and experimental metallurgists in our laboratory.

CURRENT IDEAS—FUTURE ASSETS

It is easy to look back and recognize the birth of ideas that resulted in current assets. The evidence is clear that present technology is based on earlier science. It is not so easy to look ahead and fully perceive the future assets which will accrue from present ideas. However, there are some areas of active research today in which I believe it is possible to make some speculative predictions.

We are now well advanced in a metamorphosis in world technology that is so far-reaching in its implications as to be almost revolutionary. To an ever-increasing extent, technological progress depends upon scientific advances rather than upon the availability and variety of natural products. In a primitive technology, the essential raw materials are extracted from the earth's crust and used with little or no modification. Iron, copper, gold, coal, and petroleum are examples of such primary resources.

As technology advances in scope and complexity, the value of primary resources is enhanced by modifications of the raw material. Steelmaking, for example, represents an "added value" to primary raw materials that has had an enormous influence upon technological progress. The greater the complexity of technology, the more important are the added values contributed by science.

Natural products, of course, will always be of primary importance, but their utilization to meet new technological requirements will demand much more than merely extracting and refining. In developing new assets based upon the physical properties of natural products, scientific research serves as a bountiful source of new technology. To an increasing extent, the role of industrial research will be the enhancement of these added values, which contribute to technological innovations.

IN THE AREA OF SEMICONDUCTORS

A current example of the value of fundamental research as a source of new technology is solid-state physics. Just as the elucidation of the laws of gases in the last century led to unforeseen technological developments over a broad front, so the Twentieth-century study of the properties of matter in the solid state is advancing modern technology along new lines. In particular, the development of solid-state devices employing semiconductor materials, such as transistors and rectifiers, is adding a new dimension to electronics.

This technological area provides an excellent illustration of added values contributed by research and engineering. The common semiconductor materials, germanium and silicon, while plentiful and readily obtained, are, in the natural state, wholly unsuited to electronic applications. Not only is an unprecedented degree of purity essential, but also once purified, they must be modified, or "doped," by the addition of minute and precise quantities of appropriate "impurities." An understanding of the physical role of these critical additives, as *donors* or *acceptors* of electrons, is one of the important objectives of fundamental research on this class of materials.

Ten years ago semiconductors were little known materials of scientific interest only; today semiconductor products constitute one of the most rapidly growing industries in America. An industry that was wholly nonexistent in 1950 is expected this year to gross over $140 million; and it is conservatively estimated that the figure will reach $1 billion in ten years. Thus, a whole new industry was born in the research laboratory and brought to maturity through the combined contributions of research and engineering. Moreover, the birth of this new industry has provided myriad opportunities for new business ventures—large and small.

It is likely that ten years from now the annual numerical production of transistors and other solid-state devices will exceed that of vacuum tubes. Semiconductor devices will play an important role in the growing applications of computers, information storage systems, and automation. The transistor will invade the home to a much greater extent than the vacuum tube has done—in control devices for heating, air conditioning, and electronic cooking, and possibly in communication.

It is not unrealistic to expect that the automobiles of the various members of the family will be linked to home base through a communication system involving multiple transistorized signal receivers powered by miniature solid-electrolyte batteries. As new solid-state devices like the transistor become commonplace, it should be remembered that these working gadgets are the progeny of ideas born of pure science.

Industrial Diamond

Another area of pure research that holds great portent for the advancement of future technology is the superpressure, supertemperature field. This, too, is an example of a *learning* activity motivated by curiosity about a hitherto inaccessible region in nature. When this work was initiated in our laboratory six years ago, some of the phenomena to be encountered in this region could be anticipated from the evidence of naturally occurring substances, such as diamonds; others could only be guessed at. The successful culmination of the attempt to find a reproducible method of making diamond was a confirmation of preconceived ideas based on scientific understanding. The potential value of a practical commercial process was, of course, fully recognized. The conversion of graphite into diamond is a striking example of added value derived from research and engineering. At the current market price of industrial diamond, the value of the primary raw material is multiplied several thousand times.

Only recently our Metallurgical Products Department at Detroit released the announcement that Man-Made diamonds are now on the market. Actually, they have been too modest in their announcement. The elapsed time between the date of the discovery of the basic process in the laboratory, to the first marketing of diamonds made by the process was about 2½ years, an achievement that reflects great credit upon the cooperative effort of scientists, engineers, and manufacturing specialists. In fact, this astonishing result in bringing a fundamental discovery to market must set some kind of a world's record. This prompt transition of new knowledge into practical benefits could hardly be done except by the close coupling that existed between the source of the new fundamental knowledge and the specialized engineering and manufacturing skills in the sponsor's own company—thus a quick, efficient transition from ideas to assets.

When we made the initial diamond announcement in February 1955, I expressed the opinion that scientific information gleaned from the exploration of this new region might, in the long run, prove more valuable than the ability to make diamond. If progress in the field since that time is a reliable indication, I am inclined to think I was not far wrong. The production of the entirely new material, *borazon*, is an indication of the potential riches to be found on this new frontier. In this case, a substance never observed in nature has been synthesized and found to possess properties more outstanding in some respects than diamond.

I predict that continuing learning work in this field will yield new assets more rare and more valuable than gems. With respect to Man-Made industrial diamond, it is entirely within the realm of possibility that eventually diamond, and perhaps borazon, will be commonplace in industrial abrasive applications.

And Nuclear Energy

Nuclear physics is a Twentieth-century field of investigation that will have tremendous impact upon the technology of the future. The potential value of the fusion process as a source of power presents one of the foremost technological problems of the day. Incidentally, the separation of the fissionable isotope U^{235} from natural uranium and the production of plutonium from U^{238} are notable examples of added values from research in this new and complex technology. The utilization of nuclear energy is a very broad area of investigation—one that demands the mobilization of a wide range of highly specialized skills and technical facilities.

One aspect of the problem that holds the center of the stage just now is the possibility of utilizing thermonuclear reactions in a *plasma*, a concentrated atmosphere of ionized gas atoms, as a source of controlled power. In contrast to the *fission* process, which releases energy by the splitting of heavy atoms, the thermonuclear reaction known as *fusion* involves the combination of light atoms at extremely high temperatures—of the order of stellar temperatures.

The technical obstacles confronting science in this area are formidable and the path to victory is only vaguely discernible, but the potential assets are so great that we cannot afford to do less than our utmost to overcome these obstacles. I feel confident that the time will come when nuclear energy will become an invaluable tool in the hands of man instead of a sword of Damocles hanging over his head.

It is just possible, too, that the study of plasmas will yield assets in areas that now seem remote to the immediate objective. For example, there is some interest in plasma phenomena in connection with the production of electromagnetic radiation, possibly X-rays. When the temperature of a gas is raised, it first radiates visible light, then at higher temperature ultraviolet light, and at still higher temperatures X-rays or gamma radiation. Thus a thermonuclear plasma may become an entirely new source of radiation. Another unexplored possibility in the study of plasmas is chemical synthesis —the production of complex molecules by chemical reactions in a fully ionized gas at very high temperatures. These speculations take us far beyond the horizon that now stretches out before us, but we must constantly peer into the distant future if we hope to draw continuously from the fabulous pitcher of science.

LARGE-SCALE INDUSTRIAL RESEARCH

Industrial research is an investment in the future. Like other investments, it involves risks—often long-term risks. Not all research projects bear fruit; in fact, a relatively small number of highly productive research ventures must carry the burden of the great number of unsuccessful undertakings necessary for thorough "prospecting" to insure that hidden opportunities are not missed.

The financial investment required to support a large-scale research program is very great, the risks are considerable, and the period before dividends are declared may be long. Here large industry has an important role to play. With extensive physical, economic, and human resources and with a great diversity of technological interests, the major industries are uniquely able to support the type of large-scale research effort necessary to guarantee continued technological progress. For these reasons, modern technological progress is dependent upon the research contributions of large industry—a role that is not always fully appreciated by the American public.

This is a good place to emphasize the importance of profits in large-scale industrial research. There is probably no aspect of American industry that is less understood and appreciated by the public than the vital role of profits in a free-enterprise system. The average citizen is inclined to have the somewhat vague opinion that profits are all right, provided they are not too big. On the subject of what is "too big," he is apt to be a bit hazy. He lives by the creed that the "laborer is worthy of his hire" and when he attempts to translate this creed in terms of business operations, his conclusion generally is that "some profit" is necessary and justifiable.

In our free-enterprise system, profits must be adequate to yield a reasonable return to the investors who supply the venture capital and also provide funds for the essential business investments necessary to keep pace in a progressive industry.

Research is an essential business investment—and a very costly one. The costs of carrying on a large-scale research program are subject to all the inflationary factors that are present in any business venture, exaggerated by the ever-increasing complexity of scientific work. It is unrealistic to assume that the research activity of private industry can be geared to the demands of technological progress without consideration of adequate profits with which to finance it. It is important that the American public understand that an adequate profit from the industrial enterprise will permit further investments in research and new technology with important benefits to all.

There are values in large-scale industrial research that are important to technological progress. Not the least of these is the *scope* of a large-scale research effort—the breadth and depth that result from the mobilization of a multiplicity of skills. Probably the chief characteristic of modern scientific research is the very high degree of specialization required in the various scientific areas.

The *depth* of a research effort depends upon the degree of specialization that can be focused upon the various technical aspects of the program; the *breadth* is determined by the range of specialization encompassed by the total effort. A comprehensive research program demands a multiplicity of technical skills and a diversity of scientific interests. Moreover, the productivity of the individual research worker is greatly enhanced by cross-fertilization taking place in a research environment that embraces a wide range of specialization.

One of the prime values of large-scale industrial research is the fact that the transition from new knowledge to useful practice is greatly facilitated when the entire sequence of events—from research to end product—takes place within one sponsoring organization. The coordination of effort is more efficient when all participating components are parts of the whole than it is when the research results from one or more independent sources must be correlated with engineering carried out by another organization and with manufacturing practices in yet another. Prompt utilization of research results demands integrated effort all along the line.

Although large-scale research is relatively new on the industrial scene, its role in maintaining the flow of wine from the miraculous pitcher is already well established. Its importance to future technology is clear. When you are looking ahead a decade or two in science, you must calculate among your assets, in addition to present knowledge, new discoveries that are certain to be made.

At the pace of modern scientific progress, two decades is a very long time. The last two decades, for instance, encompass the whole of the atomic age, including atomic bombs and thermonuclear weapons, as well as microwave electronics including radar, semiconductors, sulfa drugs, antibiotics, and a host of other developments that we now recognize as technological assets. There is no evidence that this pace is slowing down. Natural limits to new discovery, if they exist at all, are not discernible. Hence, we may expect the miraculous pitcher to yield everything that our scientific competence and resourcefulness can draw from it.

In conclusion, I should like to state the following beliefs.

- Future technological progress will dwarf the achievements of the past.
- As we progress technologically, appropriate and substantial emphasis must be given to *basic* research, the learning activity that constantly adds to our greatest asset—knowledge.
- Science and technology will become an ever-increasing basic source of industrial growth.
- To an ever-increasing degree, America and the world at large will become a science-oriented society.

Large Business as a Source of Technical Assistance for Small Business

A talk extracted from a paper on "Industrial Research in a Free Society" submitted to the President's Conference on Technical and Distribution Research for the Benefit of Small Business, Washington, D.C., September 24, 1957.

In my hand I am holding a small jar of material worth, at current prices, about one thousand dollars and a small silvery cylinder that is more valuable than an equivalent amount of gold. The jar contains industrial diamonds, but not diamonds dug from the earth. These are diamonds made by man and the raw materials cost less than a penny. This silvery cylinder is not silver, it is silicon, the most abundant solid element on the earth. There is more silicon in a worthless handful of sand than there is in this small cylinder, but what I hold is worth several hundred dollars because it is almost entirely silicon, prepared in a scientific laboratory so that it is 99.99999999 per cent pure.

The *practical* value of these industrial diamonds is in cutting, grinding, and machining our toughest metals, glass, and ceramics. The *practical* value of this small cylinder of silicon is that from it 5,000 transistors can be made. These materials symbolize the modern scientific "alchemy" that literally is producing a "technological revolution" in America.

Primitive man dug raw materials from the earth and used them "as is" or with little modification. An easy measure of technological progress is the extent to which industry gives added value to raw materials. Think what the art of steel-making adds to the value of iron—and what this added value has meant to technological progress over the last century.

Making diamonds and electronic devices from the simplest of elements typifies the ability of modern science to add value to raw materials. This jar and this cylinder symbolize the fact that our nation's frontiers are no longer geographical but scientific, and our prospectors and explorers are not seeking new sources of gold in mountain streams but rather they are searching for new sources of knowledge about nature in the highlands of modern research. Technological progress now depends far more on new scientific advances than upon the availability of natural products. This is the true meaning of the technological revolution.

In less than ten years, discoveries of new properties and new phenomena in so-called "semiconductor" materials like this pure silicon have resulted in the creation of hundreds of new small businesses and have given new opportunities to hundreds of existing businesses. The semiconductor industry, which was nonexistent in 1950, grossed $140 million in 1957 and is expected to reach the billion-a-year figure by 1967. It will be discussed later on today's program.

THE ROLE OF RESEARCH

Large-scale industrial research is playing a major role in the technological revolution. Because it is "large-scale" and because it is "industrial," the research I will discuss today has several *unique values* that produce benefits for all of us.

Industrial research is an investment in the future involving *long-term risks*. A small number of highly productive research ventures must carry the burden of the many unsuccessful scientific explorations that are necessary to insure that hidden opportunities are not overlooked. The financial investment required to support a large-scale research program is very substantial, the risks are considerable, and the period before dividends are declared is frequently long. Here large companies have an important role to play. Possessing the necessary physical, economic, and human resources, and with a great diversity of technological interest, the major industries have the unique opportunity to support the risk-type of large-scale research effort necessary to guarantee continued technological progress. *All* companies get very substantial benefits from the research contributions made possible by the long-term risks of *large* industry—which is a fact of industrial life that is not always fully understood by the American public.

A second unique value in large-scale industrial research is the *scope* of this research effort—the *breadth* and *depth* that result from the mobilization of a multiplicity of skills. Going hand in hand with the high order of specialization demanded by our increasingly complex technology is the need for *team* effort. An all-out attack on a major research problem—such as these Man-Made diamonds, for example—demands teamwork involving many highly specialized skills: in this case, physicists skilled in experimental work at extremely high temperatures and pressures, chemists familiar with the kinetics of phase transformation, specialists in crystal structure, and engineers experienced in the design and construction of unusual mechanical equipment. Industrial research today differs from that of a generation or so ago. The lone inventor in his solitary laboratory, once the mainstay of technological progress, is now supplemented by the scientific task force armed with the elaborate technical gear required for an all-out attack on the complex problems of modern science. Many schoolboys know that Alfred Nobel invented dynamite, but it is not so easy to answer the question, "Who invented the atomic bomb?"

SCIENTIFIC CROSS-FERTILIZATION

Another reason for the effectiveness of large-scale research, covering a wide range of interests, is the *cross-fertilization* that takes place among scientists who are experts in various fields. Each scientist, working in his own field, is stimulated by his contacts with the development of concepts, techniques, and instrumentation in neighboring fields, with important benefits to his own progress as an individual researcher. Thus, team activity triggers a chain-reaction of ideas. The psychological effect of meeting each day with fellow members in the broad scientific fraternity—in an atmosphere charged with the adventurous spirit of research—breeds creative progress. In such an environment, the scientist has readily available to him the advice and assistance of specialists from many fields. An electronics research man needs a special alloy or a new ceramic to meet certain critical requirements; he makes his wants known and the specialized skills of metallurgists and ceramists are at his disposal, together with the appropriate technical facilities. A solid-state physicist encounters a problem that calls for the specialized knowledge of the physical chemist; a nuclear physicist has a problem in high-frequency circuits; an electrical engineer encounters problems in gas discharges—examples of the specialization and multiple skills which characterize modern scientific research.

FACILITIES FOR SPECIALIZED VENTURES

Also among the assets to be found in large industry is the ability to make heavy investments in research *facilities* for specialized research ventures. At our laboratory we have a building that is more like a factory than a laboratory except that its production-size equipment is devoted to research. Here large arc furnaces, extrusion presses, and rolling mills have a low use factor that could never be justified in a manufacturing plant, but their availability is vital to a research program aimed at finding better metal and ceramic materials for application throughout our company.

The ability of a large industrial organization to take long-term financial risks assures a *continuity* of the research program. Research is not like a water faucet that can be turned off and on at will to satisfy an immediate need; the stream of new knowledge must be kept flowing in times of feast and famine or we will lose contact with the source just when we most need nourishment.

A key function of a research organization is *maintaining contact* with the world of science. By contributing on a significant scale to the body of science, over a wide range of scientific areas, the staff of a large industrial laboratory can keep its lines of communication active and be in a position to obtain, appraise, and use new knowledge developed anywhere in the world.

Industrial scientific research is a primary source of growth and profitability in modern industry, and it is rapidly assuming its appropriate functional role along with the older industrial functions such as engineering, marketing, and manufacturing. As such, the scope and character of research are determined and justified by the business objectives of the company it serves. In what way then does the comprehensive research program of the large company benefit the small business enterprise? I will show in the following discussion that this benefit consists in the creation of new products, new markets, and new industries with widespread industrial participation. (A particularly striking example of the latter will be presented in the case history to follow this paper.)

I have discussed the unique values of large-scale industrial research in terms of risk-taking, scope, team effort, cross-fertilization of ideas, facilities, continuity, communication with the world of science, and business incentives. All of these values are attained without sacrificing the advantage of operating in a free and competitive manner. Unlike scientists behind the Iron Curtain, scientists working in a free society have the reward that comes with the realization that their work is providing people with more of the things they want in this free society.

RESPONSIBILITY TO OWNERS

Industrial research must obviously be geared to the business it serves; we are past the time when it was pursued for the glamour of science, or to keep up with one's industrial neighbors. In our own company, expenditures for research *and development* in some years have been greater than the company's net profits, and thus it should surprise no one that our 380,000 share owners expect, deserve, and get substantial answers to the "why" and "how much" questions, as they apply to research.

A friend of mine who is the research director of a well-known company, having heard that I was to talk here today about large business as a source of technical assistance for small business, told me—no doubt facetiously—that he had a sure-fire illustration for me to use. He said that every time he presents his research budget to the owners and managers of his company, somebody comments, "Sure research is a good thing, but doesn't everybody—including our competitors—get as much advantage from our research as we do?"

This is an important comment and it concerns a key aspect of industrial scientific research. It is an admitted fact that *others* do benefit from *your* research. The sponsor, however, expects his research contribution to enable him to be "first" with new and improved products and thus to broaden his opportunity to serve his customers and to deserve an adequate profit. The new markets thus created will not be served by a single company. Typically, a new market is an opportunity for widespread participation by industrial units large and small, giving full play to the specialized skills of the supplier of materials, components, machinery, and equipment as well as the manufacturer of the complete product or product line. This broadening of the market is generally of substantial benefit not only to the many participants but also to the entrepreneur whose research initiated the opportunity.

Fortunately, science is a plentiful well; if pumped energetically—and steadily—it can fill the cup of the pumper, thereby enabling him to continue his efforts, and still yield abundant benefits for others.

BETTER VALUES FOR CUSTOMERS

The results of research sustain the performing company by providing the innovations that make possible more attractive values for customers. Of course, the principal objective of our research program is to "keep ahead." The fact of the matter is, however, that we must worry about *keeping up* as well as *keeping ahead.* Technological progress begins in many sources, and no single laboratory or group of laboratories can expect always to be "first." At best, a large-scale research program can give the company it serves a series of transient opportunities for product leadership. But even at the very least, the supporting company should expect that its research program will appraise new scientific discoveries reported throughout the world of science and thus help the company better serve the wants of its customers in a competitive market.

Professor Sumner Slichter of Harvard University wrote recently, "The strength of competition in American industry will, to an increasing extent, be determined by the scale of technological research and development. Large enterprises, far from being a menace, will to a growing extent be the instruments by which the country is given the benefit of large-scale technological research and of increasingly vigorous competition."

The results of industrial research are widely distributed not only to the supporting company, but also directly and indirectly to many others—in the case of our company to 280,000 employees, 380,000 share owners, 42,000 vendors, 400,000 retailers, countless other businesses, and millions of customers.

There are many reasons why the same research and development effort that benefits our own company and our customers also benefits other businesses, particularly small businesses. As I have indicated, among the most important of these reasons is the fact that new products created by large-scale industrial research almost invariably create new markets, not only for the company that supported the research, but also for many other companies as well. I would like now to cite some interesting case histories from General Electric experience.

IMPACT OF SEMICONDUCTORS

Earlier in my remarks, I mentioned the semiconductor business, which was born in 1950 and will probably become a billion dollar business by 1967. Actually the industrial impact of this new industry is not adequately measured by $1 billion. Semiconductor devices, as new and versatile electronic components, have stimulated the growth of many older businesses and have created entirely new ones. Portable radios and communication systems, improved hearing aids, compact computers, automation equipment, improved industrial power conversion equipment, and numerous important applications to weapon technology are among the products which have been rejuvenated or have been born by virtue of semiconductor components.

I hasten to add that in this case, as in many others, no one company is solely responsible for the new technology which created the new market opportunities. The invention of the transistor by the Bell Telephone Laboratories was in this case a *key* contribution, but many other developments—including high purity germanium and silicon, the semiconductor diode, and the proliferation of transistor types—have been contributed by other companies including my own.

The new market created by semiconductors and related products thus benefits the sponsor and in our case well justifies our research and development expenditures in this field. However, the *broad* participation in the new market is a corollary benefit which accrues to many participants and is a particularly attractive opportunity for the small company, many of which have been born or have matured in this new market.

Another good example is the fluorescent lamp introduced less than twenty years ago. Obviously, fluorescent lamps themselves are now made by many manufacturers, and as in the case of semiconductors, there also have resulted many other opportunities for other companies in related fields: our company does not manufacture fluorescent fixtures, for instance, but literally scores of companies do, and most of these are small, with less than 500 employees. In the manufacture of fluorescent lamps, we purchase services and materials from hundreds of vendors. Fluorescent lamps also have meant opportunities for thousands of independent dealers, electrical contractors, and others in electrical services trades.

SHARED DEVELOPMENT

The electrical home appliance field is one in which the development of a new market based on new General Electric technology has been widely shared and extended by others. I recall the hermetically sealed refrigerator unit, which, in the early days of this industry, added basically to the reliability of this device and certainly promoted its acceptability in the home. This hermetic technology, developed by General Electric in the late 1920's, is presently standard practice in the trade. The early electric range did not achieve acceptance in the home because of the open coil heaters which were both dangerous and short-lived. In 1918, the development, by our Research Laboratory, of the Calrod heater—a completely sheathed, long-lived unit—was a basic stimulant to this industry in which so many companies now share. Sheathed heaters are now standard in home ranges of all manufacturers.

Food waste disposers, developed first by our engineers in 1935 as a pioneering home appliance, are now being made by eleven other companies and are a source of new business for thousands of local retailers. The electric blanket, developed at the General Electric Research Laboratory in 1936, is now being made by many companies, and a closely related product, the heating pad, is being made by an even larger number of other companies, most fitting the definition of "small business." For our own manufacture of these products, we purchase materials and services from a large number of other companies. And we do more than merely buy from them. We also interchange assistance and technical knowledge, produced through research, with our suppliers—to our mutual benefit.

POLICY OF BUYING LOCALLY

A policy of being a good "community neighbor" allows a large plant to contribute to the business of other companies in the same area. General Electric's Appliance Control Department at Morrison, Illinois—a town of 4,000 people—buys from 1400 firms, and 780 of these are within 150 miles of Morrison. In Rome, Georgia—a town of 35,000—our Medium Transformer plant buys from 900 different companies, 540 of them nearby.

Our company as a whole spends about $2 billion a year buying the goods and skills of 42,000 suppliers across the country. Some of these are big companies, furnishing us with complete transportation service or vast quantities of giant machinery, but 90 per cent of them are small businesses offering the special skills we need.

Once again, the large volume of our purchases from small companies can be traced directly to the new products and product innovations wrought by large-scale research and development, and once again we should remember that many of the fruits of our research must—of necessity—be given to our suppliers so that we, and they, can profit by putting the results of research into actual practice, improving the value of what is being produced.

There are many other ways in which large-scale industrial research aids small business. Patents and the licensing of ideas generated in large laboratories give profit opportunities to thousands of businesses—and jobs to millions of men and women.

Of all the ways in which large-scale research benefits small business, perhaps the most important is through the creation and diffusion of scientific knowledge—the contributions of our scientists to the "world of science" from which everyone now, and in the future, can benefit.

Two-thirds of the research and development in America today is being undertaken by private industry—a major shift in the balance of research effort since the days a few decades ago when university laboratories and laboratories of individual scientists accounted for most of it. The National Science Foundation has observed that only 10 per cent of small manufacturers (with less than 500 employees) have a research program, but nearly 95 per cent of manufacturers with over 5,000 employees have a research program. The large companies, although they account for only 40 per cent of manufacturing employment, do 72 per cent of all industrial research and development.

A COMMON POOL OF SCIENCE

All scientific progress is based on published scientific knowledge assembled from sources throughout the world—"world science." Thus all scientists and nearly all of their sponsors feel an obligation to contribute to the pool from which so much sustenance is drawn. Most industries today recognize the reciprocal advantages of prompt publication of research results.

Scientists and engineers at our company, for example, have published some 4,000 papers during the past ten years. Large-scale industrial research also makes important contributions to the world of science through the participation of its representatives in technical meetings, by training personnel for scientific work in many places—including small business—and in countless informal ways.

The existence of the huge pool of knowledge that I have called the "world of science" offers great hope and great challenge to small business. Here is a foundation of "free" information on which to extend present technology and build innovations. Frankly, the principal cost of this information is the cost of learning how to understand and use it. Small business must make an investment in this world of science, if not by contributing to it, at least by participating in it, and, at the very least, by keeping in touch with it and appraising its opportunities.

PROBLEMS OF SMALL BUSINESS

Not long ago, the typical businessman kept his own books and got along nicely, but as our economy has become more complex, the accountant and the tax expert have become invaluable to even the smallest business. Increasing complexities are also apparent in our technology, and the time has come when small business must see the need for making certain that it is fully tapping the resources of the world of science.

Part of the research and development problem, as it affects the owner of a small business, is that many such owners do *not* recognize that research and development is one of their major problems. A recent survey (made specially for this conference) showed that most people in small business believe that *finance* is their most important current problem. Small business itself views research and development as a lesser problem than *finance*, *personnel*, and *sales*, but representatives of large businesses–interviewed in the same survey–generally placed "research and development" as the primary problem facing small business, ranking it above personnel, finance, and sales. Thus it is not surprising that the financial and tax advisor is commonplace in the operation of small business. And it also follows that, as a representative of large industry, I would like to emphasize the importance to small business of becoming expert in keeping up with modern technology. The cup of research opportunity truly runneth over, and there is nourishment for all, large and small, who will take the initiative to go after it.

WHAT SMALL BUSINESSES CAN DO

"Going after it" is admittedly easier to say than to do. For the small business there are several available methods of tapping the world of science, methods that have proved highly useful to various small companies under varying circumstances. Where its technological needs are limited to a well-defined area of science, the small business should consider having its own scientific staff. Even a one-man staff, with the right man trained in the right area, can make a significant contribution through its own work, by keeping up with the pertinent technical literature, and by attending and interpreting the important work being presented at technical meetings and conferences. Where its technical needs are in several well-defined areas of science, and where there is a degree of coherence between them, a small business can consider creating a suitable research staff of its own. Alternatively, the small company may employ part-time technical and scientific consultants from competent nearby colleges, and may supplement such consultation by actual project research work on a contract basis. Industrial *research institutes* provide a particularly good opportunity for the small company to contract for its research program or supplement one already in place. The fact that these research institutes are growing in size, number, diversity, and competence is evidence of the increasing use that business, large and small, is making of this method of pursuing its research needs. The research institute can be of special value to the company whose products or services are related to widely diversified areas of science. Research institutes offer to small companies their most efficient and least expensive means of keeping up with the world of modern science, including the extensive contributions to the overall pool of knowledge that are made by large-scale industrial research.

And, of course, there is the technical know-how which one business can get directly from another business. Large companies that sell materials can make available—and *want* to make available—extensive scientific knowledge about the application of these materials. Large companies which buy parts and services from small companies can make available—and *want* to make available—technical help, scientific advice, and value-analysis programs that will be good business for both buyer and seller.

We may as well face up to reality and admit that many times these sources of technical know-how go tragically unused because the "mutual help" aspects are misunderstood and so-called "pride" keeps human beings from asking for technical advice. There are normal and established business channels through which businesses —large and small—can help each other. These channels are not being used to the fullest. In many cases, a businessman needing technical assistance could get it by calling—today—the technical service representative of a materials supplier or the appropriate purchasing agent of a large-company customer. The simple establishment of these productive relationships can do a great deal to distribute the new facts and techniques being produced by large-scale research and development, as they apply to the materials, components, and processes of interest to the company.

EFFECT ON NATIONAL SECURITY

No discussion of the benefits of large-scale research and development would be complete without a reference to what this effort means to our national security. In these times, when national security depends increasingly on advanced scientific and productive capacity and versatility, all industry continues to have a deeply important public responsibility to participate in the national defense program—even though commercial business offers a generally more profitable utilization of technical skills and facilities.

Ralph J. Cordiner, president of General Electric, has summarized the national defense picture in these words:

> Today's defense challenges are of such complexity that they require marshaling the full range of American business enterprise. Large companies with extensive technical skills and resources must provide overall co-ordination of complex projects which *they* can do best, while at the same time creating opportunities for the thousands of smaller or more specialized companies to contribute what *they* can do best.

Finally, in our discussion of the flow of benefits from large-scale industrial research, we must re-emphasize the positive and direct correlation between technological advances and a rising standard of living. The vitality of the expanding American economy has come in large measure from the flood of new products and industries that have been the result of the research and development activities of American business operating in a free society. If we continue to make long-range investments in our research, we will have greater profits in terms of enriched lives for ourselves and for our children.

REQUIREMENTS FOR FUTURE PROGRESS

A variety of forces and circumstances are limiting—even endangering—our nation's technological progress. It seems to me this conference should consider seven specific things that need to be done to insure the continuation of the benefits of large-scale industrial research. I believe that all seven of these items are at least as important to small business as they are to large business.

1. TO INSURE CONTINUATION OF THE BENEFITS OF RESEARCH, WE MUST BUILD UP THE NATION'S SUPPLY OF ADEQUATE TECHNICAL MANPOWER. Engineers and scientists obviously hold the key to our nation's progress in this swift-moving age. Industry, large and small, can do much to solve the shortage of technically trained people. Working with schools, we businessmen can take an active role in encouraging young people to further their studies, and to explore their technical aptitudes and interests. We can awaken students to future opportunities, provide tangible career-guidance material, help teachers become even more effective, and provide financial assistance to colleges in several ways, including that of helping our employees repay their alma maters for their education. Within our own companies—regardless of size—we can give our men and women opportunities to acquire new skills, encourage them to continue their education by establishing co-operative programs with nearby schools and colleges, and provide opportunities for professional training and development.

2. TO INSURE CONTINUATION OF THE BENEFITS OF RESEARCH—PARTICULARLY THE BENEFITS TO SMALL BUSINESS—WE MUST REJUVENATE OUR PATENT SYSTEM. There is need for a more sympathetic attitude on the part of the courts toward the original concepts and philosophy upon which the patent system is based. We must recognize that the inventor who provides previously unknown ideas, machines, and processes definitely contributes to the progress of our social structure. There also is need for improving the facilities of the United States Patent Office for determining whether the inventions submitted for patenting are new or old. It is apparent that the Patent Office, in common with other components of our technical society, threatens to be overwhelmed by its inability to accumulate, classify, and search—as its responsibilities require—the

growing mass of technical literature, including its own, within a reasonable time.

3. TO INSURE CONTINUATION OF THE BENEFITS OF RESEARCH WE MUST DEVELOP NEW IDEAS ABOUT SCIENTIFIC ADVANCES AND HOW THEY ARE PRODUCED. We need greater understanding of scientific and technical specialization and their relationship to team effort. We must recognize, in our developing management concepts, the basic character of creative technical work, and must provide the incentives which stimulate productive effort.

4. WE MUST CREATE A BETTER UNDERSTANDING AMONG OUR CITIZENS OF THE NECESSARY RELATIONSHIP BETWEEN PROFIT AND PROGRESS. Research is a high-risk business investment and a very costly one. It is unrealistic to assume that the research activity of private industry can be geared to the demands of technological progress without consideration of adequate profits—demands which provide the incentive for research and literally make it possible. Adequate profits give us the opportunity to maintain the cycle of technological progress: research—new products—jobs—sales—and the profits that permit the cycle to start again. We must reaffirm to people everywhere the basic American belief that profit, instead of being something "left over," is an integral element of the growth of our nation and our society.

5. TO INSURE CONTINUATION OF THE BENEFITS OF RESEARCH WE MUST CREATE A GREATER UNDERSTANDING, PRINCIPALLY ON THE PART OF MANAGEMENT AND EMPLOYEES, OF THE SIGNIFICANCE OF NEW PROCESSES AND NEW PRODUCTS. The human trait of "resistance to change" has kept people in some parts of the world literally in the dark ages. Even in America, individual unwillingness to accept the temporary "inconvenience" associated with innovation is a major deterrent to insuring our defense and improving our living standards. Obviously, I am not suggesting that everything that is new is necessarily good, or that all of the results of modern scientific research will yield immediate improvements in our products or our way of living. However, it is essential that we all contribute to an atmosphere in which new things can be evaluated objectively, without fear of changes simply because they *are* changes.

6. TO ASSURE CONTINUATION OF THE BENEFITS OF RESEARCH WE MUST CONTINUE TO URGE THAT A GREATER PROPORTION OF THE

NATION'S TOTAL RESEARCH EFFORT BE DONE BY INDUSTRY—LARGE AND SMALL—AND THAT THE GOVERNMENT'S RESEARCH BE ADMINISTERED IN A WAY THAT STIMULATES THE PROGRESS, VITALITY, AND INDEPENDENCE OF BUSINESS. Even in the national defense area, private industry would be willing and able to finance a greater proportion of the needed research and development—*if* the procurement policies were to provide the same profit incentives that over the years have been so magnificently productive in the civilian economy. With research and innovation assuming such a tremendously important role in the nation's economy, these elements should most certainly be kept within the framework of the free enterprise system.

7. TO INSURE CONTINUATION OF THE BENEFITS OF RESEARCH WE MUST ESTABLISH A CLIMATE OF INFORMED PUBLIC OPINION WHICH RECOGNIZES THE POLITICAL AND ECONOMIC CONDITIONS NECESSARY FOR ENCOURAGING THE EXPANSION OF INDUSTRIAL TECHNOLOGY.

The public as a whole must recognize the importance of large-scale industrial research and its proper role in our economy. Government must not let an emotional fear of "big business" reduce or destroy the contributions that such research can make to the nation's welfare and defense. The management of all American industry must recognize the importance of supporting the kind of fundamental science that provides us with basic new understanding about the world in which we live. And small business must recognize and take advantage of the countless opportunities being made available by large-scale industrial research operating in a free society.

The opportunities presented by today's "technological revolution" are so challenging that America needs the full range of her industrial resources, large and small, each company doing what it can do best, and each drawing strength from the other.

The Appraisal of Technological Change

An address at the Stanford Research Institute's Third Annual Industrial Economics Conference, Los Angeles, California, January 13, 1958.

As you have heard, and will continue to hear throughout this conference, there are many sources of industrial growth. I wish to focus on one of these sources—technological change. This factor in industrial growth is of dual interest; technology is having a growing impact on business—and business is having an ever-increasing impact on technology. In this area we have an opportunity not merely to *watch* growth take place but to *do* something about it.

Our technological opportunity is quite different, for example, from the opportunity provided by the study of population growth as an important factor in long-range economic forecasting and business planning. In technology, it is frequently easier to *do* than to *measure* what has been done. In regard to population, it is easier to *measure*. This may not be an appropriate occasion to discuss the extent to which individuals can influence population growth, but it may be safe to say that—historically, at least—this is a field in which long-range planning has not been notably successful.

As a basis for a discussion of the appraisal of technological change, we should first take a look at some of the present relationships between technology and modern industry. This look can be brief since it undoubtedly will not be necessary to review for this audience all of the complexities which intertwine scientific research and development with today's business enterprises.

There is a growing suspicion in many quarters that new technology, which originates in scientific research, may be the source of new business enterprise, and I will confess to a strong personal conviction that this is the case.

Furthermore, this belief must be rather widely held because of the following statistical information. Although our nation's gross national product (GNP), expressed in dollars, is increasing currently at the rate of 6 to 7 per cent per year, the technical activity included in the term *research and development* (R & D) is increasing at the rate of 13½ per cent per year. This increase holds for what might be called the "gross national R & D," including government, as well as for the industrially supported R & D. These numerical values are significantly different, even when we take into account the difference in the inflationary factor in each case, and the different rates cannot, obviously, continue indefinitely.

We are apparently increasing R & D as an industrial activity at an abnormally high rate in an effort to bring it into better balance with the other functions of manufacturing businesses. Although such technical industries as electrical manufacturing, chemicals, electronics, communications, pharmaceuticals, and many of the specialized defense industries have long had a relatively large R & D activity (expressed in relation to gross sales), there is evidence that, even in these industries, R & D is increasing at a somewhat higher rate than the corresponding product output.

Although the broad relevance of technology to business enterprise and growth is generally recognized, there are wide differences in the way in which this activity is practiced throughout industry. The needs of specific companies and the environmental factors associated with each industry presumably dictate these differences in practice.

The differences in practice to which I am referring include the differences in the expenditure level for such work in relation to the output of goods or services, the distribution of R & D effort as between R on the one hand and D on the other, the ever-present question of "make or buy" as applied to R & D, and so forth. For example, companies that employ all of their available capital for new plant investment (for manufacturing increased quantities of the same products, or simple mechanical modifications of the same product) may be inclined to make little or no expenditure for research and little for development. Other companies, and industries, have a long-standing practice of carrying out extensive development programs, but consider that research—although it is the primary source of the work represented by development—is not of their concern. Still other companies and industries, particularly the technical industries enumerated previously, carry out extensive programs in all phases of R & D.

A basic question involved in these choices is *the balance between short-range and long-range goals*. Almost any business can look quite good on the current operating statement by eliminating all *current* expenditures for *future* business. A decision to pursue this course would make it practically certain that there would be *no* future business.

The appropriate level of *current* expenditures for *future* business involves all functions of the business, not only R & D, and its determination as a part of a business plan requires delicate balancing of the elements of risk and opportunity. There is no formula approach to this problem, and the best current practice is based on trial and error, or as I would prefer, *change and measure.*

If we cannot prove from first principles how much R & D work should be done for the most profitable conduct of the business—we must rely upon our experience or the experience of others—we can at least change the expenditure level for such work, and observe what happens. In other words, we may evoke experience, or generate experience and evoke it for future guidance.

To illustrate "change and measure," I would like to use as an example a case involving marketing. Since this is a field in which I have no knowledge or competence, the case appeals to me as being simple and straightforward. The relationship between the sales of a consumer product and the advertising budget is one which can, I presume, be determined by trial. If increased advertising brings about increased sales of the product, we might further increase our advertising until the incremental gain in profit is offset by the incremental cost; this is the point of balance which is sought.

Similarly, the process of change and measure may in some cases be applied to the determining of the best relationship between R & D and the other functions of the business. If the principal function of the R & D is to provide product innovation, and if the product line is not too complex, it should not be very difficult to arrive at a suitable level for the R & D work. Variations in the results of such trial and error, over a long period of time, have presumably led to the wide differences noted in industrial R & D practice. It is not quite safe to conclude, however, that recent past experience in R & D is the best guide for future practice, for there are some built-in complexities. For one thing, measurement of the result of R & D work, and even the result of changes in the level of such work, frequently is very difficult. In the first place, the marketplace does not react promptly to R & D, and this is particularly true in the case of research.

Research programs, if they are designed to have a resounding impact on the business, may do their resounding only after a long period of years has passed and a large expenditure has been placed at risk. Furthermore, in important cases, the response of the business to R & D effort may be a complex chain reaction in which cause and effect are difficult to separate. For example, a new material may be introduced which results in a product improvement; but, in addition, the improved material may completely change manufacturing methods and even product design and design methods. The improved product as a component of other products may affect the market served by these additional products, and so forth.

As a relatively simple example of a new and rapidly growing business which originated in research in the electrical industry, consider semiconductors, including transistors and related products. Gross annual sales are expected to be a billion dollars by 1967, and everyone now agrees that semiconductor technology has been and will continue to be a "good thing." However, ten years of effort in many laboratories, and corresponding expenditures, preceded the appearance of semiconductors on the market, and during much of this period the venture was a highly speculative investment.

The period required to bring semiconductors from the laboratory to market was long, but not abnormally long for a real blue-chip venture in R & D. There are, of course, many other complexities which could be outlined in determining the best *level* of R & D and the best proportion between R & D in a given business situation.

Since the appearance of the Sputniks, all of the nation's R & D effort has become a public issue, but in particular the term "basic research" has become suddenly popular in the press. It seems appropriate to take a quick look at how basic research fits into our summary of the present relationships between technology and modern industry.

Traditionally, the first step in any discussion of basic research is to define it, since there seems to be little general agreement as to just what the term means. Most definitions try to imply a motivation—or lack of it—for performing the research. Thus it becomes easy to say that basic research is "looking for new scientific knowledge without knowing what it will be used for." Actually, this is not a bad answer if you are talking about the way many good scientists visualize their work. Much of our technological progress is based on scientific discoveries by men who never visualized what the "practical result" of their work would be. Their motivation was simply an intense curiosity about nature.

Although the satisfaction of curiosity about the world in which we live provides a tremendous human drive for scientific progress, the sponsor or the supporter of the research—whether government or industry—should have a purpose which is expressed ultimately in terms of human welfare.

Since curiosity provides ample motivation for the scientist, but is not a sufficient motive for the supporter of research, the question of "why you do it" tends only to confuse the definition of basic research. What is important, to the scientist and to his sponsor, is the search for new knowledge and understanding of nature. Thus, in simplest terms: *basic research is the process of learning new scientific knowledge.*

With respect to basic research—we might well call it "learning work"—let me express the conviction that in American industry as a whole we are by no means exhausting the business opportunities available in this activity. There are many ways of explaining this lack of enterprise in our free enterprise system, but one reason is the fact that basic research, like some other creative fields, has grown from unbusinesslike traditions and backgrounds. Another reason is the long time between investment and payoff; some companies cannot wait that long. For these and other reasons, partly inherent in the nature of the work, it has been difficult to be businesslike about an activity that is so vital to the future of technical businesses.

The time has now come, I feel, to dispose of some of the trappings of witchcraft that have surrounded this trade. To start with, we can be sensible about why we, in industry, do basic research. I can state the interest of General Electric in basic research by saying that *we* are engaged in this creative activity because we regard it as a primary source of new knowledge and understanding which, translated to new or improved technologies, presents important opportunities for new business enterprise. Practically all of the products of the electrical industry, and of course of many other industries, had their origins in basic research, and it is no secret that many future products will be developed on the foundation of recent or future discoveries in science.

. . . .

I have discussed some elements of the relationships—*now*—between technology and modern industry. That we are in a period of "technological change" is obvious, and in pursuing my assigned topic—"the appraisal of technological change"—the easiest course might be simply to appraise what is happening now. But today's technological change is based on yesterday's research, and today's research will produce the technological changes of the future. Therefore, it would appear that my task, as a spokesman for industrial research, should be to make some *specific suggestions* for appraising the technological changes that can be anticipated in the future.

Although I would be hard pressed to present a rigorous proof of the thesis, I believe that our facility for predicting scientific developments *is* becoming progressively better and that *many scientific and related technological developments can be anticipated in part or in whole.* In almost the same breath, I want to make it clear that there will be plenty of room for completely unanticipated discoveries in science in the future. But it seems to me that various attributes of modern research contrive to enhance our ability to anticipate progress. I believe that in the future many scientific discoveries will be seen in outline before their actual arrival, perhaps basically because *human needs and wants will react with increasing effectiveness with scientific methodology.*

As an example, it is a fact that a nuclear chain reaction was discussed in scientific meetings and technical journals five years before the discovery of the fission chain reaction. Perhaps this example and others that could be cited are evidence that the "shadow of the future" may become clearer in the future.

In early historical times, the practice of soothsaying was held in high repute. In ancient Canaan, the soothsayer was a high priest, and no sensible person would make long-range plans without the benefit of his advice. Perhaps soothsaying will come into its own again in forecasting scientific and technological progress.

Economic forecasting inherently includes a forecast of technological progress as a gross trend, since this is an important factor in economic growth. However, such forecasting is not generally in sufficient detail to include a consideration of specific fields of technology of concern to individual business enterprises. This more detailed consideration of technological change is desirable to permit an appraisal of (a) new business opportunity, and (b) the rate of obsolescence of present products or businesses, both of which are required to permit the orderly planning of business progress.

It is not possible to provide a single "set" pattern for making appraisals of future technology. In some cases there are enough facts available—and sufficient precedents based on related achievements of the past—to permit an actual *forecast* incorporating a rough time-scale in years. An example here is our recent attempt to forecast the future of power production by fusion. In a few moments I will present the basic elements of this problem—and what we believe to be a reasonable conclusion.

Another technique for appraising future technological change is to make a detailed analysis of the scientific problems that must be solved and weigh these against an appropriate set of economic considerations. In the example I will use now, the application of solar energy, it will be seen that although there are insufficient precedents—and too many imponderables—to permit an actual forecast in terms of years, an appraisal of scientific and economic considerations can be extremely useful in planning research projects, and in measuring research progress in relevant fields.

We will consider first a *forecast* of importance to the atomic power business. In this case we seek answers to the questions: If *fusion power* can be developed, what is its probable timetable and how will this affect present plans for an atomic power business based on fission?

Energy can be produced from nuclear processes by splitting heavy atoms, as in the *fission* of uranium, or by *fusing* light atoms, as in the fusion reaction of the hydrogen bomb. All of the present plans of the atomic power industry are based on the practical development of the *fission* process for stationary as well as mobile power plants.

The atomic power industry is in an early stage of development, and like most new industries is faced with the necessity of high initial investments in additional R & D and engineering work, as well as in the facilities for R & D, engineering, and manufacturing operations. The corresponding high starting costs will defer, until a long time in the future, the attainment of a net profit from the atomic power venture. It is characteristic of the pace of modern technology that, at this inconvenient period of rapid growth and no profit, the atomic power industry must carefully study the problem of the possible obsolescence of its principal basic process—fission. This appraisal is, however, a current necessity. In judging the feasibility and timetable of fusion as a possible competitive process, very helpful perspective is provided by a study of the chronology of fission, which includes the following events:

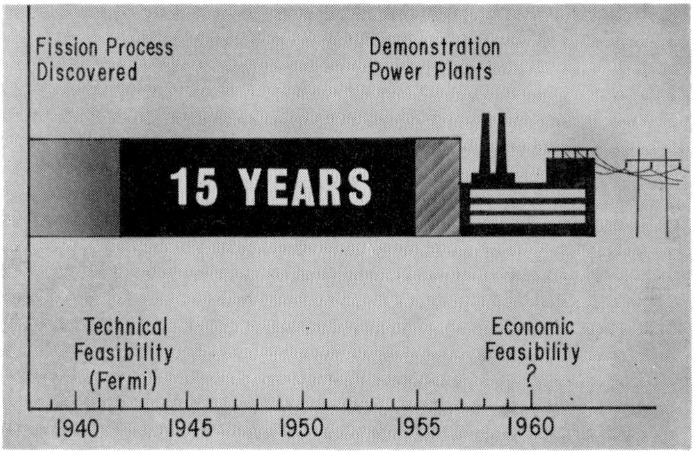

Note that four years of intensive effort separate the discovery of the basic fission process and the demonstration of the self-sustaining chain reaction. We regard the point of *technical feasibility* and the final step of *economic feasibility* as the most significant indices in a technological development following a scientific discovery. In the case of fission, the technical feasibility may be clearly identified as the demonstration by Fermi of the self-sustaining chain reaction which proved the possibility of net energy production from the fission process. A period of fifteen years of intensive scientific and technological development was then required to attain the present stage of "demonstration power plants."

The attainment of economic feasibility is not easily defined or identified, particularly in this case where geographic and defense factors enter the consideration. However, for our present purpose we may focus upon this objective: the general application of fission-power reactor plants to public utility use.

We are not yet at the point of economic feasibility in this sense, but are entering a period of progress which promises to lead to this objective soon. We may take 1960 as a somewhat optimistic date to use for this important event. The whole chronology of fission, in terms of these benchmarks, thus encompasses more than two decades.

With the history of fission in mind, we may now look at the potentially competitive process of fusion. In this process, light atoms "fuse" to form heavier ones, with a net production of energy. There are many combinations of light atoms which have been shown, in particle accelerators, to do this. In what is probably the best-known example, two atoms of deuterium, or heavy hydrogen (actually their nuclei), fuse to form an atom of helium 3 and a neutron, with the release of some energy. This particular reaction has been known since 1934, prior to the discovery of the fission of uranium. For fusion-power production it is proposed to take some of this heavy hydrogen gas and heat it to a very high temperature under controlled conditions, whereupon the fusion will take place, creating helium 3 and releasing large amounts of energy.

The immediate goal of all present fusion research is to see if this high-temperature reaction can be observed at all. The difficulty is that, to merely detect the start of the reaction requires a temperature of many millions of degrees. At these temperatures a few neutrons should leak out if the thermonuclear reaction occurs. This is the point in the chronology of fusion where we now are—recent reports indicate that the thermonuclear reaction may have been produced in the laboratory.

However, this demonstration is *not* the required *technical feasibility* for fusion power. To produce net power, it is necessary, as in the case of fission, to achieve a self-sustained reaction, and this far more difficult goal will require the production and containment of gas temperatures of many *hundreds of million degrees.* The attainment of technical feasibility in terms of a self-sustained fusion reaction lies well in the future. The technical difficulties here are well nigh overwhelming; in the view of some competent scientists they are impossible.

For perspective, however, we must keep in mind an important fact of scientific life: we do not have to solve all of tomorrow's problems with today's tools; tomorrow we will have some new tools. These tools will come from the new discoveries, the new knowledge, the new techniques, materials, and processes that we will acquire en route. We are here contemplating a development which in aggregate will require decades, and we must calculate among our assets the new knowledge we will acquire during the period of study, as well as the knowledge available today. As evidence of the pace of modern science, consider the fact that the past two decades encompass the whole of the "atomic age," including atomic bombs and thermonuclear weapons, as well as microwave electronics including radar, the semiconductors, sulfa drugs, antibiotics, and a host of other developments. Hence, in appraising the problem of fusion, we must keep in balance the extreme difficulty of the task, characterized as ". . . one of the great scientific challenges of the century," [1] and the rapid pace of modern science.

If, for forecasting purposes, we now examine the problems being faced in the proposed development of fusion power plants, we may readily identify many points which bear a striking similarity to those met in the fission power history just outlined.

These similarities, at corresponding points in the developments, include:

(a) *New Science*
The basic scientific phenomena are almost entirely new and require comprehensive theoretical and experimental development.

(b) *New Technology*
The development, engineering, and manufacturing work will all deal with problems which are unique in the new field, and will require original solutions at nearly every point.

(c) *Large-scale Equipment*
The final power plants, based on either process, will be very large; hence correspondingly large-scale pilot plants and demonstration power plants will precede full-scale production plants.

These three important considerations help to explain why it is taking two decades or more to develop "practical" fission power plants, and why, with equal effort and luck, it will take at least that long to traverse the long trail to an economic fusion power plant.

[1] Richard F. Post, *Rev. Mod. Phys.*, Vol. 28, No. 3, p. 338, 1956.

If significant differences in these two tasks can be identified, they may reside in items (a) and (c). Under item (a), it is presently questionable whether we are yet (in the case of fusion) at the point of being able to demonstrate the basic thermal process—the controlled thermonuclear reaction, although announcements of recent British research progress give substance to this hope. Under (c), it now appears somewhat likely that to operate at all, fusion reactors may have to be not large, but *very large.*

Because of the ready availability of the history of fission, and the striking similarities between the two cases, we believe that we can propose a chronology of the fusion development which has excellent validity. There can be no doubt that the development of fusion power, if indeed it can be accomplished, will take a long time. Each of the steps of developing a new science, developing a new technology, and carrying out very large-scale experimental and pilot plant studies will be time-consuming, even under forced draft.

The contemplated chronology of the fusion power development includes these events:

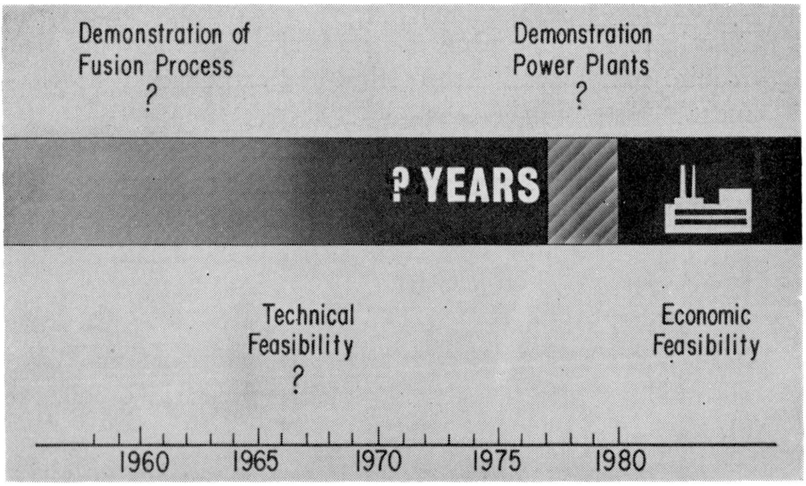

The question marks throughout this table indicate that no firm assurances can be given that the events enumerated will take place. The dates shown suggest, however, that if these events are attainable, this is a reasonable timetable for their occurrence.

Hence the final conclusion: *There is time to develop an atomic power industry based on fission, and to justify the heavy investment requirements.* However, a profitable operating basis, to recover heavy starting costs, cannot be deferred too long into the future.

The difficulty of the task of accomplishing fusion power must not in any degree relax the effort directed to this objective. In fact, the long-term projection of energy requirements and energy sources make the ultimate success of this venture a command performance.

. . . .

As promised, I would now like to consider another type of appraisal of future technology. How can we appraise the potential of solar energy? What are the prospects for converting the sun's energy into useful electricity to light our homes and run our factories?

In my discussion of this subject, I shall draw heavily upon the work of many associates, particularly Drs. K. H. Kingdon, Bruno Zimm, Ferd E. Williams, G. M. Branch, J. F. Flagg, D. L. Douglas, R. E. Halsted, and L. R. Koller.

Our interest in this area stems from many origins, but particularly from the fact that with solar energy one may generate electricity by a variety of methods. Let us consider only one of these methods—the conversion of sunlight to electricity by the photovoltaic cell, which we may call the solar cell. This is not a new phenomenon—it has been utilized for many years in photographic exposure meters.

Solar cells may be made today with a conversion efficiency of 10 to 12 per cent, and a future value of 20 per cent may be a possibility.

These cells produce direct, rather than alternating, current, but there is a large industrial market for low-voltage direct current—principally in metallurgical and chemical process applications. However, these particular uses require electric power 24 hours a day, every day, and cannot be served by eight-hour-a-day solar energy power. Bulk energy storage by a simple, reversible process would solve this problem, but no process is available that would remotely meet the necessary costs; hence, the industrial application of solar-electric power in this country awaits a new breakthrough in energy storage systems. Incidentally, for this and numerous other important prospective uses, a simple, inexpensive, and efficient reversible energy-storage process looms as an outstanding need of modern technology.

There remains, in the case of solar-electric power, the question of its possible use in individual consumer plants; for example, as an electric power supply in the home. Several inherent features of solar cells favor this possibility. For one thing, the conversion efficiency of solar cells does not depend upon size—the unit efficiency is the same, regardless of the number of square feet employed; hence, there is no advantage in efficiency in a large central-station type solar plant. This is an important difference from a heat engine, such as a turbine, which is always more efficient in larger sizes and hence is advantageously used in large central-station plants. Furthermore, in comparison to fossil fuels, which are geographically localized, solar energy is already distributed rather uniformly over the populated areas of the world. Hence, a local plant would save distribution costs and justify a higher equivalent local power rate. The local plant would, on the other hand, lose the advantage of load diversity, and would entail some other disadvantages of possible consequence. These disadvantages, however, are subordinate to the major economic factors.

The preceding considerations favor *individual* consumer plants for solar electricity, and since the *technical feasibility* of such plants is certain, even with available technology, a careful study of the *economics* of such systems is mandatory.

Several study designs of solar-electric home systems have provided the basis for an economic study. For example, we have envisioned a solar home in which the south wall would be allocated to the solar collector and provide the necessary total area of about 800 square feet. With a solar cell conversion efficiency of only 5 per cent, this plant would be able to supply 12 kilowatt-hours of electricity a day, which is somewhat marginal in comparison to the present average consumption in the United States of eight kilowatt-hours per day. Of course, in California you could plan to get substantially higher output from the solar plant, but we have had to consider averages. As a matter of fact, it is surprising to find that there is only a 25 per cent variation in the amount of solar energy received annually in eighteen principal cities lying between north and south 40th parallels of latitude. California climate would provide a much smaller seasonal variation in solar energy than many other parts of the country, and this would be significant in designing the storage element which is a necessary part of this plant. A study of numerous alternatives indicates quite clearly that the lead plate storage battery is the principal contender for this particular function, and design exploration has been based on this choice.

Throughout the systems studies, numerous technically feasible alternatives have been explored for their economics.

It will be clear that balancing the elements of cost between various system components—for example, between the solar cells and solar collectors—permits numerous choices that could be optimized in detailed designs. We have assumed a storage capacity of four days at full output in the case to be described. This would be adequate for most of the continental United States for the plant design contemplated.

I doubt that people in California would be willing to allocate the south side of their homes to a solar cell bank, in which event the cell bank might be placed on the back of the lot remote from the dwelling. In any case, we assume that between home-owner, architect, and landscape gardener some acceptable place can be found for the solar cells, without which the solar-electric plant becomes definitely unattractive.

There are many interesting aspects of solar power that might have been worth dwelling upon if the economics had turned out to be more encouraging. However, it is clear that much technical and technological progress needs to be made before we can achieve economic feasibility in the home solar plant. The two principal difficulties occur in the two principal components—the battery and the solar cell. To be economically competitive in our balanced plant of the stated output, with four-day, full-power storage, we would require a cost reduction in the battery of *sixfold*, in comparison to best present technology. In the solar cell, the unit cost must be reduced by a factor of *100-fold* on the same basis.

We regard the factor of *six* as a far more difficult hurdle than the factor of 100, because storage batteries represent an old art that has had the benefit of much R & D over a long period of years, including World War II when it received concentrated attention as a weapons component.

A factor of 100 in solar cells represents, of course, a not inconsiderable hurdle as well. However, the whole art and science of radiation conversion is new and rapidly developing, and in particular the areas of semiconduction and solid-state science and technology are both experiencing rapid growth. It therefore appears not impossible that the required solar cell progress might be accomplished in the long term. Manufacturing concepts and techniques radically different from any now in use are clearly required in this area.

In this application, as well as in completely different contexts, inexpensive energy storage looms as a technological factor of tremendous potential leverage. Lack of progress in this quantity, more than in solar cells, stands in the way of the practical realization of such systems as the one under discussion. The net conclusion from our solar energy study is that bulk—as distinguished from specialty use—electric power generation by this method is not economically feasible at present. The limiting factors which have been determined therefore provide goals and indices of progress for future R & D.

. . . .

Although our two examples of technological appraisal—fusion and solar energy—obviously employed different techniques adapted to different problems, you will note that in both cases the element of *economics* played a significant role. In light of the purposes of this Industrial Economics Conference, it seems appropriate to emphasize the growing importance of economics in many decisions affecting modern research and development programs. The question of "Can it be done?" is becoming increasingly subordinate to the questions "What will it cost to do—how long will it take—is it worth doing?"

And, finally, it must be emphasized that mere appraisal is not enough. Technology can have great impact on the direction of industrial growth, but—equally—industry can direct its technology along the paths that will lead to the fulfillment of human needs and wants. Having *appraised* the potential of science, we must *act.*

Our economic destiny—indeed our national security and our entire way of life—will be influenced significantly by how well we *appraise*—and then *pursue*—the opportunities offered by modern technology.

Today's Research and Tomorrow's Power

An address presented at the Southeastern Electric Exchange meeting in Boca Raton, Florida, March 31, 1958.

During the final few years of the Nineteenth century it became apparent, even to the most persistent pessimists, that the electrical industry had come to stay. The makers of kerosene lamps were still enjoying a good business, but they had some misgivings about the future. The situation had developed so rapidly that even arc-lamp manufacturers were worried about Thomas A. Edison and his carbon lamps.

As an interesting and authentic case history of these early days of the electrical business, I would like to refer to the steam-engine industry; it was sharing the electricity boom. Much of this industry was concentrated in and around Erie, Pennsylvania, where an even dozen companies made boilers and steam engines to drive the "dynamos" in the electric power stations which were springing up everywhere. In Erie—then called "The Steam Engine Capital of the Nation"—the future looked rosy for this business around 1890.

Of course, as usual, there were some troublemakers around suggesting that there *might* be better ways to provide power to those big dynamos which people were just beginning to call "generators." A few wild-eyed fellows, observing that the steam engine was about to reach its maximum practical size, had the temerity to suggest that the electrical industry *must* find new prime movers. But in "The Steam Engine Capital of the Nation" there was little interest in investigating the possibilities of the steam *turbine;* it was devoutly hoped that this idea would go away.

I will not belabor the by-now-obvious point: today, with one exception, those twelve steam-engine manufacturers in Erie are either out of business or making some other product. The single exception is a company which had some bright young men who adapted the uniflow principle to steam engines, and then began marketing their improved product not for central-station power plants but for the kinds of jobs for which it was particularly suited.

It appears that not one of the twelve companies realized the potential of the steam turbine and took the trouble to investigate its possibilities. Certainly from the standpoint of manufacturing know-how and marketing contacts they had a great opportunity to move quickly in this new business. But a preoccupation with the *status quo* stood in the way. There was nothing new then—or now—in this situation. The human trait of "resistance to change" has kept people in some parts of the world literally in the dark ages. Even in America, individual unwillingness to accept the temporary "inconvenience" associated with innovation has been, and continues to be, a major deterrent to national progress.

PLANNING FOR INNOVATION

For all of us in the electrical industry, there are several apparent "facts of technical life" that are most reassuring. The most pleasant of these I will postpone discussing until the end of my talk. In the meantime, I want to talk about a fact that is reassuring to *me*, as a representative of research, and another fact that should be reassuring to *you*, as leaders in the electric utility industry. What pleases and reassures *me* is that our advance parties exploring the frontiers of science report that there appear to be many fine prospects for future progress in the generation, transmission, and utilization of electric energy. *You* will be particularly reassured, I believe, when I suggest that these technological changes can be foreseen and forecast to a degree that should permit orderly planning for their arrival and integration in the technology of the industry.

With research and development rampant in the land, and particularly in the electrical industry, our exploring parties are constantly alert to and are continuously appraising scientific progress, scientific opportunity, and the timetable of technological change. But I do not regard overnight obsolescence as a serious threat in the electrical industry, although one should not confuse "overnight" with long periods of sleepy hibernation. To cite examples from the recent past, *there was time* for electrical manufacturers to appraise the capabilities of and opportunities in electronics—and to put electronics to use. *There was time* for radio manufacturers to learn how to make television sets in time for the demand for them. *There was time* for the makers of vacuum tubes to learn how to make transistors. *There will be time* for people now in the electrical business to adjust themselves to fission power—and still time to learn about, and eventually apply, the tremendous potential of fusion power.

There will be time to plan for technological progress—*if* we use time to best advantage. If we call upon alertness, analysis, and appraisal, and above all a willingness to contemplate change, technological innovation will present many opportunities and few unpleasant surprises.

THE "SCALE-UP" FACTOR

There is a basic scaling factor that provides inertia to any change in the large power equipment of the utility industry. The physical size of such equipment is in itself a major factor in extending the time required to develop and introduce fundamental innovations.

For example, the steady upward progress of the temperatures and pressures in steam turbines has been made in necessarily small increments because of the size and complexity of the machinery involved. Any basically new power production or conversion process which might become available in the future will have a long, slow, and expensive trail of "scale-up" to traverse before it can become available for large-scale use in any practical system.

The scale-up problem, with its associated time and cost, is one of several reasons why it is taking about twenty years to move from technically feasible atomic power generation, to economically feasible atomic power. This important objective has not yet been reached, but it is at long last in sight. The scale-up problem is also an important element of the several similar reasons why fusion power generation will require a long R & D history to bring it to the point of practical use. Because the history of fission power is such an important reference case history, and because the proposed future history of fusion power has so many important possible consequences, perhaps I should review some conclusions we have come to on this subject. An interval of fifteen to twenty years lies between the point of *technical feasibility* of fission power, proven in 1942 by Fermi, and the *economic feasibility* of fission power, which we are now approaching but have not yet reached. This distressingly long period of time is explained to a substantial degree by the problems of *scale-up.* Fermi's reactor proved a self-sustaining reaction was possible, and produced net power, but only a tiny amount. In typical small-scale, test-tube-size experimental research in the laboratory, many experiments can be completed in a short time, and as a result much exploration can be accomplished quickly. However, as soon as the test tube becomes the pilot plant or the semiworks plant—or the demonstration power plant—progress is much slower and costs become much higher. This is a fact of technological life.

With the perspective provided by fission-power history, and a detailed consideration of similarities and differences in fission and fusion, it is easy to conclude that the earliest timetable for fusion power will not be shorter than the timetable for fission power. Important in this consideration is the fact that the *technical feasibility* of fusion power has not yet been proven; to pass this significant milestone is the objective of tremendous R & D effort of the period immediately ahead. From our studies we have proposed that practical fusion power should not be anticipated until at least twenty years have passed. With reasonable diligence and forethought it will be possible for the utility industry to conform its future plans to this timetable of technological progress.

Fission and *fusion* are undoubtedly the two principal long-term prospects for basic change in the technology of the utility industry. Fortunately, and for the reasons stated, both of these fundamental and revolutionary changes will become operational and practical at an *evolutionary* rate.

SOME FUTURE POSSIBILITIES

In addition, however, there are some interesting further technical possibilities which, though presently remote from large-scale use, may in the future create an appreciable perturbation on electric power production. These possibilities include new methods for storing energy, superconductivity, the fuel cell, semiconductor converters of various kinds, and the thermionic power converter. I will say a few words about each of these subjects and indicate how we presently appraise their future possibilities.

ENERGY STORAGE

One of the really basic problems of today's electrical industry is energy storage. When the housewife wants to bake a cake she has plenty of time to prepare the ingredients, but the power she uses in her electric oven must be generated the very instant she pushes the button. The absence of a truly economical system for storing electrical energy has been, is, and will continue to be a fundamental limiting factor on the usefulness of electricity. Economical, reversible energy storage would not only solve the electric utilities' peak-load problems but also would be a major step toward the utilization of solar energy and might—in the meantime—revolutionize the transportation industry.

To explain the fundamental nature of the problem, let me oversimplify the case in considering an actual example—the storage battery. The energy which we store in a battery appears at an atomic level as a slight rearrangement of the electrons in the chemicals used in the battery. To use an energy unit which is appropriate for such cases, the rearrangement is measured by a few electron-volts. To accomplish large-scale energy storage in matter we must reversibly accomplish far more drastic molecular or atomic rearrangements, such as might be measured in tens or hundreds or more electron-volts. We must eventually solve this problem, but at present no one has the foggiest notion of where to start. Until we get a foot in the door, the intriguing fact remains that—if we only knew *how* to do it—we would have enough room in a cubic foot of matter to store hundreds of times more electrical energy than the 1.3 kilowatt-hours achieved with today's storage batteries.

SUPERCONDUCTIVITY

Superconductivity is one of the most startling phenomena of present-day science. About fifty years ago the Dutch physicist Kamerlingh Onnes observed that when the temperature of certain materials approaches absolute zero their electrical resistance disappears completely—they become literally *perfect* conductors of electricity. Superconductivity has been a fascinating research puzzle ever since, but only very specialized practical applications have been made.

The appraisal of the future of superconductivity has, until recently, been hampered by the lack of a satisfactory understanding of the phenomena, preferably expressed by an analytical theory. However, a far-reaching analysis of superconductivity is contained in a theoretical study published last year by Professor John Bardeen and his associates at the University of Illinois. Of course, the $64 question is: Can we hope to develop superconducting phenomena at temperatures appreciably above absolute zero—say, at room temperature? The latest analysis does not rule out "room-temperature" superconductivity, but it clearly indicates that this possibility is extremely remote. One of our own physicists recently told me he would demand odds of "about a million to one" before betting that his grandchildren will live to see the day when uncooled wires can transmit electricity at perfect efficiency. There must be similar skepticism about the possibility of "superconducting windings" for generators, even though our design engineers tell us hopefully that this would solve a lot of their problems.

Originally, superconductivity could only be demonstrated at temperatures of 1 or 2 degrees above absolute zero. However, the present "high-temperature record" for superconductivity has been established at 18 degrees above absolute zero—if you can call minus 255 degrees Centigrade a high temperature. There is some hope among scientists and engineers that superconductivity might be used to advantage in power transmission lines if it could be achieved at liquid-nitrogen temperatures of about 80 degrees above absolute zero, possibly even at liquid-hydrogen temperatures of 20 degrees absolute. However, the technical problems look formidable, at best, and the economics are marginal unless we can work at substantially higher temperatures. That slim possibility cannot be ruled out, however, so we are keeping one eye on this problem.

A more immediate improvement in the transmission of power should come from the higher transformer efficiency promised by new core materials. One recent result of our intensive research program in the field of magnetism has been "four-square" silicon-iron, a remarkable new kind of magnetic sheet material. "Four-square" is a doubly oriented material which, in experimental transformers, significantly decreases losses and permits a higher operating flux density. If we can learn how to produce this material in quantity, as we hope, it will add importantly to motor, generator, and transformer technology.

THE FUEL CELL

A fuel cell is a device that directly converts chemical energy—the oxidation of a fuel—into electrical energy. In simplest terms, a fuel cell is merely a primary battery which consumes cheap fuel materials instead of such relatively expensive raw materials as zinc, cadmium, or lead. The fuel cell idea is attributed to the English scientist, Grove, who first published on the subject in 1839.

Theoretically, a fuel cell conceivably could be 100 per cent efficient, but in the laboratory its efficiency has never reached any such figure. After World War II there was a resurgence of interest in fuel cells, and currently there are scientific efforts reported in Germany, England, the Netherlands, and at several laboratories in the United States. Last year the National Carbon Company announced the development of a fuel cell in which hydrogen and oxygen are combined to produce electrical energy. This "electrolysis-in-reverse" system yields about one volt per cell and about one kilowatt per cubic foot. National Carbon reports efficiencies of 65 to 80 per cent, and the U. S. Army is testing the new device for use in small radar sets. For bulk rather than specialty power generation, however, the fuel cell must burn cheaper conventional fuels, and it cannot presently do so with satisfactory efficiency and life.

That hydrogen is an expensive fuel seems paradoxical in view of the abundance of hydrogen and oxygen in a familiar combination that covers 71 per cent of the earth's surface. But pure hydrogen—by itself—remains an expensive item. It appears, then, that the use of fuel cells will be limited in the near future to military applications, stand-by plants, and other systems where convenience or portability are more important than economy. Some of our scientists are optimistic about the long-term possibility of fuel cells, although they are very much aware of the difficult technical problems that remain unsolved.

It will not be news to you if I recall that well over 99 per cent of the world's electrical energy is produced by rotating machinery—the dynamo—the electric generator. Primary batteries—electrochemical generators—produce most of the remaining fraction of 1 per cent. You may be surprised, however, to learn that in addition to rotating machinery, batteries, and the fuel cell, there are at least fifteen fundamentally different methods for converting nonelectrical energy into electrical energy. This does not include the electric eel which, in the absence of information to the contrary, I will assume is another example of the electrochemical principle.

Most of these methods have been known for many years and are not adaptable to large-scale power generation. Under the heading of mechanical-electric converters are: (a) the *piezoelectric effect*, used in vibration detectors; (b) *oscillating magnetic induc-*

tion, most familiarly observed in the phonograph pickup; (c) *electrostatic generation*, which is produced by small children scuffing across rugs and by huge Van de Graaff machines; (d) the *magnetostriction effect* achieved when a magnetized material is stretched or compressed near a coil; and (e) the *streaming potential*, which is sometimes used to measure pressure fluctuations in oil pipelines.

Nuclear energy is being converted to electrical energy at several power installations throughout the world, but in all cases the conversion employs a heat cycle. Direct conversion of nuclear-to-electric energy may occur in a variety of ways: for example, (a) when high-energy electrons emitted by a radioisotope are collected on a conducting plate; (b) when the field produced by two unlike electrodes collects charged particles produced in a gas by nuclear radiation; (c) when a semiconductor junction is exposed to beta or gamma rays—the so-called "atomic battery"; and (d) when nuclear energy is used to make a phosphor surface generate light, this light in turn being converted into electricity by a photovoltaic cell. In all cases, the efficiencies of known nuclear-to-electric conversion techniques are fantastically low. The power obtained must be measured in terms of small fractions of a watt, and the handling costs are high.

SOLAR ENERGY

The direct conversion of light energy to electric energy is most commonly observed in photographic exposure meters, but a more glamorous application is in what is popularly called "the harnessing of solar energy." Solar energy also can be achieved by thermal conversion, which I will discuss in a moment, but most of the recent interest has centered around the photovoltaic process which transforms light energy into electricity.

We have made a thorough study of the economic and technological aspects of solar energy. Without going into detail, I will summarize by saying that before solar energy can become competitive, we must improve energy-storage techniques by about six times in comparison to best present practice, and must develop solar cells that are 100 times cheaper than anything now available.

Parenthetically, I might mention another approach to solar energy even though it is obviously not a direct-conversion process.

It has been suggested for such areas as the southwestern part of our country that the sun's heat might provide "free" preheating of boiler feed water. Apparently this heat is not quite "free," because studies based on readily available materials and components indicate that it fails to be economical by a factor of about two.

In summary, it seems safe to predict that solar energy may have a number of important specialty applications—including power for satellite transmitters—during the next few years, but in this country it will not be economically feasible for bulk electric power generation for many years to come.

The most exciting of all the "unconventional" power sources comes under the heading of direct heat-to-electricity conversion. Here there are at least five possibilities beginning with: (a) *thermoelectric generation*—the familiar thermocouple. Thermoelectric generators have always had small power output, negligible efficiency, and large weight and volume per watt produced. A thermoelectric generator employing semiconductor materials is somewhat more promising, however, and it may also be used in reverse to provide thermoelectric cooling. We are investigating a variety of approaches here, all of which are scientifically fascinating and at the same time a long way from serious employment in power applications.

Among heat-to-electricity converters we also should mention: (b) *thermomagnetic generation*, which produces low-frequency alternating current; (c) *phase transition potentials* observed between ice and its surrounding salted liquid, a phenomenon which, in conjunction with the previously mentioned streaming potential, produces electricity in thunderstorms; and (d) *pyroelectric phenomena* in which certain crystals such as tartaric acid generate a current when heated. Some of the phenomena have interesting possibilities for measurement devices, but once again the efficiencies are infinitesimal.

THE THERMIONIC CONVERTER

In spite of the small efficiencies previously observed, the direct conversion of heat into electricity has continued to be a hopeful dream of people in the electrical industry. Recently one of our scientists invented a *thermionic converter*, an electronic device that takes advantage of the simple fact that electrons can be "boiled out" of a hot metal surface and used to produce an electric current directly. But this device also, I hasten to assure you, is not now a threat to the steam-turbine generator. It is a very interesting and potentially important scientific achievement. Small laboratory versions of the thermionic converter have turned more than 8 per cent of the applied heat energy into useful electric power. The inventor, Dr. Volney C. Wilson, believes an efficiency of 30 per cent or more may be theoretically possible. But application of the thermionic converter as a major power producer is a long way off. Its actual efficiency must be greatly improved, and it must be built in very large sizes; both of these problems are formidable. When and if this progress can be made, the thermionic converter may become a *supplement* to the steam turbine rather than a *replacement* for the steam turbine.

New ways of doing things in the electrical industry are inevitable and desirable; the chief concern is the *timetable* of progress. Our best and only hope for staying in business is to keep in close touch with those "advance parties" of scientists which I mentioned —those who are exploring the frontiers of science and providing us with the hints and hopes of things to come in our business. With each passing year, there are more of these exploring parties to keep in touch with, and the job becomes increasingly complex and expensive.

"ACROSS-THE-BOARD" RESEARCH

There is time to keep in step with the march of progress, but only if we realize the magnitude of the job and act accordingly. Particularly, we must realize the importance of an "across-the-board" approach to scientific research. Large research efforts, delving into a multiplicity of scientific fields, give the advantages of *breadth* and *depth;* large-scale research offers the advantages of *supporting skills* and *specialized facilities,* which make it easier for modern scientists, who are necessarily specialized, to do their work.

It should not be suggested that everything that is new is necessarily good, or that all of the results of modern scientific research will yield immediate or even eventual improvements in our products or services—or our way of living. However, progress requires new technology, and the thoughtful appraisal of its economic opportunity. In the future, your business will be doing things differently—

and so will mine. If we are thoughtful about it, we will do things not only *differently* but also *better,* better in terms that will be persuasive to our customers at the marketplace.

To anyone in a currently profitable business, innovation can be a downright nuisance. In our free enterprise system, however, this nuisance is the lesser of two evils, the greater evil being business extinction. In the electrical industry it is clear that there is *no substitute for enlightened technical work as a base for technological innovation.*

In our own company, in some years, the expenditure for *research and development* has exceeded the net profits. There is no more difficult problem in modern management than determining the best balance between short-range and long-range goals—for example, the proper balance between today's profits and research for tomorrow's business.

It is difficult to determine both how much is *too much* and how much is *too little,* and there are risks in both directions. However, one is hard put to assemble a very long list of businesses that have failed because they did *too much* research. On the other hand, one can cite a long list of businesses that have disappeared or have experienced serious loss of position because of obsolescence of principal production processes or services.

"Innovation," you may say, "*is* a damn nuisance." That is true, but business *obsolescence* is a disaster.

The "nuisance aspects" of innovation are more than offset by the *abundance* of the payoff provided by solid progress based on innovation. In addition, while seeking long-range goals, we stumble across countless unexpected treasures, and frequently these are directly applicable to the improvement of our present way of doing things. In solving the problems of materials for atomic-powered aircraft, we will solve metallurgical and mechanical problems applicable to other types of heat engines. Fundamental studies of new heating and cooling techniques—required by space missiles and space flight—may be expected to have additional far less glamorous and more workaday applications. The search for fusion power probably will provide by-products in circuit technology or radiation generation long before the energy of the hydrogen bomb has been controlled for peaceful use. Scientific research produces *improvements* as well as *innovations.* I firmly believe that the cost of research can be justified to the sponsoring company and its customers on the basis of *unexpected payoffs* alone. But without the long-range goals and challenges, we might never set out on the explorations at all.

The steam turbine of the early 1900's was an *innovation,* and certainly a *nuisance* to the steam-engine business. The impact—on our whole economy and way of life—of the steam turbine would be hard to measure. Its fantastic success story is not, however, confined to the first chapter—the invention of the turbine. Successive chapters have added exciting installments to the dramatic history of constant *improvement.* In more than fifty years no one has produced a fundamental method of generating bulk electrical energy that is more practical than the turbine-generator. But during this half century, research and engineering have combined to produce turbine-generator systems that, on the average, use only about one-eighth as much coal per kilowatt-hour as did those first turbines which edged reciprocating steam engines right out of the central-station picture. These improvements, toward which the entire electrical industry has contributed, have made electricity the American consumer's biggest bargain.

Nothing on the technological horizon gives promise of replacing rotating machinery for generating electricity—not on a big scale or in the near future. The tremendous size and complexity of the power industry would make overnight obsolescence *technologically* impossible even if by some unexpected discovery it became *scientifically* feasible. For all of these reasons, it appears obvious that the steam turbine, although nearing retirement age by human and corporate standards, is far from being "over the hill." We have discussed some of the more exotic long-range goals of science that may affect the electrical industry, so now it seems appropriate to discuss some current and possible future developments in turbine power plants.

Such a discussion is particularly important because your industry is now facing its greatest expansion program; you are being challenged by growing population, a growing economy, growing defense needs, and growing use of electricity to install as much generating capacity in the next ten years as has been built and installed in the entire seventy-five-year history of the electrical industry. In the face of what seems to be continuing inflation, the industry will be hard-pressed to maintain its record of selling electricity for the same or less dollars, while the average price index of other commodities has multiplied again and again since 1900.

I will remind you that, at your meeting here last year, General Electric's Glenn Warren said: "If we are to continue to make strides toward holding the cost of electric power down, we must investigate and develop *all possible avenues*—for not only the further potential improvements in fuel consumption but also for further reductions in capital expenditures per unit of output. We must develop plants that will require still less operating cost. In this latter area, great strides have been made in the past two decades. However, we face the fact that there is probably no big single development in sight which will give us these further economies of operation and installation cost. There are, on the other hand, many individual items which will give us a gain here and a gain there which, when added up, will give us continued improvements in the direction in which we are seeking to go."

You will recall that Mr. Warren emphasized the need for investigating and developing "all possible avenues." Here the electrical industry is, fortunately, in an excellent position, because more than any other industry it has supported "across-the-board" research programs in a wide variety of scientific fields. In our own laboratory we have continuing fundamental projects in many areas that may be helpful in further improving and reducing the cost of turbines. Among these are studies of gas dynamics, corrosion, friction, fluid mechanics, superpressure, heat storage, lubricant synthesis, vibration and balancing, combustion, reaction kinetics, shock phenomena, surface phenomena, basic fracture phenomena, diffusion kinetics, metal surface phenomena, metal fatigue, thermal stresses, high-temperature alloys—and literally dozens of projects in solid-state physics, metallurgy, and fundamental chemistry which give the greatest hope of achieving the kind of basic improvements in materials that will further improve our turbines, motors, transformers, and other equipment.

RELIABLE ROTORS

Projects of this type are basic, fundamental approaches to improving steam turbines, but I should also like to emphasize the extensive *developmental* and *engineering* efforts being sponsored by electrical manufacturers. A recent and excellent example of a combined scientific-engineering approach to a difficult problem could be called the "Case of the Bursting Rotors." This mystery story had several tragic chapters, but now we have come upon what we believe to be a happy ending.

Rotor bursts over the years prompted the development of inspection tools that made it possible to "see through" forgings many feet thick. But it became apparent about four years ago that even the best of these techniques was not good enough to explain failures at stresses only one-third as great as those at which rotors *should* have failed.

Three large turbine manufacturers and five principal forging manufacturers then joined together under the sponsorship of technical engineering societies to conduct a co-operative study of the problem. One result of this work was the ordering by our own company of more than thirty oversize rotor forgings, each made of a slightly different material or the subject of some variation in heat treatment. These king-size forgings gave our turbine engineers the opportunity to slice off dozens of two-inch- to six-inch-thick disks, each with the full rotor diameter. These disks proved to be the means by which for the first time it has become possible for fractures of the same type observed catastrophically in the field to be related to carefully measured experimental conditions.

The analysis of these results recently has been published in a series of technical papers so that the entire industry can benefit from what has been learned. We are confident that major additions have been made to the knowledge of large rotors under centrifugal stress, the effect of alloying elements and heat treatment as related to varying degrees of brittleness, and the benefits of the newly developed vacuum-pouring techniques. Each month we are receiving improved forgings from the steel companies, and you, as a result, are receiving more reliable machinery.

This increased reliability should provide the happy ending to what started as a mystery story, but it will be obvious that such happy endings are achieved at a considerable R & D expense. Furthermore, the improved materials have a substantially higher cost than the ones we used formerly. However, we have achieved a firm technological foundation for future progress in large electrical machinery.

Our turbine people have specifically listed some of the potential turbine-plant gains that promise lower overall installation costs and reduced operating costs per kilowatt-hour produced. Most of these gains will be based on improved engineering techniques—and, of course, on enlightened management taking full advantage of economic factors. These are gains that can be made by using existing knowledge to the greatest possible extent. The gains from new knowledge, produced by scientific research, are the type that can be predicted only in general terms—gains that will be made but we cannot say just where. These "new-research-discovery" gains should not be confused with the more specific suggestions I am about to enumerate.

LARGER UNITS FOR ECONOMY

First, among the gains available today in generating practice is that to be had from increasing the size of individual plants. These gains are available in units extending up to at least 300,000 kw per shaft, while the average size of units now being installed in the United States is currently at 120,000 kw. Moving this average higher seems to be the *single most substantial opportunity for increasing the economy of central station power generation.* Although the problem of reserve system capacity enters this picture, the steadily increasing overall size of most systems permits an increase in the size of individual units. Thus the use of a greater number of multiple ties with surrounding systems is offering many companies the opportunity to put "more eggs in one basket" and enjoy the attending economies of larger generating units.

Two new re-re-superheat 450,000-kw cross-compound 3600-rpm units now being built for the American Gas and Electric Company are expected to establish new records for fuel economy, kilowatt-capacity per cubic foot or square foot, and low operating costs.

In addition to increasing the size of units to increase their efficiency, great attention has been given over the years to using higher pressures and temperatures. Inlet pressures have been doubling every twelve years during the past half century, and initial temperatures have been increasing at about 12 degrees per year during the same period. As things now stand, it appears that the law of diminishing returns will certainly begin to set in when inlet pressures reach 5,000 to 6,000 psig and when initial temperatures exceed 1050°F. You will note that I said, "As things now stand."

Here we must discuss a problem that is perhaps best described as the "austenitic barrier." As inlet temperatures have gradually increased up to today's 1050°F, the materials used for rotors, large shell castings, and steam piping have remained fundamentally the same: low alloy ferritic steels. Beyond 1050 degrees these ferritic materials, by their fundamental nature, begin to weaken rapidly.

To achieve equivalent strength at higher temperatures, we must turn to alloys with a basically different composition and structure: the austenitic, or stainless, alloys. Thus I have suggested the phrase "austenitic barrier."

AUSTENITIC PROBLEMS

Using austenitic materials—or, if you prefer, "stainless steels"—creates several problems, the greatest being the cost of the raw materials. Higher alloying, and the requirement for using more critical alloying materials such as nickel, actually triple the cost of bar stock for turbine blades.

Cost is a serious problem, but not the only problem. Austenitic materials are more difficult to fabricate in any size; and there is no experience with the large ingots required for huge turbine parts. Further, the twenty-year life and performance of such components is completely unknown. To say the least, the "austenitic barrier" presents some formidable technical problems, many falling in the category of "scale-up" mentioned previously.

Are ferritic materials up against a real limit imposed by nature? Our metallurgists are pretty flatfooted about it—they say *yes*. So we have got to face this "austenitic barrier" for further performance gains through higher temperature.

In the past, increases in inlet temperatures have been made in small increments—from 900 to 950, from 950 to 1000, and so on. In leaping the austenitic barrier, one might consider going suddenly from 1050 to 1400, and indeed such a jump might, in theory, produce enough added efficiency to justify the higher first-cost of materials. But the idea of suddenly redesigning everything at once—from boiler to bucket—on a system as big, complex, and expensive as a turbine entails risks that are best described as overwhelming. It is very likely that the austenitic barrier will postpone for some time the kind of substantial improvements in efficiency-from-higher-temperature which the industry has enjoyed for so long. The scientists tend to concur with the turbine experts in saying that, *until we discover new basic knowledge we do not now have,* it will be best for us to forgo temperature increase as a source of turbine performance in the near future. Although the austenitic barrier presently looks difficult, it is well to bear in mind that the history of turbine development includes instances of equally tough barriers that have since been overcome. In the 1930's, for example, the creep of steel—a very fundamental property of matter—imposed a limit of 750 degrees which looked very formidable at the time. Research and development which led to new knowledge, and the resultant technology, solved the "creep barrier."

OTHER AVAILABLE GAINS

If we cannot go to higher temperatures *now,* let us look at some of the things we *can* do with technology available today. The older practice of re-superheating has been revived and has been demonstrated to be a good method of achieving the advantages of a 175-degree rise in operating temperature—without actually exceeding the 1050-degree limit. *Re*-re-superheating has been shown to add the equivalent of another 50-degree rise in operating temperature. But that is about the limit by this route.

Some new cooling techniques also are being used to lick the high-temperature problem in turbines. For generators, in addition to the use of hydrogen and insulating liquids as coolants, our engineers are now experimenting with the use of pure water for cooling the inside of pipe-like electrical conductors. New cooling techniques have done much to erase some previous notions that we might be approaching the theoretical ceiling on the size of generating units.

Because of the greater glamour of high-pressure–high-temperature approaches, we may have neglected some efficiency opportunities in the low-pressure end of the turbine.

There are frequently clear and calculable advantages in going to longer last-stage buckets, larger condensers, and supplementary cooling towers for summertime cooling; and there is no reason why the available two or three points in efficiency should not be taken.

Obviously, such economic calculations should be weighed against possible *future* price changes in fuels.

Another unglamorous but practical advantage can be gained from the use of high-speed steam turbines instead of electrical drives for boiler feed pumps. The capacity of the boiler feed pump has increased to the point where a steam-turbine boiler pump will have lower first-cost per installed kilowatt and can also give a fraction of 1 per cent increase in overall efficiency.

Engineering teams are working hard to coordinate the operation of new high-pressure boilers and turbines so that variable initial pressures can be used as load demand changes. They foresee simpler designs that will combine full-load efficiency with economic light-load operation. Another idea, available to the user of natural gas supplied at high pressure, is to get rid of the throttles now widely used, heat the gas with steam from the main unit or by flue gases, and put it through a pressure-reducing turbine-generator. Such systems can add from 2 to 3 per cent to the total capacity of a station through the utilization of presently wasted energy.

FUTURE FOR GAS TURBINES

And here a word on the operation of *gas turbines* is appropriate. Three things are happening that make it essential for utilities to take a new look at the use of gas turbines for peak-load or end-of-line stations. First, there are no longer enough old, inefficient stations around to meet peak requirements. Second, the substantial and rapid improvements in efficiency, which in the past have made it uneconomical to delay installing new base-load machines, are almost certain to be less rapid in the future. A system saturated with generators nearly as efficient as the best available can—and should—look toward accepting slightly less efficient peak-load units. Third, gas turbines are now being built in sufficient size—we are making units up to 21,000 kw and designing units of up to 30,000 kw—that they have a definitely lower first-cost when compared to other types of peak-load installations. Efficiencies are increasing, also; some of the larger gas turbines, incorporating regenerative units, now have a thermal efficiency of 25 to 27 per cent.

There are also a number of interesting plans for gas-turbine–steam-turbine combinations. In most instances, the practicality of such systems will depend on adapting the gas turbine to use coal fuel—a possibility, by the way, about which we are very encouraged and toward which a substantial effort is being made. The exceptionally large investment which we have made in developing the gas turbine as a new prime mover for the public utilities is an outstanding example of the technological strength available within our great electrical industry.

THE OUTLOOK SUMMARIZED

The outlook for the steam turbine-generator can be summarized about as follows:

1. FIRST COSTS. Large gains in the cost per kilowatt of turbine-generator units can be made by going from smaller units to units in the range of 150,000 to 250,000 kw. Additional, but less spectacular, gains in the turbine unit are available beyond this range. However, there are very significant gains in overall plant costs which continue to give direction to the trend to even larger units up to at least 500,000 kw.

2. OPERATING LABOR COSTS. Savings here are real as units become larger. While the savings are less per kilowatt in the larger sizes, they are important.

3. FUEL CONSUMPTION. The past six years have seen the greatest gains in history, thanks to the move toward larger, more efficient plants. But now the fuel economies which are achieved when inlet temperatures exceed 1050°F. are offset by the much higher first costs. The opportunities for extending the operation of turbines above 1050 degrees are substantially blocked by the "austenitic barrier."

Past technological improvements have offset inflationary trends to the extent that the utilities have been able to maintain—or reduce—the cost of electricity while the cost of nearly every other commodity has multiplied again and again.

To minimize the anticipated leveling off of the improvement curve, we must take full advantage of the opportunities described for gains in system, component, and auxiliary areas.

An *educated guess* about the future is that improvements will be noted in small increments; for example, we can anticipate, overall, a reduction in fuel consumption per kilowatt-hour of perhaps 1 to 1½ per cent a year for the next ten years. We cannot anticipate anything matching the three-per-cent-per-year averaged over the past fifty years, or the fantastic five-per-cent-per-year the industry has enjoyed during the past six years. Considering all of these things, it would seem to me that it will be necessary for the utilities and rate-making bodies to face up to the realities of probable future increased electric power costs in terms of then current dollars.

I speak as a representative of science, not as an economist, but two conclusions seem self-evident:

1. *Unless the basic inflationary monetary trend is halted, the health of your business and our business—indeed, the health of the nation's economy—may depend on our making a realistic reappraisal of the appropriate selling price for electricity.*

2. *As never before, the electrical industry must insure tomorrow's progress by continuing its vigorous exploration of scientific opportunity and exploitation of technological gains.*

. . . .

You will recall that early in my remarks today I mentioned that there were, for all of us in the electrical industry, several "facts of life" that are reassuring. I postponed the most pleasant of these until the end, and here it is: The news these days is filled with new ideas for *energy sources* to generate electricity, and new ways to *convert* this energy into electricity, and improved ways to *transmit* and *distribute* electricity, and new and more efficient ways to *utilize* electricity in homes and in factories. But even in the wildest, most blue-sky prophecies we find no indication, nor even hope, that there is a more useful *form* of energy than electricity. This fact provides the immovable, permanent foundation stone upon which the future progress of the industry will be built.

The Manageability of Scientific Research

An address at a conference sponsored by the American Management Association in Pasadena, Calif., May 5, 1958.

Science and management began to associate with each other approximately fifty years ago. At first the association was only tenuous. Some pioneers in the profession of management decided in 1910 that a convenient name for the new philosophy they were developing would be Scientific Management. At about the same time scientific research was beginning to be performed on an organized basis. Although scientists, by their nature and tradition, tended to resist the idea of managing or being managed, the fact remains that the "managing of science" became as inevitable as "scientific management."

Relations between these two disciplines, or, if you prefer, professions, have not always been as close as the names might indicate. Scientists have been skeptical about how much science is actually involved in "scientific management," and managers have openly doubted that scientific research *could* be managed. Many scientists have been less than enthusiastic over the idea that "the scientific method"—involving, particularly, the reproducible experiment—*can* be applied to a subject like management. On the other side, it has been said that "a good professional manager can manage anything" —except research, one usually adds. There has been some agreement, particularly on the part of the scientist, that in research, good management, like good government, is that which manages the least.

But, in spite of some prejudices arising from professional prerogatives, science and management are becoming mutually indispensable. I think most scientists would agree that modern management, with its intensive analysis of its functions, and with its computers and cybernetics, is becoming more "scientific." Last October *Fortune* magazine, in an article on "Research Management: The New Executive Job," reviewed the steady progress in the management of industrial research. I think we can conclude that research—like any multibillion-dollar business—will have to develop some business management, even at some professional discomfort on the part of both managers and researchers. Research is, of course, not a "business" in the ordinary sense—it is usually a functional part of many businesses. There is no doubt, however, that the aggregate expenditures for this work are in the billion-dollar range and growing rapidly. Thus, my theme is that management and research must get together to study mutually important problems. If we can make some progress here—and I am sure that we can—it will expand the usefulness of this highly creative work in industry and will provide increasing opportunities for the engineering of tomorrow's products, processes, and services.

Research is a broad term that includes a great variety of technical work, and I now wish to focus more closely on industrial scientific research including, hopefully, some basic research. It is in this variety of research that the greatest challenges and opportunities for management lie. I think it is timely to discuss the management aspects of scientific research because a great many people in industry and in government are re-examining their interest in this kind of work. Oddly enough, the Sputniks stimulated much of this interest, just as they stimulated a lot of interest in other things that we have not been thinking enough about recently.

I have observed frequently that businessmen who feel entirely comfortable about the management aspects of engineering, manufacturing, and marketing may feel baffled when it comes to scientific research. Much of this discomfort arises from the difficulty of putting numbers on the quantities with which we deal in the management of research. This is a serious difficulty, and I feel that we will not develop the full potential of this industrial activity until we acquire more facility in measurement. The only thing about research that we measure very well at present is the expense input. That particular index we measure with great competence. I will return to this point shortly.

In spite of some management difficulties—and to make a point I have exaggerated a bit—research, including basic research, is rapidly growing in American industry. The statistics, which are probably familiar to you, are very impressive. It *must* be that research expenditures turn out well in a significant number of cases if businessmen are allocating increasing sums for this work. I can think of many instances in which it has been a good thing in our company.

Actually, scientific research may be thought of as the starting expense of an investment in technological innovation. Whether it is a good investment depends upon a lot of things, including the kind of industry in which the business is conducted. Some businesses, and even whole industries, seem to require very little technological innovation. In some cases that come to mind, innovation is achieved annually by the rearrangement of chrome strips. This is a very powerful factor at the marketplace, because people are very reluctant to be seen with their chrome strips in the wrong place.

The other extreme is presented by the technical industries, including electrical products, electronics, chemicals, instruments, communications, semiconductors, pharmaceuticals, aviation, the defense industries, and many others. The customers for these products are generally not greatly impressed by chrome strips, or whether the handle is right side up or upside down. It is here that technological innovation has the greatest leverage, and scientific research its greatest opportunity.

PLANNING BASIC RESEARCH

Under the heading of *planning,* I wish to discuss here only a few of the many important aspects of this subject. It is a too commonly accepted notion that you cannot plan basic research. That is, you cannot plan on what will come from the research—with any certainty. It is also too obvious that, if you are exploring in a strange land, you cannot predict what you are going to discover. Both of these things are too true and too obvious, and they are also not quite correct. A little thought will show that, although a prediction of specifically *what* is going to be discovered and precisely *when* may be highly speculative—or even impossible—a good deal *can* be said in advance about expected discoveries. In fact, it is important in business planning to look forward as far as possible, and to do so effectively, one *must* speculate about scientific progress. One should, to the extent possible, anticipate its probable outlines and timing and probable impact on industrial technology. Important in the present context is the fact that a great deal can be done by research management to plan the organization and the operation of its scientific research programs to attain the research objectives which *would* have the desired impact on important technology. To illustrate the approach, consider this question: How many metallurgists have you heard of who, in the course of their research, discovered a new antibiotic drug? Or, how many solid-state physicists do you recall who have discovered a new polymer synthesis? This leads to Rule 1: The results of research will be heavily weighted in favor of the field of specialization of the investigator.

Now let us take a closer look at the work of the research specialist in relation to the objective of the research. In the course of his work, the researcher is presented daily with many alternative choices and directions. He can go to a higher pressure, a lower temperature, a more dilute solution, a narrower band pass, a higher Q, a lower frequency, and so forth, *ad infinitum*. If his research objective is clear and if he is determined to reach it, he frequently can and does make the choice that favors his objective. Because this is so, we regard the attainment of the objective as a very probable result of his investigation.

Successful researchers, whether in industry or elsewhere, do not plod through their scientific careers stumbling upon things accidentally. That is, they do not constantly stumble upon *important* things. Some strikingly successful research careers, like that of Irving Langmuir, have been characterized by a constant succession of important research results. Powerful innate curiosity and many other unusual aptitudes are involved in such cases, but one cannot avoid the conclusion that some careful sorting out of trivia and the selection of important objectives have played a decisive role in such careers. I conclude that a knowledge of what to look for, and what would be important if it could be found, is a source of motivation and guidance that significantly influences the results of research. This leads to Rule 2: Motivation is a very important management practice in planning and operating research programs.

OPERATING CONSIDERATIONS

Because there is in industry far more experience in the management of engineering, manufacturing, and marketing, we frequently employ practices in managing research that are more suited to these older fields. For example, it is appropriate in an engineering development activity that is geared to a product plan to require weekly or monthly reports of progress. In an overall business plan, these reports are very valuable because they make possible the coordination of the various functions involved and permit a measurement of progress in comparison to the plan. In a scientific research program, we have found that the requirement of reports on a calendar basis is unrealistic. The results of scientific research investigations do not emerge with chronological regularity. There may be no result for five months, and a jackpot a day later. Any ingenious researcher can produce a one-inch-thick report each week, if required. But he may have time for little else, and most of his reports will be extremely dull reading. A better plan, we feel, is to require that reports be written only when there is something to report, regardless of the calendar. The reports that are written on this plan are worth reading. Furthermore, the scientist who issues reports opportunely is on your team, because he knows that you understand the nature of the work in which he is engaged and have adapted your management procedure to this understanding.

Communications are a very important activity in industrial scientific research, and some intensive attention to the subject is required to attain satisfaction at all points in an industrial organization. Let us consider first the problem of internal communications associated with technical work programs. We have to keep in mind that the transition of an idea from scientific research to the point of utilization by an ultimate consumer may require many *steps* in technological, engineering, and manufacturing development. The applied scientist, the development engineer, the designer, the test engineer, the manufacturing engineer, and a host of other specialists may each contribute essential elements to the successful transition. Communication from one step to the next is a minimum requirement in a transition of this kind, and the steps must not be so large that people have difficulty of communicating, even by shouting. Thus, the problem of effective communications cannot be solved without reference to organization concepts, which I would like to discuss next.

ORGANIZATION OF TECHNICAL WORK

Organization, as an element of management, has been experiencing growth and evolution as our economy has been growing, and this evolution has extended to the organization components for doing technical work. The location of these components—laboratories, for example—in a decentralized company is a problem which I once thought was confined to a few large companies. This seems not to be the case, for there is a lively interest in the subject in companies large and small. A decentralized operating plan may be an operating necessity for a large company for reasons of size, but may be equally necessary or desirable in medium and small companies because of product or market diversity or geographical dispersion.

In all of these cases, the matter of organizational location of laboratories is involved, and from the variety of different plans being tried by different companies, it appears that no single best solution is in prospect. Again, companies differ, and solutions have to be designed to meet all of the industry and company environmental circumstances. It is important to point out that one of the significant circumstances that must be taken into account specifically in the planning of organizations is the nature of the technical work itself. For example, the character of the work may be such that the choice between one central laboratory and five decentralized laboratories may not be a free choice at all. Or, it may be that the cost of doing the work will be substantially different in the two cases. That this should be so in the case of research work is not surprising, because similar considerations apply to other industrial operations. For example, the large manufacturing plant required for the production of steam turbines might produce very uneconomically, or not at all, if divided into five small plants geographically dispersed. Some scientific and technical operations have a "critical size" imposed by the requirement of a great variety of technical skills or facilities. Fragmentation may reduce the effectiveness of such units nearly to zero. Complex and expensive gear and equipment characterize much modern scientific research, so that the economy, utilization, and duplication of research investment must be taken into account explicitly in considering organization alternatives.

In the company with which I am associated, where we regard a decentralized plan as an operating necessity, we have been able to successfully sort out these interrelated problems of organization and technical work. Our *product oriented* laboratories are located in the organization of the decentralized operating or business components of the company. Our *science oriented* laboratory is one of three "service" components serving the operating components and the company as a whole. Its objective is not products, but *opportunities* for new *technology*, embodied in ideas, phenomena, materials, and processes that originate in scientific research. Whether the "opportunity" is translated into a product or a business is a decentralized decision, in which Research Services has a great interest, but no authority.

MEASUREMENTS

In the professional work of the research scientist, *measurement* is an essential activity. A theory of nature is not truly descriptive until it becomes quantitative, or is a scientific observation fully meaningful until numbers can be associated with the phenomena under study. This being the case, it is surprising that scientists themselves have shown little interest in measurements of their functional work. Such measurements are a necessity if scientific research is to grow in magnitude and in contribution to American industry.

The measurements required include: (a) the individual achievement and productivity of the research scientist, and (b) the overall result of a research project activity on company operations. I will discuss these items separately.

Some progress is being made in identifying the elements of output of a research organization, and, by extension of those methods, we may learn how to measure the performance and productivity of the individual research scientist. That result is earnestly desired because we wish to recognize promptly his achievement and to reward him equitably for his work. The recognition I have in mind is not recognition in the world of science. This recognition, too, is very important to a scientist, but it may take place a distressingly long time after his immediate associates, his colleagues, and his organization are aware of his ability and accomplishment.

Some elements of output of a researcher are readily apparent, and include leadership, scientific papers, reports, patents, and consultation. Dr. William Shockley [1] published an interesting analysis some time ago of the "paper product" of numerous research groups and explored some correlations with salary levels. More analyses of this kind and quality are much needed, and could greatly extend our knowledge of scientific productivity.

The direction of progress for the measurement of individual scientific productivity seems clear. We must identify *all* of the elements of output, weigh them, and measure them. For example, patents undoubtedly provide an index of output, but this index is

[1] *Proceedings of the IRE*, 1957, Volume 45, #3, page 273.

currently of more value as a statistical, rather than as an individual, measurement. Patents differ greatly in value. Furthermore, the value of a patent may depend upon time in an unpredictable manner. A patent that appears to have limited promise when issued may at a later date acquire great value because of additional developments in related fields. Or the contrary may be true. Although we do not know how to measure the value of scientific papers, it is certain that they differ greatly in their contribution to science. Indices over and above the "paper product" are clearly required. For example, some very valuable members of a research staff publish few papers and apply for few patents. Their contribution lies in leadership, which they impart through counsel, advice, and consultation. No tangible index of this output lends itself readily to the measurement of this valuable work.

Now let us consider the value *to the company* of the research output we have been discussing. The appraisal of the overall company benefits from a venture which originated in research is a useful endeavor. If such studies could be projected forward, they would have greater usefulness, but even as an appraisal of the results of a completed venture, they provide valuable perspective which may be brought to bear on future decisions concerning research.

The manager of a business making a profit is constantly considering a great many alternative routes for expenditures for new or expanded sales, markets, and profits. He may, for example, build *new plant capacity* for an increased volume of presently manufactured products, or he may invest in *new machinery*, including increased mechanization and automation in present manufacturing plants. On the other hand, a greater opportunity might appear to be in the *marketing* area where a new approach and program may pay off handsomely. Alternatively he may decide on the *integration of* his manufacturing, by extending his operations in the direction of basic raw materials, and so forth. Or, he may invest in *research.*

In all but the last of these alternatives, it is generally possible to forecast the intended result with sufficient accuracy to permit decision making. To take a simple case, if the question concerns the purchase of improved machinery for reducing manufacturing costs for a present line of products, it is usually a straightforward matter to calculate the return on the intended investment. On the other hand, the calculation of the future return on present research expenditures involves more intangibles, and hence is generally much more speculative. Unless the management is farsighted, the research proposal may lose out in competition with alternative investment opportunities. Too often, when the partisan of a research proposal says that its possibilities are incalculable, he is probably exactly right—they cannot be calculated.

We must learn more about this problem by analysis and study. I am greatly encouraged by the positive usefulness of project case studies, or "model" studies, which we have undertaken in our organization. They have clearly justified the effort involved.

I have endeavored to show that the extension of scientific management to the management of scientific research is an increasingly important feature of modern industry. Although research directors are constantly confronted with—and occasionally confounded by—the numerous intangibles that always thwart their efforts to find simple formulae for operating procedures, it is possible to find a rationale for guidance in a specific situation. It is through progress in all phases of scientific management that today's research will have increased impact upon the growth of technology, upon the opportunities for engineering tomorrow's products and services, and hence, upon the growth and vitality of the economy as a whole.

The Opportunity for Basic Research in Industry

An address at a meeting of the National Science Foundation, Washington, D. C., May 20, 1958.

Interest in basic research in our country has been greatly stimulated by the Sputniks. There has been some substance to this interest, too, for many agencies, including the National Science Foundation, have found increased sources of support for such work. At first sight it is odd that the Sputniks should have had this effect. Basic research could not possibly produce a quick antidote for Sputniks. However, a more mature consideration shows that the Sputniks are merely a symptom and not a disease. Sputniks are seldom fatal, in themselves. But the underlying malady can have most serious consequences, and basic research may be the required wonder drug, or an essential ingredient of it.

This resurgence of interest in basic research extends to industry as well, and many leaders in diverse fields are now taking another look at this subject. This is an encouraging sign. There is a big stake here, and we must be right in our appraisal of the role of basic research in industry. There are at least two primary competitive pressures that will force a thoughtful consideration of this issue. The first is the pressure of economics. Basic research is costly, but it is an increasingly important source of the technological opportunity that provides the innovation which is the keystone of competitive industrial progress. The second reason is almost the same reason, expressed in a context of national survival: basic research is a key source of new defense technology. Both of these reasons are so important that we must have the right answers to the many proper questions about basic research in industry. Before getting down to some of these issues, let us take a look at the nature of the work itself. What do we mean by basic research?

Almost all of the leaders in modern American industry would acknowledge an interest and concern in "research" as an important factor in their business. Many would state an interest in *basic* research. A further inquiry would, however, show a great spectrum of opinion on the kind of work included in the terms *research* and *basic research.* Many people believe in basic research, provided it is the kind of basic research in which they believe. Or, to paraphrase the words of the song—"I'm true to research, but in my way." Opinions differ on the amount of such work that can and should be done in a given industry and company, and on the methods of organizing and managing this work, as well as on the methods of utilizing the results. These differences of opinion are real; they are based on real differences in industries and companies.

The National Science Foundation studies of research and development have revealed some of the first statistical information on industrial activity in the basic research classification. A new approach to this question has been made recently by a group of people under Dr. John Fisher of our Research Laboratory. They have developed a correlation between scientific publication and basic research activity that has permitted a study of industry patterns. The relationship of basic research to company size and to product obsolescence shows some striking and unexpected features. This work is in the process of publication.

In the universities where traditionally much of the world's basic research has been accomplished, there is a continuing interest in this kind of research and its relationships with teaching as well as its relationship with industry. Here, too, some differences of opinion may be detected. A few educators hold to the view that basic research is the exclusive domain of the universities, while others believe that this tradition has outlived its usefulness and should be abandoned as unrealistic. The heavy influx of sponsored applied research and development into university activities poses a further complexity to the question of the role of the university in this area.

I propose to examine some of these questions and to expose both the areas of opportunity and the roadblocks to the expansion of basic research, particularly in industry.

KINDS OF TECHNICAL WORK

But we have to start with some agreement on terminology. We must have a common understanding of the nature of the different kinds of technical work. We have long had a very great need in our own company for common understanding of these matters, in connection with the organization, operation, and evaluation of technical work. In recent years we have evolved and presently use a set of four technical work categories and corresponding definitions. We are now starting to use them in our accounting practice. For my present purpose, I will group these categories under two broad terms whose meanings are almost self-evident. We speak of *learning work*, in which the immediate objective is new knowledge and understanding of nature, and *applying work*, in which the immediate objective is the output of goods and services. Obviously, in the pursuit of applying work, new scientific knowledge may result; this, however, is not the focus of the work. We find these terms useful, and I will employ them in my discussion.

I would like to add one more word on terminology. We commonly refer to research and development, or R & D, as, for example, in the National Science Foundation surveys of this subject. Since this statistical study and others like it typically include the engineering category of technical work, a more appropriate term, I suggest, is Research, Development, and Engineering–RD & E. If in our abbreviation, RD & E, we indicated the relative industrial magnitudes of these activities, the R would be very small, the D of intermediate size, and the E would be the largest by far. There is obviously a considerable opportunity here for self-deception; we frequently delude ourselves into believing that we must be doing a lot of *research* because we are obviously doing a lot of research, development, and engineering–RD & E; the line between these elements has been conveniently unsharp.

There is no lack of interest in *applying work* in American industry, since the bulk of technical work in industry has this objective. Such work includes the development of materials, processes, and prototypes, and the subsequent extensive work in the engineering category. The objective of all of this work is the solution of practical technical problems required for the business. In the pursuit of a business plan that involves innovation of product, process, or service, it is usually possible to analyze the proposed results in terms of markets, output, and profitability and to relate this to the cost of the required technical work. On the other hand, the corresponding relationships for a proposed basic research program are far more difficult to determine; they are more speculative with respect to the results sought and especially as to when results will be obtained. Hence, business decisions concerning such work have traditionally relied upon a substantial amount of faith, belief, and hope, rather than upon forecasts and schedules against which performance could be measured. This particular deficiency is an important and fundamental difference between learning work and applying work. I shall return to this point shortly.

In spite of the difficulty of pursuing this work within a framework of business objectives, there is no doubt that basic research is a primary source of new industrial technology required for growth and progress. Many examples from actual practice could be cited, but to make certain that a current illustration is before us, I would like to review the case of the semiconductors. This area, as a phase of the study of the solid state of matter, has been the object of intensive research for a long period of years, particularly in industrial laboratories where many basic research contributions to the knowledge and understanding of this subject have been made. Technological opportunities based on this knowledge have made possible a new, already large, and rapidly growing industry of far-reaching consequence. In addition to the companies that invested in the basic research on this subject, many new companies have found opportunity for growth in the new industry, and hundreds of suppliers have found a new market for their specialized goods and services. More important, semiconductor products and components with unique properties have stimulated many other product areas, and, in particular, are providing new and important technical capabilities for our national defense. This example illustrates why a competitive progressive industry must necessarily interest itself in the kind of business opportunity provided by basic research, and, hence, why management is increasingly concerned with the management aspects of research in this category. In the electrical industry with which I am associated, everyone knows that much of the technology of the trade originated in the basic researches of the early scientists like Coulomb, Ampere, Faraday, Henry, Hertz, Maxwell, and many others. The discovery of the principles of electromagnetic induction by Faraday ranks with the great discoveries of all time. It led to the technological foundations upon which much of the traditional electrical business rests.

THE OPPORTUNITIES FOR BASIC RESEARCH ARE NOT EQUAL IN ALL INDUSTRIES

Industry is a very broad term covering a vast array of human activity, and not all segments of industry afford equal opportunities for exploiting the results of scientific research. In addition to the electrical industries, the chemical, communication, electronic industries, and many others have all had their origins in the laboratory. But some industries have grown to great size and profitability with very little expenditure for or benefit from scientific research. For example, the exploitation of inventiveness has in some cases provided the initial business opportunity and the continuing source of product innovation. However, all goods and services are becoming more complex, and the technology of materials, processes, and manufacturing methods is becoming correspondingly more complex. This trend would suggest that the traditional differences between industries—with respect to their employment of industrial research—will in the long term tend to disappear.

However, that point is somewhat in the future, and today the opportunity for the *application* of new technology varies widely in different companies as well as in different industries. At one extreme are the technical industries, some of which I have mentioned, in which opportunities for the development of new technology and for its prompt application are abundant; and at the other extreme are the industries making simple mechanical products which seem neither to employ nor to need much sophisticated science, nor even engineering. I received a rude shock some years ago when I discussed research with the head of a highly successful company in the metal fabricating industry. I asked about his laboratories; he said he had none. I suggested that of course he must be doing a good deal of engineering work. He said that he was doing some, but that he actually employed no engineers. I asked how he kept his products up to date and competitive. He replied that he watched his competitors carefully and copied their innovations. As I learned more about his company and the industrial environment in which it operates, I became convinced that his policy with respect to RD & E was sound. There are many different ways of skinning a cat, so I am told, and it is important to find the best way—the best way for you.

PLANNING BASIC RESEARCH

One of the most serious roadblocks to the planning of basic research is tradition, but the tradition is based upon some real difficulties, including numerous unsolved problems. Some hold that the difficulties are insurmountable and hopeless. The view is all too prevalent that you cannot tell what is going to come out of basic research, and that it may be better not to try. It is not uncommon in this field for witchcraft, superstition, sleight-of-hand, and hunches to compete on a basis of parity. This intolerable situation must change if we are to do more basic research in American industry. These problems of planning are difficult, but they are not insoluble. If we would put to this task a small fraction of the courage and determination we use to solve scientific riddles, the results would be astounding. I believe that the steps which we must take here are becoming clear, and that some real progress is in the making.

Are the results of basic research completely unpredictable? Is the explorer in science entirely at the mercy of a capricious nature? Is it hopeless to inquire as to possible or even probable areas of scientific progress? I think the answer to each of these questions is *no*. Let us consider some typical cases. To be realistic, we must admit at the outset that there will always be room for the completely unpredictable discovery in scientific research. But it does not follow that all results of basic research are completely unpredictable. Consider, for example, the biochemist engaged in a study of the effect of ionizing radiation on enzymes. Is it likely that his research will lead to a new understanding of enzymatic reactions, or is it equally likely that he will discover a new ceramic process? Alternately, consider the physical chemist studying the structure of ceramics. Is he equally likely to discover new knowledge about enzymes? I think not, and that the area of scientific specialization of the researcher in itself provides a strong orientation to the results of his research. The selection of scientific skills is thus a first and fundamental step in planning a basic research result.

A more detailed look at the working method of the scientist shows that nature presents him with many possible choices in the pursuit of his investigation. His decision as to the experimental or analytical approaches to be pursued, the techniques or tools to be employed, and particularly his appraisal of areas of opportunity, promise, and importance all influence significantly the outcome of his inquiry. I think we should discourage the perhaps popular conception that the most successful scientists follow a career of stumbling upon important discoveries. I think that the contrary is true and that the scientists who always happen to be present when important discoveries are made are, in fact, strongly motivated by an understanding of what would be important to discover. They must constantly discriminate between the trivial and the significant. The objective of scientific inquiry—that is, what the scientist is looking for specifically, and especially his conviction of its importance—thus provides further detailed orientation to his work.

The thought I am trying to express is that, although the prediction of specific discoveries and inventions is generally impossible, we can do a great deal to anticipate and encourage a given area of scientific progress. We do this by the specific research skills we apply to an investigation and by the individual interest and motivation we elicit in the researcher. Knowing what to look for and what would be important to find are of course significant attributes of a successful researcher. Mother Nature may not co-operate with his plans and hopes, but she should be given every opportunity to do so. The discovery at the Bell Telephone Laboratories of the semiconductor phenomena which made possible the transistor was in my view not an accident, but the result of the kind of orientation which a creative environment can impart to a basic research inquiry. The long and distinguished tradition of this laboratory in the fields of communications and electronics thus provided an altogether fitting and probable birthplace for this important discovery.

THE CHALLENGE IN INDUSTRIAL RESEARCH MANAGEMENT

The task and challenge with respect to basic research in industry are clear. We must foster a creative environment for learning work which is consonant with the character and objectives of the business. To do so we need to learn a great deal more about the environment and climate, and the motivations for creative scientific work. Recognition and reward are fundamental elements of this climate, but there are other important factors directly related to the operating work of scientific research. These considerations include:

1. The role of specialization and the corollary role of multiple skills in the pursuit of research objectives;
2. The increasing complexity of the tools of research, and the required supporting services;
3. The basic functional necessity for communication with scientific colleagues, associates, and with counterparts elsewhere;
4. The functional and economic relationships between learning work and applying work;
5. The identification, appraisal, and measurement of the output of research.

In all of these subjects, present knowledge and capability leave much to be desired. Because management practices are better developed and more widely understood in older industrial functions, such as engineering, manufacturing, and marketing, we tend to employ procedures in the management of research that are more appropriate to these older functions. In the management of creative learning work, we face an outstanding need and a tremendous opportunity. We must develop practices that are more appropriate to this kind of work. Enlightened concepts are called for here if basic research is to grow in extent and in contribution in American industry. I will comment briefly on each of these areas of need.

Specialization is an inevitable result of the increasing complexity of science and technology. The specialization applies not only to skills, but also to services, equipment, and facilities, and this factor must be taken into account, explicitly in the planning, organizing, and operating of scientific research. Some kinds of research investigations can be launched profitably with relatively few scientists and a minimum of services and facilities, while investigations in other fields will require the full treatment if useful results are to be obtained. In other words, there is a "critical size" for research effort, and the size is related to the specific field of work.

We must study the role of communications in scientific research in industry, and understand its important bearing on our research programs. Intramural communications problems currently limit our ability to utilize new results of research programs. Extramural communications are a functional necessity for a scientist engaged in learning work, and the lack of them may impose a limit to his progress. Modern researchers depend upon colleagues and counterparts elsewhere to *appraise and recognize* the results of their work. Prompt publication of scientific results is a working necessity and has almost nothing to do with the loss of proprietary rights. The publication of *applied research* and *industrial know-how* is, however, an entirely different matter.

I referred to the relationship between learning work and applying work. I had in mind such factors as the size of a basic research program in relation to effort in applied research, development, and engineering. This question of relationships also enters into the problem of transition of research results to practical use. We need much clearer guidelines here, and I will suggest some approaches that promise to lead in this direction.

In the field of measurements, almost everything remains to be done. In any economic activity, it is necessary to measure both the input and the output. Although we have exceptional ability to measure the expense input of industrial research, we still have to learn a great deal about measuring the output.

HOW MUCH?

How much learning work should American industry be doing? To answer this question, we must develop broad guidelines from the study and analysis of current and recent past practice in our own and in related industries. From such "model" studies, we must acquire meaningful relationships between the amount and timing of learning work and other industrial technical work, and the bearing of both on the objectives and progress of the business. To have specific usefulness, such studies should be undertaken by each industrial unit through the detailed study of its own areas of opportunity and need. The technique of making such studies is becoming clearer, and guidelines of much specific usefulness are coming out of significant case histories or "model" studies.

One of the toughest jobs is trying to find the balance between short-range and long-range goals that is most appropriate to a particular business in a given industry, and this includes the decisions on research. It is a constant temptation to improve the current operating statement by eliminating *current* expenditures for *future* business. Too much preoccupation with this kind of shenanigan will make it practically certain that there will be *no* future business. However, when you buy a piece of the future with today's research it has to be paid for today, out of today's profits, if any. So the industrial leader deciding on research expenditures, and particularly expenditures for *basic research*, is necessarily engaged in a juggling act in which the balance must be constantly re-examined in the light of the latest information. This leads to the "on again, off again" problem which is particularly difficult to accommodate in a basic research activity. It may also be quite costly and wasteful. Fortunately or unfortunately, depending upon your point of view, the penalty for the wrong decision does not fall due until some time later. The time from a new scientific discovery to the cash register is seldom less than five years, and is not infrequently ten years; seven or eight years is a fairly representative figure. Hence, one industrial research decision is easy to make: If you do not expect to be in business five years from now, there is no need for expenditures for scientific research, especially in the learning category.

LIMITATIONS TO INDUSTRIAL RESEARCH

For the industries that make more or less complex technical products, or employ more or less complex process or manufacturing technology, what factors limit the amount of scientific research, and especially learning work which it is feasible and economic to do? There are a lot of intangibles and unknowns here, and again specific answers must be sought for specific industries and companies. In the case of a given company, for example, perhaps all available cash that can be retained from earnings must be allocated to new manufacturing facilities for currently competitive products. This might make it necessary to forgo research expenditures until later. In all cases, the question of "make or buy" needs to be considered. The opportunity presented by the support of a basic research program in a university may be a solution, or a part of a solution, of a specific need. Many similar considerations, or combinations of them, may prevail in an actual case. In the circumstances of the company with which I am associated, I am most conscious of two limits. The first is the problem of the manageability of the research function within a framework of business objectives, many aspects of which I have outlined briefly. The second limitation, which looms large, is the difficulty of quickly and economically accomplishing the transition from science to technology—from technical feasibility to economic feasibility. This is the problem of the *interface.* It is axiomatic among my associates, and among my counterparts in other industrial laboratories, that the transition from research to engineering is the most difficult problem with which we deal. This vast problem encompasses organizational concepts, communications, an array of questions concerning skills and facilities, and finally considerations of psychology and emotion which bear on the problem of human resistance to change. The scientist is almost invariably too indulgent of the faults of his offspring, the *New Idea.* On the other hand, the engineer has by experience acquired a gimlet-eyed perception to the frailties and ineptitudes of new technology. We are gaining greater facility in dealing with these multiple problems of the interface between laboratory and practice, but our gains are partly offset by the increasing complexity of the transition projects. I believe, however, that net progress can be observed.

WHAT SHOULD THE GOVERNMENT DO TO ENCOURAGE BASIC RESEARCH IN INDUSTRY?

A report by the National Science Foundation said: "Industry is well advised to emphasize basic research increasingly in its research-and-development programs, and the Federal Government might well offer positive encouragement to this end." The meeting today is tangible evidence of a determination to explore avenues of possible approach.

The NSF report suggests some plans for the dissemination and exchange of research information, and also discusses the possibility of making some specific R & D costs, such as training of scientists at universities, tax deductible as business expenses. Actually, the Federal Government encourages industrial research now, to the extent that it is allowed as a current operating expense for tax purposes. There is no doubt that the tax structure is an important factor influencing the level of research expenditures in industry. From an industry viewpoint, the industrial expenditures for all functions of a business, including research, are determined by an economic balance, and this balance is, of course, influenced by tax practices. I feel reasonably sure that various segments of industry would prefer to determine, from purely economic considerations, a balance between industrial functions that is most appropriate to the pursuit of their business. This consideration, however, does not take into account fully such essentially uneconomic questions as national survival in weapons competition with the Soviets. This question is of substantial importance and in need of much thoughtful study.

Some of the bills before Congress or in preparation, which have the stated objective of encouraging more basic research in industry by tax credits, include some additional features that would largely neutralize the proposed tax advantage. In one case, the provision for reporting research results to a government agency as a condition for tax credit is an unfortunate proposal, which reveals an alarming lack of understanding of scientific research and how it is pursued. Industrial laboratories that do scientific research in learning categories permit and encourage prompt publication of research results, because they know that their scientists rely heavily upon the scientific community to appraise, evaluate, and recognize their scientific contributions. Encouraging basic research in industry by the tax route may be feasible and desirable, but, if so, better specific proposals are required to bring this about.

One particular thing that government must do to insure a continuation of the benefits of research is to rejuvenate our patent system. I am aware that this problem has been recognized, and that some steps toward a solution are in progress. I think, however, that much more fundamental and extensive changes are needed. There is need for a more sympathetic attitude on the part of the courts toward the original concepts and philosophy upon which the patent system is based. We must recognize that the inventor who provides previously unknown ideas, machines, and processes contributes significantly to the progress of our social structure. There is also need for improving the facilities of the United States Patent Office for determining whether the inventions submitted for patenting are new or old. It is apparent that the Patent Office, in common with other components of our technical society, threatens to be overwhelmed by its inability to accumulate, classify, and search, as its responsibilities require, the growing mass of technical literature, including its own, within a reasonable time.

The Federal Government's interest in basic research has a strong point of focus in the Department of Defense. Each of the armed services is carrying on extensive RD & E programs in support of its objectives, partly as an "in-house" activity and partly through contractors. Included in the totals is a substantial amount of basic research which conforms to the learning work category. This relatively small expenditure can develop sources of very important leverage in future DOD activities and deserves more detailed study and consideration than it has had to date. The problems of management of basic research in the DOD are entirely analogous to the ones I have been describing, and they require a similar consideration. Through its Naval Research Advisory Committee, the Navy Department is undertaking a study of the relationships of its basic research program to the activities and objectives of the department. It is expected that clearer concepts of objectives sought, criteria for the work, and relationships in magnitude and timing will come from this study and will provide more objective bases for the future planning of this work.

SOME PROBLEMS IN THE EDUCATIONAL AREA

Obviously, the future of industrial research will depend on the availability of adequate technical manpower. Industry, large and small, can do much to help solve the shortage of technically trained people. Businessmen, working with schools, can take an active role in encouraging young people to further their studies and to explore their technical aptitudes and interests. They can awaken students to future opportunities and provide tangible career-guidance material. Most important, they can help the teachers of high school science and mathematics to be alert to the aptitude of their students and the careers available to them in science. My company sponsored an original high school science fellowships plan in this area, and the National Science Foundation now carries out a most effective program with this objective. Finally, we can help our employees to identify an obligation and to repay their alma maters for their education.

However, let us not go overboard on the question of the shortage of scientists and engineers. Industry simply could not use twice as many this year, or next year, even if they were available. This year, in fact, the supply in both categories appears to be quite adequate, and a little stretch may be required to place all of the technical graduates. The fact that Soviet Russia is training engineers at three times our present rate does not automatically mean that we should equal or exceed her rate. In our system, we require an economic balance, and to use more technical people we have to learn how to employ them effectively. Although there is an economic limit to the *numbers* of technical people we can employ in industry there is no limit to the *quality* of the required skills. In the present public discussion of this subject, I think we should focus more attention on quality and less on quantity. If, as I hope, we increase the basic research in industry, and I have outlined some of the homework that must be done to make that economically feasible, then we will have an increasing number of opportunities for capable industrial scientists. In the universities, there will be an increasing number of opportunities in science teaching and in scientific research because of the great increase in student enrollment. And, of course, we must anticipate both of these needs to some extent, since the incubation period for a scientist is quite long.

I mentioned earlier that a few educators feel that all basic research should be done in the universities, and I believe that this subject deserves some frank discussion in the present context. This feeling may be based on the importance of the coupling between basic research and science teaching. This coupling *is* truly important and is a source of vitality and strength in the science faculties, and it should be preserved. It is, however, not the only coupling, or the only consideration bearing on this question. Another important coupling is the one between basic research and applying work, and as I have pointed out, in industry this is the most difficult task with which we deal. Until this interface is successfully bridged, the ultimate material benefits of scientific research—in the form of the satisfaction of human needs and wants—are not attained. Furthermore, the coupling between science and the national defense is important also, and we must make certain that we build no interface here that would impede the flow of new scientific knowledge toward defense technology. So in this case too we must keep in mind a balance, a balance between teaching effectiveness on the one hand and applying effectiveness on the other. Since both are important and neither one can be forsaken for the other, I suggest that both the universities *and* industry have a proper role in basic research. Not only are these respective roles proper, but also our interests are quite different in many areas of science. In the world of science, fields of science are susceptible to fashionable trends, which may or may not coincide with a given industry's area of technological interest and hope. Currently, for example, in the physical sciences there is an overwhelming interest in high energy nuclear physics, which does not have a counterpart interest in industry. In industry, we have had a prevailing interest in solid-state phenomena and semiconduction, which has not been balanced elsewhere. In the recent past, few universities even offered course work in semiconduction, although many are now starting to do so. As another example of many that could be mentioned, in our laboratory we have recently undertaken a research program in the physics of incandescent solids, which has been neglected scientifically for at least three decades. The *need* arises from the fact that incandescence is a basic phenomenon underlying important segments of our business. Since it has not been studied scientifically in recent years, we feel the subject may present an *opportunity* for new scientific knowledge. Hence a consideration of both *need* and *opportunity* bear heavily on a specific decision to undertake basic research in industry.

I conclude that in basic research there is room and need for all of us—in the universities, in industry, and in the government. In the Twentieth century and in the years of the Sputniks, we in the western world must make sure that the best creative minds we can challenge, wherever they are located, must have an opportunity to contribute in science.

Education and Science

An address before the Scientific Research Society of America on the occasion of receiving the William Procter Prize, at Washington, D. C., December 29, 1958.

It is easy to assume that a bond between *education* and *science* is completely natural and would exist even if no one gave the matter any particular attention. We take for granted the ideas that only education can provide scientists, and that science, as a primary force reshaping the modern world, is automatically worthy of the educator's interest. However, as I will indicate, the bond between the two is somewhat frayed.

Historically, a mutual interest between education and science has *not* always existed. We have not yet celebrated the hundredth anniversary of the first granting in America of a Ph.D. in science. This latter event took place at Yale University in 1861. Many more years passed before science really became an integral part of American education, if indeed this has actually come to pass. It is a fact that many people graduating from college today have little or no understanding of science, and that not many decades ago it was customary—and necessary—for young Americans bent on a scientific career to take their advanced research training in Europe.

The mutual interaction of education and *industrial* research is new if for no other reason than the fact that industrially-sponsored science is, of itself, a comparatively recent addition to the American scene. It seems apparent that a good educational system and a thriving pattern of industrial research have some common interests. It all sounds very pat: the schools provide training for the bright young men; industrial research provides opportunities for careers in science for the bright young men, who make discoveries, which make possible new industrial enterprise, which increases the nation's standard of living and thus provides the source of support for more and better educational institutions.

Of course, education and science—particularly industrial science—should work together because of these common interests. And, thanks to organizations like RESA, they do. But are we, as a nation, doing the job as well as we should? Have both scientists and educators fully identified the joint obligations, responsibilities, needs, and, particularly, the *opportunities* in this relationship between education and industrial science? Has industry fully recognized the problems of modern education, including its financial support, and determined industry's appropriate role in this issue?

Has education understood the motivation of industrial science and its tremendous impact on the economy, the standard of living of all of the citizens, and the national defense? Both fields abhor the thought of "losing," as we put it, our most skilled people to each other, yet our common interests would probably be served best if we had a much greater exchange of experienced people between the campus and industrial research centers. Most important of all—have educational administrators truly understood the need to teach science as a part of our culture? Have any of us—educator or scientist—really come to grips with the fact that in the most scientific age in the history of the world, science itself has not become a matter of real interest and understanding to the individual citizen?

Frequently I have been amazed to find so many similarities between *individual education* and industrial research. For example, both are popularly regarded as "good things." One trouble with a good thing is that it is so easy to believe that, if it's good, twice as much would be twice as good. We might, for example, conclude that to meet the Soviet threat we should train scientists and engineers at twice our present rate. Many of my waking hours are spent trying to determine how much scientific research is appropriately done by my company. That question is about as complicated, and has some of the same elements, as the question of how much education is optimum for an individual. Perhaps an individual's investment in education should be in proportion to his intellectual capacity or his ability to absorb it, and perhaps a company's investment in scientific research should be in proportion to its technological capacity and hence its ability to absorb the results of research in furthering its economic objectives.

Passing from the education of the individual to education in a broader context, I note some similarities to industrial scientific research, also. The college president evaluating the relative merits of strengthening the Physics Department versus adding courses in Russian must feel some kinship with the industrial research director trying to divide a given amount of money between Information Theory and Polymer Chemistry. College presidents and corporation presidents could, with only a few changes in the words, interchange speeches on the subject of "growing costs"—in education and in research. There are some discussions in my business about the appropriate role of the Federal Government—and government money. It is reliably reported that educators also have varying opinions on this issue as it applies to their institutions. The management of research and the administration of education have some common denominators. In particular, they both deal with many intangibles, with strongly intrenched practices, and inadequate measurement. Both have problems in productivity that we are reluctant to face realistically. Both have basic problems of balance, with only intuitive methods for determining the desired answer.

MUTUAL MEASUREMENT PROBLEMS

In scientific research, we have superlative ability to measure natural phenomena and almost no ability to measure any of the economic attributes related to the work, except cost—let alone the political and sociological consequences. This lack was not serious when scientific research was only a ripple on the economic pond, but this activity now makes a big splash and it may become a tidal wave. The results of research have great leverage on the economy; they are a prime factor in the national defense; they affect the life of every citizen. Most important, with an $8 billion annual bill for research, and its resultant development and engineering, this activity has leverage on Mr. Citizen's pocketbook.

In industry, research and related technical activity is becoming a big component of cost, and it promises, or perhaps I should say threatens, to become much larger in the future. Hence we must be more businesslike in conducting this big business, and we are becoming dissatisfied with traditional intuitive methods of administration and management.

I am no expert on educational matters, but as an outside observer I would guess that there may be some need for measurement in education, too, and for similar reasons. For education has become a big business and business administration rests on measurement. The techniques of measurement are difficult enough, but the problem is a lot more serious if we do not know what to measure; and I am not sure any of us, in our groping for measurement, are really measuring the right things. For example, when educators stamp their product magna cum laude, have they measured the right things? I can think of some specific instances that would at least support doubts. When we measure the output of an industrial research venture, we frequently have our fingers crossed, too. I am not sure that we are measuring the right things, either. I would like to leave the distinct impression that we both have a great deal of unfinished business in the area of measurements.

PRODUCTIVITY IN EDUCATION

Let us focus a bit of attention on a word that is too much of a stranger in our midst—productivity. This word has all of the elements of irritation built into it, for its mere mention implies that there may be some dissatisfaction with the subject. Yet, in any economic system, this is a basic concept, and preoccupation with productivity in the economy as a whole increases each year, as we add new and more efficient machinery, manufacturing processes, better engineering methods, more automation, and more power to lengthen and strengthen the arm of each worker. Let us look at the result purely on the basis of power alone, which is by no means the whole story. In American industry, the installed horsepower per worker is now about 15. It is easy to determine that the sustained rate of effort for an individual is in the range of $\frac{1}{10}$ to $\frac{1}{20}$ hp, and, for a small side bet, I could identify a few individuals who work at a still lower rate. But let us say that one horsepower equals ten manpower and go on from there. The average worker in industry thus has at his disposal about 150 manpower.

This basic relationship of power and mechanization to individual productivity is fundamental to the high standard of living in the industrialized countries of the world. Productivity paces the economy; not, however, without creating a few problems in the cultural areas of the same economy. Cultural pursuits may, by their very nature, not be adapted to mechanization, and may thereby price themselves out of reach of all but a few of our citizens. This appears to be a serious problem in mass higher education, and one in need of thoughtful study. The economic status of the teacher, especially the college teacher, has not been keeping up with the economy as a whole, and productivity is one of the factors. There has been little or no increase in teaching productivity to match the progressive increase in productivity in the manufacturing industries, in agriculture, in transportation, communications, distribution, and in the service industries.

The class of fifteen to twenty students has been a constant

denominator of college education for about two centuries with good cultural results, but with obvious economic consequences. One of the most serious problems here is tradition and habit, and, until recently, no one had the courage to raise this question and to examine it carefully for possible solutions. The Fund for the Advancement of Education of the Ford Foundation has for a number of years, however, sponsored a series of studies of this subject. This work now appears to be yielding some exciting and very significant results, some of which have been incorporated into the teaching plan of the sponsoring institutions. I recommend that educators look into this work.

The rising costs of scientific research and the resultant and related technical work have focused attention on the productivity factor in this kind of activity also, with some distress in certain quarters. Since, in most cases, we cannot satisfactorily measure the output of scientific or technical effort, we are not entirely sure what is happening to productivity in this kind of work. But in many typical laboratory research operations, it is difficult to identify elements of organization, mechanization, services and facilities which would support firm claims to substantial and progressive increases in productivity in research operations. So industrial researchers also have some unfinished business and consequent opportunities in this area.

While I have been discussing some difficult problems of education and science, I would not leave with you the impression that negligible progress is being made in their solution. I might digress for a moment to describe the work of one of my associates, Ned Landon, whose tongue fits neatly in his cheek, and who recently produced what he calls "The Chief Executive's Utterly Exact Method for Measuring Scientific Research"—which goes as follows:

I multiply your projects by the words I can't pronounce,
And weigh your published papers to the nearest half an ounce;
I add a healthy bonus for research that's really pure
(And if it's also useful, your job will be secure).

I integrate your patent-rate upon a monthly basis
And I figure what your place in the race to conquer space is;
Your scientific stature I weigh upon some scales
Whose final calibration is the company's net-to-sales.

And so I create numbers where there were none before;
And thus have facts and figures and formulae galore—

And these volumes of statistics make the whole thing very clear:
Our research should cost exactly what we've budgeted this year!

Now, more seriously, in the spirit of this interchange of ideas, I would like to discuss some additional relationships between education and science.

Science training in our schools must, first of all, be based on a better understanding of science and its practitioners. A primary motivation of the true scientist is a desire to understand nature. He believes that nature is orderly and that by properly formulating questions, he may learn the pattern and organization of natural phenomena. The starting point in technological progress is often provided by scientists of this type, working at the growing edge of science, pioneering in areas never before explored.

The number of persons with aptitudes for this truly exploratory work is relatively small. Generally, they are found among the brightest students. It is essential, then, that they be identified at an early age, and encouraged and motivated so that they will realize their great potential. This, however, is *not* to say that all bright students should be encouraged to become scientists. Nothing could be more wrong—or more dangerous. Every field of human progress needs its share of the very best minds. Scientific progress in the future is not likely to be limited by nature, but rather by our progress in the social sciences, the humanities, the political and cultural order—in short, by the total environment. We need to make *balanced* progress in many fields, not in one.

I have already mentioned that the greatest single objective in teaching science in the schools—and the one that has largely been overlooked—is the matter of making science a part of our culture. It is a fact that there are still in existence "liberal" education programs that virtually ignore science. In a time when science is shaping the lives of all men, everywhere, this subject *must* be a matter of real interest and understanding to the individual citizen. For the individual citizen is the key element in the total environment, and he can no longer be isolated from science. Of course, no citizen can be expert in every field of science, and most citizens not in any of them. But it is vitally important that the level of public comprehension of science and its impact on people be raised materially.

Today's student will be tomorrow's citizen. As a citizen, he will have responsibilities as well as rights. As a responsible citizen he should know something about the history, dimensions, and trends of our technological civilization so that he can intelligently help to shape our total environment tomorrow.

The increasing role of science in government is particularly apparent to us in these days—and here in this city. Our government depends on an informed citizenry; in a democracy we cannot rise to heights much greater than the levels of public enlightenment. In the long run, public opinion is a controlling factor. Each of us has been born into the most complex technological civilization the world has ever known. Unless people, individually, come to have an understanding of the methods, objectives, capabilities, and the content of science—the basis of technology—their elected representatives cannot intelligently represent the "will of the people."

Further, if the representatives themselves do not have at least a modicum of scientific knowledge, they will not be competent to select advisors nor to evaluate technical advice. Many of the decisions of government are made in highly technical areas that are fully understood only by technical specialists. Someone must choose the specialists. The future may depend upon how well the choices are made.

I will not excuse the scientist from a substantial responsibility in this task of developing a public understanding of science. Too often his contribution to the understanding of scientific progress, often based on his own work, has been less than optimum. Too often he succumbs to the temptation to mystify, when he should be bending every effort to translate and interpret and explain his progress to his fellow citizens.

Education, then, has a tremendous challenge. It must accomplish the double purpose of providing a continuing supply of scientists, and an informed citizenry, including representatives in government. This means that science must be taught not as a set of mystical facts, but as a living, breathing, and growing component of modern civilization. Much more must be done to interpret science and technology to the public.

Can I tell you how this should be done? No. But it is obvious where we must start—with the *recognition* and *identification* of this important problem. It will take the contributions of many—scientists and educators alike—to make progress here. And it would be presumptuous for scientists to claim an all-knowing role in the planning, for scientists have frequently been remiss in their duty to interpret science for the public. We often have shown ourselves ignorant of how the mind of the public works—at least as ignorant as the public has been of the workings of the scientific mind. Better communication is certainly indicated. And in the long run, we must look to the schools to do the job. Are we passing the buck? Perhaps, but we can justify our buck-passing if we give our schools the support they deserve.

PHYSICAL SCIENCE AND SOCIAL SCIENCE

The same science that is the source of so much of today's progress could also lead to mankind's destruction in the years to come. There are many things we must do in a material way for the survival of our civilization, but all of them will avail us nothing unless we somehow learn to live with each other. That is why *all* the bright young men and women must not become scientists. We need, somehow, to achieve advances in the social sciences comparable with those that have been made in the physical sciences. And I am certain that, to an increasing extent, knowledge of *physical science* will be an essential part of the education of every *social scientist.*

Alexis de Tocqueville was a Frenchman who came to America in 1831 to study democracy. Shortly thereafter he published a book called *Democracy in America.* Many are familiar with his fantastic prophecies about the coming struggle between America and Russia. Another less well-known passage predicted a future for industrial research nearly a century before industrial research really began to come into its own.

"You may be sure," Tocqueville wrote, "that the more a nation is democratic, enlightened, and free, the greater will be the number of interested promoters of scientific genius, and the more will discoveries immediately applicable to productive industry confer gain, fame, and even power, on their authors. For in democracies, the working class take a part in public affairs; and public honors, as well as pecuniary remuneration, may be awarded to those who deserve them. . . . Possessing education and freedom, men living in democratic ages cannot fail to improve the industrial part of science; and . . . henceforward all the efforts of the constituted authorities ought to be directed to support the highest branches of learning, and to foster the nobler passion for science itself."

Tocqueville understood, far better than many modern-day prophets, the proper relation between science, education, and other disciplines. Well over a century ago he recognized the growing role of science—and even industrial science. But in his writings he also made it clear that, in spite of its importance, science is just one of several essential elements of our civilization.

Am I suggesting that what we need today is another Tocqueville? Well, the trouble with individual prophetic geniuses is that we tend to believe their predictions more *after* they have come true than before. And in any case, we will be hard put to find many single individuals who can simultaneously grasp all the complexities of today's social, political, economic, and scientific problems. What we need is not so much a *universal genius* as a broad base of *mutual understanding* between *all* physical scientists and *all* social scientists. I think it is evident that education must play the key role in achieving this understanding. I speak, of course, as a physical scientist. For us it is essential to remember that man's future destiny will be achieved not *by* science alone, but *with* science as part of the whole.

Creativity and Technological Progress

An address at the Conference on Scientific Manpower, sponsored by the Civil Service Commission and Office of Naval Research, Washington, D. C., April 29, 1959.

I think we can agree at once that we are all in favor of creativity. Certainly creativity in a scientist is a desirable attribute, and certainly the creative results of a research organization constitute a principal element of output. Yet, creativity is not a universally desirable quality for all technical people and for all technical programs under all circumstances. In some cases, we may have a proper interest in the balance of effort—as between creative innovating work, on the one hand, and applying work concerned with reduction to practice on the other. I will describe some instances in which the determination of this balance is a challenging objective.

Creativity is an intensely personal quality of some individuals, and any discussion of the subject cannot avoid a consideration of the attributes of the creative human and, particularly, the environment that stimulates his effort. Since scientific creativity is basic to technological progress, and since the latter is an important driving force in modern civilization, some of the relationships between scientific creativity and technology need discussing. Also in need of discussion is the basic managerial question of whether an organization, in industry or in government, really wants creativity, or is merely paying its respects to a symbol currently in vogue.

Motherhood, for example, is a universal symbol of nobility and virtue, but thoughtful consideration will show that even this virtue has its proper place. I believe that one may identify some roadblocks which are looming up on the road to creative technological progress, and I will discuss one of them. I will express some points of view on each of these matters, but will admit at the outset a certain presumption in dealing with any one of them so briefly.

THE CREATIVE INDIVIDUAL

Creativity is an elusive element of human performance, difficult to plan for and to achieve, and especially difficult to measure and evaluate. The term "creativity" is itself a rather elusive word, not precisely defined by usage, and susceptible to delicate shades of meaning and interpretation. For example, does the physical scientist engaged in research *create*, or does he *discover*? Perhaps his role is to probe nature perceptively to determine its preordained order and beauty—as a voyage of discovery. In this view, the scientist does not create but rather discovers what has already been created. Few would deny that the fascination of discovery is, with curiosity, an important source of motivation for the scientist. Yet, even the scientist on a mission of discovery has a highly creative role that goes far beyond the skillful perception of the order in nature. His basic contribution is the creation of new knowledge and understanding where none existed before.

On the other hand, one might conclude that, in a more fundamental sense, the very "order in nature" is actually a creation of the mind of man—it is his response to the complex stimuli which originate in natural phenomena. By virtue of his creative perception and analysis he endows these stimuli with form, with order and meaning, thus adding to our understanding of nature. *This* is the much desired output of creative basic research. In industry we have great interest in this output because it is a primary nucleating source of technological progress.

The understanding of order in nature—how atoms form crystals, how organic molecules react, what holds nucleons in the nucleus—is the basic function of science. This firm foundation supports the wide activity of modern scientific endeavor. As a result of understanding natural phenomena, we open the door to vast additional creative activity in scientific invention. We make possible, for example, the creation of new crystals not found in nature, the creation of polymers never before seen, and the synthesis of atoms beyond the periodic table. I think we can safely conclude that creativity *is* an essential attribute of scientific research, and that scientists, and sponsors and administrators of research, should have a proper concern for the identification, encouragement, and recognition of this all-too-rare quality.

Although creativity is a vitally important quality, it is also an extremely rare one. Few people, if any, since Chopin have been able to assemble sounds with the unique charm of a Chopin sonata. Similarly, the top levels of creative scientific research are by no means crowded; room could undoubtedly be provided for additional talent. The similarity between art and science is not coincidental. As Bronowski[1] put it, "it is not surprising that creators of

[1] J. Bronowski, "The Creative Process," *Scientific American,* September 1958, page 58.

the first rank are scarce in both fields." It is characteristic of creative work in science that relatively few individuals make disproportionately large contributions to scientific progress. These creative sparks, by a chain reaction, frequently kindle a technical and technological fire of tremendous proportion. In a very real sense, material progress is nucleated by creative science and technology.

THE ENVIRONMENT FOR CREATIVE WORK

Research is the business of innovation. Its successful pursuit depends upon creative talents; hence, people who sponsor and administer this activity give much thought to the subject of creativity —how to generate it, and cause it to flourish. Motivation and tangible incentives for stimulating creativity are obviously important desiderata in any consideration of creative work, but there is no guaranteed system for providing such stimulation. There are, however, certain environmental factors that are clearly related to creative scientific and technical work, including motivation and incentives. I would like now to discuss certain of these.

The most important feature of this environment is the managerial climate that pervades the enterprise. People, especially ambitious young people with excellent aptitudes and training, are remarkably responsive to their environment. They will sense this climate and respond creatively if it is made very clear that creative scientific and technical work is truly an objective of the enterprise. This objective cannot be made convincingly clear, however, by posting a statement to that effect on the bulletin boards. The entire atmosphere established by organization and operation must conform to the stated objective in some detail. It must conform in a basic detail, for example, like the compensation plan, and, in an elusive detail, like the receptivity to new ideas. Let me discuss receptivity first.

It may be doubted whether anything stimulates a creative individual as much as evidence of real interest in his creation. But the converse is equally true, and, if anything, more important. A lack of interest, and especially a lack of receptivity, will cast a funereal pall of discouragement over the activity that will stifle all but the hardiest of innovators. A reasonable degree of acceptance of innovation is, in fact, a real component of the cost of innovation effort.

In a creative environment, the first reaction to a new idea should be favorable and encouraging. The new concept must be innocent until proven guilty, and if finally found guilty, this determination must be made only after due process of consideration and study. And, best of all, the innovator must be on the jury so that he may aid in detecting the problems connected with his creation, if such problems exist.

There is a vast difference, with respect to environment, between a positive attitude that is expressed by, "Let's see how we can do it," and a negative "Let's see how we can avoid doing it." It is safe to say that the barriers to innovation presented by natural phenomena are minor compared to the barrier presented by the closed mind and the all-too-human resistance to change. Similarly, when a course of action has already been firmly decided upon, the stimulation of creative thought toward alternatives—principally to prove that the original concept is the best choice—may be occasionally justified. However, this negative procedure is not likely to encourage real innovation.

Another important, but elusive, aspect of the climate of receptivity is the treatment of authorship. The spark of innovation frequently is the only absolutely indispensable element in a chain of events which produces technological progress. Without the discovery of fission, for example, no amount of refinement of conventional prime movers would have led to atomic power. In view of this, however, it is distressing to find that the innovator's creative contribution may be lost sight of in the vast welter of subsequent activity that takes place under the heading of "reduction to practice."

This activity necessarily requires much time and the participation and creative effort of many additional contributors. Frequently the subsequent applying work may show that the creative concept was deficient in one or more aspects, or was even incorrect in some important respect, while still serving as the indispensable starting point for the entire activity. It is not unusual for the creative idea to undergo extensive change and improvement as it traverses a long path to practicality.

Regrettably, the subsequent workers will not infrequently succumb to the temptation to depreciate the innovator's work in favor of their own, or worse still, to conceal the fact that his original work was the starting point upon which their significant, but later, contribution has been built. Under the worst combination of these human circumstances the unfortunate innovator is left with the distinct impression that his creative activity has been downright wicked, and possibly even sinful. Needless to say, this situation does not encourage creativity.

A compensation plan involves not only the relatively simple question of professional salary levels and forms of compensation additional to base salary, but also the more complex question of the relative compensation levels for managerial versus professional contributor skills. This very important subject needs discussing. The issue, which must be faced squarely at an appropriate management level and carefully considered, is the theory and practice of the "parallel paths of progress." The question is whether the most expert professional skills in the work that is the primary objective of the organization are to be compensated as well as the corresponding expert managerial skills.

If it becomes perfectly evident to everyone that to succeed in the organization one cannot remain in research, but must aspire to management, the net result will be a great many managers and few able researchers. A real test of organization is the extent to which the laudable theory of parallel paths of advancement is actually embraced in the operating practice of the laboratory.

It is a common platitude in any discussion of the climate for scientific research to note that "salary isn't everything." This, however, should not be confused with the somewhat similar but vastly different statement that salary is unimportant.

As someone put it, "Money may not be everything, but it's a long way ahead of whatever's in second place." I have found few highly creative individuals, even those completely immersed in exciting scientific work, who were indifferent to salary matters. It is true that some may occasionally forget to deposit a paycheck, but, in such cases, urgent advice from the home front may be depended upon to promptly restore a sense of reality. Financial compensation is "only" a foundation stone, but it must be firm if the structure which rests upon it is to have stability.

Let us now consider some of the elements of motivation beyond salary. The *Wall Street Journal* not long ago carried an article entitled "Scientists in Industry Often Baffle the Bosses." The very good article on this subject might have been equally well entitled

"Bosses in Industry Often Baffle Scientists." The solution of this problem of mutual understanding is the key to the creation of a climate for creative scientific and technical work. The whole idea of "bossing" scientists is so alien to this trade that we had better start back at the beginning and consider the sources of motivation for creative scientific research.

The thoughtfully conceived compensation plan will not in itself provide the required motivation. I doubt if any able scientist, arriving at the laboratory in the morning, probably late, plans his day's work while saying to himself, "Now my employer pays me quite a generous salary, even considering the fact that I am very bright, indeed. In view of this I will work very hard today to make a significant discovery."

The real driving force for creative research, as I have indicated, is found in an intense curiosity and in the excitement of discovery. This motivation is not induced—it is congenital. It may be encouraged and stimulated, but in the absence of a strong natural endowment, it cannot be learned, like a foreign language.

I do not subscribe to the philosophy of research that exalts the "pure" scientist as an individual who exists in an intellectual sphere completely out of touch with reality. The causal relationship between science and human welfare is a thundering fact of contemporary civilization, and is in itself a source of stimulation to research. To deny the overwhelming importance of this modern reality is, in my view, inexcusable. One may regard science solely as a source of personal esthetic stimulation, and put out of mind the intruding thoughts of utility, but to do so requires a disturbing compromise with social consciousness. Irving Langmuir is an example of an individual who accomplished scientific work of the highest professional quality. But his awareness of the material world and its problems and needs led to a tremendous technological bounty from his excursions in pure science.

The interaction with other researchers in the same, as well as in other, fields of specialization is, of course, also a source of great stimulation to a creative individual. The communication and interaction with scientific societies is an ancient tradition of science, but with modern specialization it has become a functional necessity. This communication is the basis for recognition, which is at once a source of motivation and a significant reward for creative work. I will say more about this shortly.

THE USES OF CREATIVITY

Because creative talent is rare, it is common in enterprises dedicated to innovation to wish that they had many more creative individuals.

Nature, however, seems to have had in mind some sort of balance, because she has not been overly bountiful with this talent. It may be just as well. A small amount of catalyst can trigger off a large chemical reaction, but there is an optimum amount of catalysis which determines an efficient chemical process. Similarly, in the chain of technological progress, which starts with research and leads to the satisfaction of human needs and wants, we seek an effective balance between innovation and output. This involves all of the contributors to the effort. The research is frequently the creative source of the new knowledge—the new fact, new phenomenon, or new process that catalyzes technological development. Abundant experience teaches that each of the steps that follows research typically requires more people, more investment, and much more expenditure than the research itself; and, hence, involves much greater risks.

THE INTRODUCTION OF INNOVATION

One point that is important in relation to the uses of innovation should be clearly understood. In the chain of technological development, it is better to introduce innovation early than late.

The ideal time and place for the introduction of innovation in this scheme of things is in the research phase, or early in the development stage. If at all possible, the critical questions should have been asked of nature before we build the production plant or embark upon the production of a weapons system. Experimentation, in these final stages of project work, may be very costly indeed. It must be admitted that this much-to-be-desired orderly pattern of Research and Development frequently runs into practical difficulties. If, at an advanced "hardware" stage of a development, where change is costly, a new discovery appears which will completely obsolete the development, the project must grind to a halt to permit an assessment of the damage and a study of alternatives.

We may use chemical research as an example of the advantage of introducing innovation at an early stage of technological progress. Exploratory chemical research frequently involves all of the functional elements of a chemical business venture, but in subminiature scale. The analog of the production plant is a test tube, and the output is a few cubic centimeters of product. This product may be tested in its unique market—development and application engineering—and preliminary manufacturing costs and business appraisals may be made.

At this point, the miniature businesses have a high mortality rate. Bankruptcies and plant closings occur with frightening frequency, but these miniature business failures have little impact on the community and the economy. The whole business can be swept away, and the plant site cleared, in moments. The bankruptcy report is prepared at no legal expense, filed, and forgotten, and the whole procedure starts all over again on a new idea. Of course, this idea may have originated in the unsuccessful venture. The occasionally successful miniature venture grows at a phenomenal rate to full scale, and eventually pays for the failures.

The test-tube stage of the technological process—the research laboratory—is the ideal time and place for scientific innovation and the free play of creative talents. Here competing ideas can grow and die in rapid succession and at relatively minor risk, and optimum variables and configurations can be selected.

This is not to say that we do not want innovation at an advanced engineering stage, or even in the manufacturing phase of an important project. We do, for innovation must traverse that route to the ultimate customer. But it is expensive to experiment at full scale, and the design of the new process, or material or method, is preferably done at small scale. A constant change in "final" design and performance objectives in response to the introduction of innovation, too late, causes the project completion date to recede constantly into the distance, and the expenditures to grow out of all relation to forecast costs.

This problem is met in its extreme case in the development of complex weapons and weapons systems in the Department of Defense. Because of the usually long development cycle, the typical horrible dilemma is the competition between the availability of hardware and its obsolescence rate. If the system design is frozen too early, it will be obsolete when produced, and if it is not frozen at some point in the development cycle, it will never be produced at all. The Solomon who sits in judgment in such cases must determine a certain point of no return, and at this point the innovators must be firmly ushered out of the room so that the production people can take over to manufacture, if necessary, a potentially obsolete product. Errors in judgment in this matter may be extremely costly, and, in the specific case of weapons systems, could be fatal. To make the right decision here one requires, in addition to judgment, of course, a particularly clear, unblemished crystal ball, and a rabbit's foot of proven efficacy.

Another particularly difficult and expensive example of the competition between *innovation* and *output* now occupies our national scene in the form of our atomic power plant program.

This particular technology provides an outstanding opportunity for creative innovation, because the number of permutations and combinations of fuel, moderator, and coolant in the power reactor is extremely large. A great many of these combinations are under various stages of test and development but, unfortunately, it is not possible to build a *miniature* reactor that will react, and experience shows that it costs about $100 million and takes many years to really test a reactor configuration.

No matter how many power reactors are in active development, there appears to be an irreducible backlog of attractive reactor projects, and proving that they are attractive in demonstration power plant scale is very costly and time consuming. Because we have been exploring reactor configurations for a period of fourteen years—power reactor development started in earnest in 1945—some feel, and I share the feeling, that the time has now come to reduce to practice some of the things we have learned in fourteen years of reactor development. We cannot, of course, forgo a continuing program of exploring new reactor configurations, but significant effort should now be devoted to the best distillation of our accumulated knowledge of this new power technology, so that the atomic power age may move on to a production phase. It appears that the boiling water reactor concept has exceptional promise for this objective, and it may permit a start in the generation of some useful *output* from a technology which has thus far been characterized principally by innovation.

Finally, there is a relationship between creative innovation and the economy as a whole which needs discussion. I think I detect a large roadblock in this area.

The relationship comes to focus in our patent system, which is a little-understood foundation stone in our system of free enterprise. In particular, the relationship between patents and technological progress needs to be exposed.

As background for the discussion of this subject, it is important to clearly understand a basic fact of scientific and technological advance: *All scientific and technological progress is built on a foundation of prior progress.* No modern technological progress starts at Genesis—on the contrary, the foundations of chemistry, physics, mechanics, metallurgy, and mathematics have been laid one stone at a time, and modern progress adds to this imposing edifice—one stone at a time. People who make creative advances on the scientific frontier build on these foundations, and they readily acknowledge an indebtedness to this vast accumulation of scientific knowledge. Because of this fact, communication is a necessity in modern science. As I have indicated, there are many reasons why communication is necessary, but above all there is an *obligation* to replenish the storehouse of knowledge from which every researcher draws so heavily in the course of his work. Even in industry, where proprietary rights to new phenomena create an economic impact, prompt publication of new scientific results is now a common practice.

Important in this consideration is our patent system, which makes it possible to protect proprietary information for a period of years. Since our patent system is reeling under some body blows recently received, it may be well to remind ourselves of the theory and practice of patents, and to take a look at what is happening to this particular foundation stone of a free enterprise system.

The original idea back of the patent system was quite simple: In exchange for a temporary proprietary right granted to the inventor, he is required to publish his invention for the benefit of all. This publication was originally accomplished by the publication of the patent itself, but today one or more scientific or technical publications generally precede the issuance (and publication) of the patent by many years, and are the principal published information concerning the new development. The sum total of technical publication is the primary bank of world science and technology. Creative workers will make heavy withdrawals from this bank and deposits in it as a vital necessity in the pursuit of their work. It is the foundation on which all future technological progress will rest, and without it such progress would cease. We now know that even in the absolute dictatorship of Soviet Russia, where close control of information is a matter of policy, a very high volume of scientific and technical publications is sustained. It is unfortunate, incidentally, that, until recently, we did not take the trouble to translate many of the Soviet journals.

It might not appear at first sight that our patent system is an essential element in the system of full and free interchange of scientific and technological information. It is, however, the indispensable protective device which enables all users of the great bank of world scientific information to make deposits in confidence that they will not thereby lose their invaluable property rights. These rights, after all, constitute the inducement and stimulus for exploration and creation in a free society.

We have come to the time when all of the beneficiaries of the creative process—science, government, industry—indeed all of society—must show some concern for the security and integrity of our patent system.

In particular, let me suggest that the trend in recent years on the part of the government to claim the sole rights to patents which originate in its contractor's work could result in a significant, perhaps even vital, impairment of the whole patent structure. For example, the recently enacted space legislation (NASA) has sweeping provisions for exclusive governmental title to a wide range of contractor inventions, discoveries, and information. This particular legislation even places at risk, to a degree not yet fully clarified, the contractor's prior patent equities. This is a very broad change in government policy, indeed, and the full impact of this trend may have very serious consequences. Most serious of all is the influence of our patent system upon free scientific and technological publication. Both are absolutely essential if we are to enjoy a creative dynamic technology, which is basic to our country's economic health and vital to the national defense.

Finally, let me say that the innovator is a powerful source of nucleation for human progress. The innovator's work in all fields of human endeavor, sustained by his curiosity, his vision, and his sensitivity to human needs, is the generating and stimulating force of our civilization. His work, in science, is a wellspring of technological advance which exerts a profound leverage on the economy and on the national defense. It is not only appropriate, but also vitally necessary that we make every effort to understand, to encourage, and to reward the indispensable work of the creative individual.

Creative Science in an Organizational Setting

An address at the Conference on Scientific Creativity in an Organizational Setting held at the Harvard University Summer School of Arts and Sciences and of Education, Cambridge, Massachusetts, July 22, 1959.

The dramatic advance of science in our century, when compared with the less spectacular progress of social systems, has been a constant source of concern. It is often said that we do not know enough about living together to guarantee that we will not one day misuse the power that science has given us, and blow ourselves to bits. It is true that human beings are more difficult to study and regulate than are atoms and molecules. With maddening consistency, mankind thwarts the classic techniques of analysis, measurement, hypothesis, prediction, and verification with which scientists have been able to pry into the physical world and control its forces. For this reason, it is difficult for social systems to keep pace with physical sciences. Progress in science can be viewed as a continuing replacement of uncertainty with verifiable fact, and uncertainties in human nature have proven to be far more elusive than uncertainties in physical nature.

However grave this imbalance in progress, we must not, I feel, permit it to distort the positive values of science. Of itself, science is not a threat to society; it is quite the opposite. Although many seem to think that science has turned our stable and serene planet into a powder keg, surely the productive power of science has thus far been more characteristic than its power to destroy. It is true that science has given us our most destructive weapons, but by the same token, these weapons have proven to be our strongest shield. Science has also given us our increased life span and relative freedom from disease, our increased and improved food supply, our rapid transportation and communication. By freeing us from much of our toil, it has added to our leisure; and by giving us systems for preserving and reproducing our knowledge and art, it has given us the opportunity to use that leisure productively.

But science does not extend these gifts directly or automatically. For example, a fission experiment does not of itself generate useful atomic power. In this case, as in most others, there is a vast gap between new scientific knowledge and its usefulness to society. Industry plays a very large and constantly increasing role in accomplishing this transition. Moreover, as science has become a factor of major significance in the economy, industry has increasingly assumed a participating and contributing role in the progress of science itself. It has been discovered that creative basic research may be a most useful form of applied research, and hence the scientific laboratories in American industry are extending their effort toward the fundamental end of the spectrum of science and technology. This trend has focused the attention of American industry on basic research, including the nature of creative work and the climate in which it thrives.

To begin with, creative work is pursued in areas that are largely uncharted and unknown. It is demanding work, extremely difficult and discouraging, and often lonely. Seemingly impenetrable barriers erected by nature bar the way to progress, and the search for a breakthrough may be painstaking in the extreme. A high order of determination and drive is essential to progress. The research laboratory must maintain a climate wherein creative scientific endeavor will flourish. We must not add man-made frustrations to those imposed by nature. If a creative environment is achieved, creative scientists will be attracted to the laboratory, and will there advance the important work that they are uniquely endowed to perform.

The industrial research laboratory, moreover, must not operate in a vacuum, in theoretical self-sufficiency. An important part of its function is to interpret appropriate world science to its parent organization. It would be unrealistic to suppose that all of the scientific information that any industrial organization might use could be produced in its own laboratories. It is now generally appreciated that in order to partake of world science, the research laboratory of industry must also participate in and contribute to world science. For this reason the laboratory must publish its findings promptly, and its scientists must be encouraged to take active part in the meetings of their colleagues throughout the world.

THE WORK OF THE INNOVATOR

I have stressed the importance of science in advancing the material progress of civilization, and have indicated the vital role of creativity in this process. Let us now consider the nature of the creative individual. Perhaps the most characteristic attribute of the creative individual is his driving curiosity. He seeks, rather than avoids, problems, and he is dissatisfied until he has solved them. Moreover, when faced with a number of complicated solutions, he is motivated to reduce these to one that is simple and esthetically satisfying.

Creative scientists with whom I am acquainted invariably experience deep satisfaction when their work is reduced to useful forms, and I suspect they experience the greater satisfaction when the reduction is accomplished by others. Yet altruism is not the immediate motivation for most creative work. More than anything else, I feel, it is the enchantment that comes when profound curiosity is satisfied. The creative scientist is intrigued by unsatisfactory situations, by the enigmas, puzzles, and contradictions of the physical world, and he is unhappy until he has resolved them. Often, of course, his overwhelming curiosity and his desire to engage the more formidable situations immerse him in problems that will not yield to his efforts. It is for this reason that the creative personality must be highly resistant to disappointment. The more difficult the problem, the more interesting it is to the creative mind, and the more speculative is the outcome. Consequently, we should not expect the proportion of successes to be high.

The innovator must know when to cut through the cumbersome details of routine scientific method to grasp, intuitively, the meaningful relationships in a wilderness of raw data. This intuition is, perhaps, an essential distinction of the creative individual: his ability to perceive significant order that is invisible to others. At this point, of course, analysis of the creative process breaks down and our understanding fails. We can say only that this special gift of the innovator emphasizes his individuality, and we must fall back upon such mysterious terms as "vision" to suggest this special power of the innovator to see the form and relevance to which others are blind.

THE CREATIVE ENVIRONMENT AND ORGANIZATION

There is no guaranteed system by which an organization can stimulate such an elusive attribute as creativity. But there are identifiable environmental factors that bear definite relationship to creative scientific and technical work. Broadly, these factors can be grouped under the general headings of recognition, orientation, and freedom. There is not time to discuss them all, and today I wish to address my remarks principally to a discussion of the nature of organization "pressure," to the question of creative "freedom" in relation to organization, and to the issue of individual versus group effort. Finally, I would like to identify a reverse effect—the pressure which scientific creativity exerts upon organization.

ORGANIZATION PRESSURE AND CREATIVITY

Let us acknowledge that to some degree organizational "pressure" is inevitably communicated to the members of an organization, since an important objective of the organization is to define the cooperative relationships between the individuals and the groups that constitute the enterprise. Inherent is some "give" here, and "take" there, to accommodate the complexities of co-operative human effort. The more pertinent question is whether the organization generally required for the operating framework of a research enterprise necessarily inhibits creative output. This is an important question, and continuing study must be devoted to the issues and problems involved. As I will try to indicate, creative scientific effort is polymorphic in nature. Some kinds of work are adapted to individual effort, and some creative workers prefer and work best on an individual basis. But there are other kinds of work that are better suited to joint effort, and some creative workers prefer a cooperative approach. In both cases, an "unseen collaborator," to be described, is essential to creative progress. An organizational framework is, I believe, fully appropriate to creative scientific effort. In fact, under the best of circumstances it may enhance creative output greatly. More important than the tangible structure of organization are the climate and environment created by the attitudes and expectations of management; these intangibles may exert a very significant influence on creative effort. Sound organizational objectives, therefore, and the effective communication of them, are important elements of the climate for creative work.

CREATIVITY AND FREEDOM

It is almost intuitive to associate the words creativity and freedom, and much evidence could be cited to support the thesis that this association is meaningful. In the present instance, however, we wish to examine this generality in relation to an organization environment. We wish to expose the encouragement and enhance-

"private exploratory research" (a currently popular idea which I would ridicule, if I had time) may have few takers if it is noticed that the people who keep their noses to the grindstone are the ones who progress in the organization.

From these considerations I conclude that an organizational environment is completely compatible with creative work in science. What is required, however, is the sponsor's sincere determination to engage in such work, and the effective communication of his sustained interest to the participants.

INDIVIDUAL AND GROUP CREATIVITY

In science, the idea of *individual creativity* is almost synonymous with *creativity* itself. To infer that *groups of people* can be creative is thought of as a sacrilege in some quarters. Supporting this idea, a patent, as legal evidence of inventiveness, is generally issued to one person; co-inventors are much less common. Apparently, the patent office feels that it is relatively improbable that two people will each conceive complementary halves of a single new idea at the same moment in time. If they do, they are co-inventors; if they do not, one is the inventor and the other the collaborator. This, however, should not be interpreted as a depreciation of the role of the collaborator.

On the other hand, much has been said, although generally not in the scientific community, about the "brainstorming" technique.[1] As I understand it, this procedure involves a meeting in which multiple participants attempt to lift themselves by mental bootstraps to a greater aggregate height than would be possible to any one individual participant. I am told that frothy, stimulating conversation, a thick veil of smoke, and possibly other ingredients are essential to the success of this method. The method is claimed to be a sort of *thought alchemy* in which high-grade ideas are created out of low-grade fragments by mental togetherness. I hear that it is *de rigueur* on Madison Avenue.

We see here two rather contradictory views of the creative thought process. One holds that it is essentially an individual accomplishment, the other that it is susceptible to mass attack. On an issue as important and as fundamental as this, it is necessary to

[1] *Fortune*, October 1957, p. 145.

form some clear judgments. It is a fact that some kinds of modern scientific work cannot be accomplished solely by individual workers. For example, in significant and important cases we need to call upon more than one area of specialized skill in the attainment of a complex research objective. One may indulge a nostalgia for the "good old days" when a lone scientist, armed only with sealing wax, bent paper clips, odd bits of string and wire, and possibly homemade glassware, could accomplish important scientific results. Such cases are becoming rare. Whether we like it or not, scientific research is becoming more and more complex, and we must contemplate a continuing necessity for joint as well as individual research effort.

Perhaps we should not overlook the opportunity of applying a simple probability analysis to the brainstorming technique, and I will do so. But it may as well be acknowledged at the outset that the analysis will settle nothing. However, let us not be perfectionists in a field in which uncertainty is certain, for the analysis, if I may continue to use the term, does lead to some stimulating conclusions. To illustrate, suppose that we have 10 participants, each of whom has a great many rather poor and incomplete ideas. Suppose, in fact, that each participant has $\frac{1}{10}$ of an idea per minute. If we put them together in a "brainstorming" conference and permit them to talk at random, how long does it take to combine the partial ideas in the right sequence to produce a full idea? It can be shown that the time required to produce a *full* idea is $n^n t$, where n is the number of participants, each of whom has one/n random ideas in t units of time. For the case in point, where we have 10 participants, each of whom has $\frac{1}{10}$ of an idea per minute, it will thus take 20,000 years to produce a full idea. Frankly, that is too long to wait, but interestingly enough, during the time required to produce a full idea, there will be produced a bountiful by-product of several million half-ideas. This may be a much better way to produce half-ideas than to achieve complete ones.

As I think I mentioned, this analysis of "brainstorming" is somewhat lacking in rigor, and is in some respects unfair to the concept. Specifically, we have not taken into account interaction effort—that is, mutual stimulation—and enthusiasts might say that this is the main point, and they may be right. But by the same token, however, the calculation indicates that the mere proximity of a lot of poor ideas does not help much—you *must* achieve effective interaction to make a net gain. I think one might also conclude that if we have any people around who have the aptitude for producing complete ideas, we could give them quite a bit of time for contemplation, and perhaps be quite tolerant about their working hours while they are producing that full 100 per cent idea. We could also afford to pay such individuals quite a bit more than people who have partial ideas, and it might still be a bargain—if it is ideas we want.

Real life is, of course, not as simple as this. There is a middle ground between the individual genius and multiple dunces, and a more detailed look at creative scientists at work should yield a more realistic understanding of some of the related patterns of creative thought.

In practice the scientist leans heavily upon the foundation work and current progress of many other scientists. His immediate colleagues, associates, and assistants, if any, are at once evident as joint contributors to his research progress. Less in evidence, and if anything more important, however, are his unseen professional counterparts elsewhere in the world of science. The contribution of these unseen collaborators is an absolutely essential ingredient of modern "individual" scientific progress, which is always built on the foundation of prior progress. The published works of these unseen collaborators are, of course, essential bases for scientific advance, and they comprise in fact the principal tangible embodiment of the "world of science."

Let us give a bit of thoughtful consideration to the collaborator in being the research assistant—this useful but unobtrusive individual whose duties vary all the way from fetching coffee at mid-morning to doing the boss's thinking. We have perhaps taken him too much for granted. We are prone to discount his contribution, and perhaps we should acknowledge some conceit in this matter. Some outstanding examples of research assistants that I have in mind have made major creative contributions to the output of a research venture. Perhaps the most typical case is the complementary pair—theoretiker and the experimenter. It appears to be a fact of scientific life that top analytical talents and top experimental talents are seldom found together in one individual. When a compatible team of two scientists provides these complementary skills to a high degree, the combined research ability may be exceptional; we have had many examples of this case in the history of our research laboratory. Another significant case of team effort is found in the complex research task which requires multiple skills in different fields of specialization. A striking case of this kind was the group that discovered the diamond synthesis in our laboratory; at an appropriate time I intend to describe this team effort in historical detail.

Whether the relationships in cases of the kind under discussion are those of co-conspirators, or of boss and assistant, is oftentimes a matter of semantics. There is no doubt that the junior member of the team frequently makes senior contributions to the creative output. I can recall some cases of highly skilled research assistants, where one might justifiably entertain doubts as to whether the scientist or the assistant had made the more important contribution to a new research result. Joint contribution would appear to be both an accurate and an equitable description of such effort. Consider, for example, the assistant (and I have an actual case in mind) whose somewhat negative genius consists in an uncanny skill in detecting blind alleys and boobytraps. He is devoid of originality, but once an idea is in evidence he can detect its infirmities with unerring and devastating skill. The man he is assisting, who is highly original and creative, has no choice but to converge on the sound and verifiable concepts—for his assistant annihilates all others.

One might be justified in concluding that in such cases as those described the whole research effort is somewhat greater than the sum of its individual parts.

As an interesting observation of joint effort and collaboration, Dr. John Fisher, of our laboratory, has counted the number of authors for each of the 239 articles appearing in the Physical Review, Vol. 113, Nos. 1–5. The results are:

Number of Authors	Number of Papers
1	89
2	99
3	35
4	13
5	2
6	1

The average number of authors per paper is *two*. However, if one counts the number of authors plus the number of collaborators acknowledged in the papers, the average number per paper is 4¼. On the reasonable assumption that much of this work is creative in character, one might conclude that creative people perform better when they have someone to work with and talk to.

From these considerations, I suggest that there are significant exceptions to the symbol of "individual creativeness," and that some of the wickedness which has sometimes been ascribed to joint effort and teamwork in science may not be wholly bad. It should really give offense to no one.

THE PRESSURE OF CREATIVITY ON INDUSTRIAL MANAGEMENT

Since we are examining the relationships between the organizational environment and creative work, we might properly look briefly at an inverse or reverse image. I have reference to the pressure which creative output exerts on sponsoring organizations. This is a subject of considerable consequence and importance, and I am not aware that it has been given much public consideration. To come directly to the point—the increasing flood of creative new science and technology is presenting industrial management with a somewhat overwhelming array of challenges and opportunities. We have a good deal of homework to do in learning how to manage our abundance of technological innovation. To explain, let us contemplate for a moment the "explosion" in science and technology. "Explosion" is an appropriate word, because all of the statistics of growth and rate of growth of science and technology show that most of what has happened has happened recently. Thus, growth has been so rapid that 90 per cent of all the scientists who ever lived must be alive today. Science and technological change had almost no impact on the outcome of World War I, while it was a major factor in World War II. As Sumner Slichter has pointed out, the great economist Lord Keynes did not recognize technological innovation as a factor in the economy twenty years ago, yet today it assumes major proportions. Thus technological change is an exploding growth phenomenon which is rapidly pervading vast segments of human activity, for better or for worse, but in any event, for certain.

Technological change as an economic factor is now well established. The more general symbol of change—innovation—is, in a large and growing number of businesses in the industrial economy, a primary source of competitive advantage. In industry many sources of innovation are being pursued and studied, and in the progressive industries the *rate* of innovation is being recognized as an explicit growth determinant. Technological change is an important specific source of innovation, and it may become the most important source of innovation in the near future.

As indicated, in the milieu of the American industrial scene, these explosive growth rates are having a considerable impact. We are expending ever-increasing funds for science and technology, but the guidelines for determining the right amount in relation to other industrial functions are not entirely clear. We face vast and divergent opportunities for technological innovation and we recognize that we must learn a great deal about how to manage this burgeoning factor in a competitive economy.

None of these problems is new in kind; all are new in magnitude, and the magnitudes have suddenly become large. A few basic statistics will point up the numerical framework of the problem. Gross national product (in constant dollars) is growing currently at the rate of about 3 to 4 per cent per year, corresponding to a doubling time of 20 to 30 years. "Gross industrial R & D," on the same basis, is growing at the rate of 10 to 12 per cent per year, corresponding to a doubling time of 8 to 10 years. Obviously, these relative growth rates cannot go on indefinitely, else industrial activity will eventually comprise *only* R & D, which would greatly simplify things, but one can think of numerous attendant disadvantages. Thus, the technological factor is rapidly coming to the fore as a basis for innovation in products and industrial growth. The technological factor is also becoming a more significant component of cost. For all of these reasons, management in industry has become increasingly interested in the management of technological innovation. It is safe to say that some unsolved problems have been identified. *In a few words, the proliferation of science has presented industry with a baffling array of challenges.*

Let us consider two current examples of technological innovation. A few years ago organic chemists in our research laboratory discovered a polycarbonate family of polymers which have rather unusual and useful properties. It was recently placed on the market under the trademark of *Lexan.* Since, when this discovery was made, we were already engaged in the polymer business, and had available a great deal of background in pertinent phases of this industry, we were able to appraise the resulting new technology and to arrive at appropriate business decisions promptly. In summary, the venture has moved forward rapidly and is expected to result in altogether desirable and useful new products, and to promise an appreciable positive cash flow in the foreseeable future. Everyone is quite happy about this result, and no unmanageable stress and strain seem to have been experienced anywhere.

Let us now consider a much more difficult case of technological innovation—atomic power. This particular "opportunity" so bristles with challenges that one hardly knows where to begin; almost no aspect of the development is simple. First, the technology is completely new and is unrelated to the technology of any prior industry. The manufacturing processes and materials are new and largely unknown. Long consideration has been required to determine the characteristics of the products and to appraise the markets. Basic questions of product economics cannot be considered except in relation to political issues of speculative significance. Problems of hazard and third party liability have intruded as specific components of business risk. Finally, the investment requirements for the development of the prospective business have been of major magnitude. It is no exaggeration to say that these considerations have posed difficult management problems for companies engaged in this business, and that the most painstaking and thoughtful study has been required to arrive at appropriate business decisions in this particular venture in technological innovation.

Most opportunities in technological innovation lie between the two limits illustrated by these examples. On the one hand is the development which draws heavily upon the sponsor's available experience in engineering, manufacturing, and marketing. In such cases the transition from science to useful products or services is straightforward and rapid. On the other hand is the case that baffles the experts—posing major problems and quandaries and difficult and highly speculative choices in all of the industrial functional areas. We have, however, made this progress—we recognize that technological innovation is increasingly a way of industrial life, and that we must reckon it as a major factor in our future business plans. We must learn how to manage the complexities of our flood of technological resources with vast material benefits for mankind as a goal. Moreover, by concentrating greater effort on the proper sponsorship of creative scientific research, we will add substantially to the already formidable contributions that industrial organizations have made to the fund of scientific knowledge.

Clearly, it is within the power of industrial organization to insure that the rate of scientific innovation will continue to rise, and to insure that scientific innovations will reach their proper fulfillment as advances in the material well-being of mankind and milestones in the progress of civilization. Thus can the dramatic progress of science, so characteristic of our age, become the positive force that society has every right to expect.

The Age of Energy

An address at the Annual Meeting of the New York Section of the American Chemical Society, New York City, December 2, 1959.

The present, just as every other age, has been variously described. We have heard it called, for example, the age of science and the atomic age. It has recently become the space age. Chemists have thus far been too modest to label it the age of chemistry, but from the way they are talking, I think some of the chemists in our laboratory are about to proclaim the age of polymers almost any day now. We have also heard of the advertising age and even the age of tranquilizers. Supporting the latter idea, one philosopher sees the present, rather grimly, as the age of anxiety.

Alternatively, we might take a larger view of our time, and characterize it the *age of energy*. There is no doubt that the vast energy consumption, and the exceptional growth rate of energy being generated, converted, and used are key features of modern civilization. There is also no doubt that the technical aspects of energy will occupy the attention of scientists and engineers for a long time into the future.

Of course, we do not really "produce," or "generate" energy—that is merely a manner of speaking. It would be accurate to say that we *convert* energy from one form to another. Typically we convert chemical energy to heat, or heat to mechanical energy and then to electrical energy, which is obviously the most useful form of energy. We perform this energy conversion in order to make otherwise inaccessible energy available and convenient; in that limited sense only, do we generate energy.

For the conversion to electrical energy, generally from a source of heat, we have long had a theoretical choice of a great many processes, but experience has shown that the choice for economical production of bulk electric power has been reduced to one principal contender—the turbine-generator system. This system produces 80 per cent of the world's electric power, the remainder being principally hydroelectric. The same proportion holds true for the United States. I will have more to say about turbine-generator systems later as a background for a discussion of other types of energy converters.

MAJOR ENERGY SOURCES

Let us give a moment's consideration to the sources of energy, and since the turbine-generator dominates the field, to its energy source, the fossil fuels. We are, really, borrowers and spenders in the grand manner when it comes to our use of the fossil fuels—coal, oil, and natural gas. Frugality, historians tell us, was a cardinal virtue in the early days of this country. You may recall some of Ben Franklin's thoughts on the subject: "If you would be wealthy," he said, "think of saving as well as getting," and "When the well's dry, they know the worth of water." One wonders how Franklin and his contemporaries would have regarded our pronounced talent for spending our fuel reserves.

The energy stored during the prehistoric past by photosynthesis —that happy chemical process to which we owe our very existence —is vast, but it is limited. Moreover, when we consider the ever-increasing rate at which we invade these reserves—our capital, if you will—the drying up of the well appears not improbable. (Our knack for extravagance, incidentally, is not confined to coal and oil. We do equally well with dollars. Our combined national and personal debt, in excess of $434 billion, puts every man, woman, and child in the country $2500 in the red.) Undoubtedly, Poor Richard would have thrown up his hands at such a state of affairs.

The question of whether we are, or should be, concerned with the depletion of our stores of fossil fuels has both philosophical and analytical aspects. Analytically, it seems obvious that we should run out of fossil fuels in the foreseeable future. But several things raise doubts about that too self-evident conclusion. Perhaps our so-called fuel reserves are simply a measure of our geological prospecting effort; in which case, how do we estimate our true reserves? In any event, the available reserves depend upon our methods of extracting them from the earth; and who is willing to forecast with confidence what these methods will be five or six decades ahead?

The foreseeable future, when it extends much beyond human lifetimes, becomes much less foreseeable. We do not have much perspective for geological time, if that is what we are dealing with. But there is no doubt that we will have a requirement for the conversion of energy to useful forms at a rapidly growing rate for a great many years. In the case of that most useful form of energy, which I think I have mentioned, the production level is forecast to increase five to six times in the next three decades.

In view of these issues, we should give some thought to the

spending spiral in the department of energy reserves—and for one thing, we might mount some defense of our extravagance. Although we must plead guilty to looting and pirating this ancient gold, we have, after all, much to show for our prodigality. The very fact that we have been able to come to this pleasant room with speed and comfort, that we can enjoy this companionship, this exchange of ideas, this food, this light, and this warmth derives from the use we have made of the fossil fuels. More than half of us, moreover, owe our very existence to these same sources; during the brief advent of the extensive use of coal and petroleum it has been possible for the population of the world to more than double. (This, however, may be a phenomenon with diminishing returns, for the eventual consequences of population growth pose some of the most critical problems facing the human race.)

Only a short while ago, in this region where many millions thrive today, life was meager, marginal, and savage, for man had access to only a negligible fraction of the energy that is now available. Game and the scant harvests that he wrested from small fields worked with wooden implements supplied his only food, and smoky, little wood fires his only heat. It is not difficult to see how human life multiplied as the use of stored energy and increased technical knowledge provided the warmth needed to range farther north, to manufacture and power machines that could do the work of many men—especially the machines that could produce food by opening ever-increasing areas to cultivation.

This, of course, explains the spiral effect that I mentioned earlier. The more energy we harness, the greater the population that we can support; and the greater the population, the greater the demand for energy. The spiral is also advanced by the psychological phenomenon by which each improvement in man's lot suggests another. When shall we satisfy the desire for faster transportation and for more extensive communication? The plane suggests the rocket, just as radio suggests television. And when, I wonder, shall we have satisfied our desire for freedom from toil? One man can already do more work, by virtue of the power machinery at his disposal, than five men could do in 1900. But this is only the beginning, for freedom from routine physical labor suggests freedom from routine mental work. The lathe leads to the fully automated plant, and the abacus to the high-speed computer. Computers with discrimination and elementary judgment are already appearing in the laboratory, and technology will undoubtedly develop rapidly in this area.

With regard to our increasing freedom from work, the ladies may be interested in a phrase that I came across the other day in a study of power production. The author was discussing the principal source of power for mechanical work among civilized nations in former times and among primitive peoples today. This mechanical work was done, he said—and please remember that I am quoting—by the muscular work of "labourers, housewives, slaves, and draught animals." [1] You see now, perhaps, how much the scientists and engineers have done for you by discovering how to convert energy from coal and oil. And be assured that chemists, physicists, and engineers continue to have the plight of the housewife very much in mind. Before long she may have sufficient time at her disposal to pursue the advanced engineering degree that she will need to understand, and possibly maintain, the functioning of her fully automated household. She will be able to do so, that is, if we can support the rising spiral by which we harness more energy, expand our population, and advance our civilization.

We have been most fortunate in this effort so far, but we have had many warnings and a few close calls. During the Seventeenth century, for example, primitive English industry faced the cold fact of depletion of its major energy source, which at that time was the forests. This depletion was inevitable, for the iron, copper, and glass furnaces had been consuming far more timber each year than annual growth could replace. Happily for the English smelters—and for the shipbuilders as well—the New England colonies intervened with their combination of excellent harbors and abundant forests that grew to the very edge of the sea.

But all the forests of the world could not satisfy the increasing demands for fuel. In 1880—around the beginning of the machine age—coal replaced wood as the major source of energy. Today, coal, petroleum, and natural gas provide more than 80 per cent of the total energy used throughout the world. Thus coal saved the forests from extinction, making possible the economic harvesting of valuable wood products that we practice, with the aid of chemistry, today.

[1] Hans Thirring, *Power Production*, London, 1956.

But how much longer can the fossil fuels themselves last?

You are probably familiar with the study of this subject which P. C. Putnam conducted for the Atomic Energy Commission, in which he concluded that the depletion of the economically recoverable fossil fuel reserves will occur between 40 and 65 years from now.[2] This is probably the most complete analysis of the subject which has been made thus far, but as I have indicated, the problem has some highly speculative aspects.

Quite a different question, of course, involves the strategic wisdom of using these valuable substances as fuel. Upgrading to polymers, for example, would appear to be economically preferable to downgrading to heat. Just as the availability of coal and oil saved the forests during an earlier period, it may be that the development of atomic energy will save our fossil carbon supplies so that they will continue to be available as a primary raw material for the chemical industry of the future.

The precious materials that the science of chemistry can and will draw from fossil fuels surpass even those that can be extracted from wood, but unlike the forests, the fossil fuel reserves can never be replaced. Obviously, atomic energy sources must be exploited soon to save coal and petroleum from the sacrificial fires.

One might naturally think of increased utilization of water power as a solution. Here is clearly an economical source of energy —one that is happily free of the waste that is so serious a threat in the combustion of coal and oil. But even after allowing for the development of water power at an accelerated rate, hydro power is so limited that it is destined to become a decreasing fraction of our energy sources.

Moreover, no really effective technique has yet been devised to tap the tremendous energy of the sun directly. Nor have we learned to make extensive use of such major energy sources as the wind, the tides, and the earth's heat. These not-inconsiderable sources may yield to research and development in the future.

Speaking of special sources of energy, by the way, I am reminded of a rather exotic source of both raw material and energy that a chemist in our laboratory told me about just the other day. Faced with shortages of wood pulp during the Civil War, it appears,

[2] P. C. Putnam, *Energy in the Future*, New York, 1953, p. 169ff.

ingenious paper makers in Maine and upper New York State imported shiploads of Egyptian mummies. The cloth wrappings provided material for a coarse wrapping paper that eventually found its way into the shops of grocers and butchers. The ancient cloth so unceremoniously removed, incidentally, caused an epidemic of cholera among the rag-pickers. The historical record indicates, surprisingly, that there was stiff competition for the mummies, for during a ten-year period Egyptian railroad engineers used no other fuel than that supplied by the dried, well-wrapped bodies of their ancestors.[3] This source of energy, of course, finally suffered depletion.

Now let us return to our contemporary problem of energy conversion. Since the turbine-generator is firmly established as the primary energy converter, with electricity as the available and desirable output, why all of the present excitement about fuel cells, thermionic converters, thermojunctions, and the like? The answer is that the study of these systems is quite interesting, and it *may be* useful. Whether or not any of these new converters will ever compete with the turbine-generator is at least questionable; but this matter requires a detailed answer.

In any event, a new energy converter may be quite useful even if it cannot compete with the firmly established methods. To provide the perspective against which to judge the progress of competing methods, we should be reminded that the most efficient turbine-generating plants going into production today operate with a conversion efficiency of about 40 per cent, but to do so they must be very large—in the range above 150 megawatts. In addition to being large, turbines are also quite heavy, weighing about ten pounds per kilowatt. They cost approximately $35 per rated kilowatt. As indicated, their performance depends upon their size; a small turbine-generator would have much poorer performance in every parameter.

Some of the unconventional energy converters that are being investigated today are new, and some of them are very old indeed. This latter fact does not in any sense prove that they will continue to be unpromising. Fields that were plowed years ago and found to be unproductive are not necessarily so today, for we have some

[3] The origin of this account can be found in Dard Hunter, *Papermaking*, New York, 1957, pp. 382–384.

new agricultural methods. For similar reasons, the old hopes in science and technology need to be re-examined every five to ten years in the light of new knowledge, new materials, and new processes. At the pace of modern science, a decade is a long time and it produces a host of completely new capabilities which need to be weighed in relation to old goals.

THE THERMOJUNCTION

Let us take the thermojunction as an example of an old discovery whose promise has been enhanced by recent developments. In 1821, Thomas Seebeck observed the basic phenomenon, which was that heat applied to a junction of two conductors would produce a small amount of electricity in a connected circuit.

In 1834, Peltier discovered the inverse effect. He found that passing a current through a junction would heat it, or cool it, depending upon the direction of current flow. With the aid of a theory which simply related the coefficients of both effects, the two discoveries added up to an intellectually satisfying, symmetrical set of heat-electricity energy converters. Thus, for over 100 years, the thermojunction remained a bright little facet of classical science which was the delight of the lecturer in college physics. Except for its use in measuring instruments, the phenomenon has had very limited practical utility or application.

Then came the semiconductors. If you make a thermojunction out of two appropriate dissimilar semiconductors, rather than two different metals, the result is striking. A Seebeck-semiconductor junction will convert quite a bit of heat to electricity, and a Peltier-semiconductor junction will do quite a bit of refrigerating. Here is a good example of new knowledge rejuvenating old science. The semiconductor "vitamin" has completely revived a long-dead horse, and he is galloping off to the races. Just how far and how fast he will go is a very important question. Shortly, I shall estimate the odds to date.

SEEBECK JUNCTION

Let us consider first the Seebeck junction for producing electricity from heat. These junctions are made typically from compounds and mixtures of lead, zinc, tellurium, antimony, and similar elements. The overall conversion efficiency depends upon the temperature-voltage performance of the junction itself, and upon the temperature of the hot junction and the temperature difference. Conversion efficiencies of 8 to 10 per cent have been obtained for the best combinations and for rather high temperatures. This performance is much better than that obtainable from pure metals, but it is still a long way from turbine-generator efficiency.

However, these semiconductor generators have some other advantages that are quite important. They are silent, for one thing, and they may be lightweight for a given power rating. Also, their efficiency does not depend upon size, so that you can obtain the observed efficiency in a small device if you wish. I do not hear of anyone proposing to put a steam turbine-generator in a space vehicle or a missile, but one might use a semiconductor generator for space applications.

The Russians have a novel application of the semiconductor generator which is very much down-to-earth. A small unit mounted in a lamp chimney provides power for radio receivers in rural areas, and has been rather widely used in Russia. Unless kerosene lamps come back into vogue in our own country—an abhorrent idea to some of my associates—the U. S. market for this ingenious product appears unpromising.

The current data on the semiconductor thermojunction are given in the following table.

Semiconductor (Seebeck) Junction Performance

Materials: Lead, Zinc, Antimony, Tellurium, Arsenic, and Similar Compounds: Typical Junction-Lead Telluride, p and n Type.
Overall Conversion Efficiency: 8 to 10% (for 700° to 800°K Max. Temp.)
Weight-Power Ratio: 10 to 50 lb./kw, Depending Upon Size. (?)
Cost-Power Ratio: (?)

Undoubtedly, the available efficiencies of semiconductor thermojunctions are sufficient to develop application opportunities for semiconductor power generators. Whether the efficiency can be brought into the more competitive range of 20 to 30 per cent is presently speculative. This possibility cannot be ruled out, however; there is a very long list of binary and ternary compounds and mixtures, each of which needs detailed study before a firm answer to this question of the ultimate efficiency of the thermo-junction will be possible.

PELTIER JUNCTIONS

The Peltier, or cooling, junction may be made from similar semiconductor materials, in which case its effectiveness as a refrigeration element is vastly greater than the classical Peltier junction made from metals. With the latter, one could demonstrate perceptible cooling, while the semiconductor junction will develop a large temperature difference.

The rapid rate of recent progress in cooling junctions can be seen in the accompanying figure.

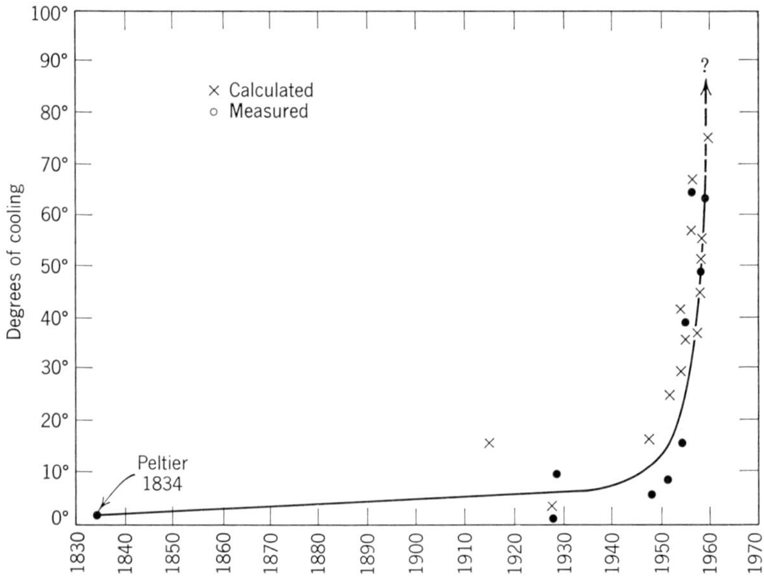

The quantity plotted in this figure is the maximum temperature difference which the junction has produced under no-load conditions; of course, here the junction is at room temperature. When the junction carries a refrigeration load, the working temperature difference will be smaller. Note that a temperature difference of nearly 80°C is obtainable, and that on the selected time scale, which includes the date of Peltier's discovery, this quantity is about to go off the chart, almost at once.

How this will finally turn out I do not know. The curve may actually stop at the last point shown, but I think this is rather unlikely. For one thing, we have a long list of compounds in this case too, and each requires careful and painstaking work to determine its usefulness in cooling junctions. This takes quite a bit of time and effort, and is costing quite a bit of money. When and if the no-load temperature difference climbs up into the range 110° to 120°C, the corresponding coefficient of performance of cooling junctions will equal the performance of mechanical refrigeration systems. Obviously, the jury is still out on semiconductor refrigeration.

In the meantime, I keep a neat little semiconductor water cooler on my desk at the Research Laboratory. It requires only 25 watts and does not disturb me at all with compressor noise.

Now I would like to get back to energy converters that produce electricity.

THERMIONIC CONVERTERS

Recently, Dr. Volney C. Wilson of our laboratory has made some remarkable progress in the direct conversion of heat into electricity by thermionic emission.

The principle of the thermionic converter is simplicity itself. Electrons are "boiled" out of the cathode, carrying away the heat of vaporization. They are then condensed at the anode, giving up the heat of condensation. Because the heat of vaporization is greater than the heat of condensation, the electrons retain energy which is available to do work in an external circuit. A unit of this kind will put out about a volt, at a current which depends upon the area (and other features) of the electron emitter, and at a conversion efficiency which depends upon a great many things.

The space between the emitter and collector may be a high vacuum, in which case very close spacing is required for efficient conversion. Alternatively, an easily ionized gas may be used to fill the space. In both cases, neutralizing the effect of space charge is an objective and a requirement for satisfactory performance.

Wilson's best conversion efficiency thus far is 17 per cent, with a gas-filled device. This is a good deal more than the 8 per cent which was available when we originally announced this development in November of 1957. The vacuum and gas-filled thermionic converter can be compared in the following table of presently available performance data.

Thermionic Converter Performance

Type	Efficiency	Weight Per Unit Of Power	Emitter Temp.
Vacuum	4½%	25 lb./kw.	1100°C
Gas Filled	17%	4 lb./kw.	1900°C

Note that a thermionic converter may be a lightweight generator, and hence adaptable to portable applications. It is also true that its efficiency does not depend upon its size, so small converters can be built without loss of efficiency. The little, wafer-like, vacuum thermionic converters that Dr. James E. Beggs of our laboratory has developed show how remarkably simple and effective a way this is to produce electricity from heat.

The ultimate efficiency of the thermionic converter is certainly going to be greater than the presently available 17 per cent. How much greater, is not so much a question of theoretical limits, which are much higher than 17 per cent, as it is a question of the practical limits which are imposed by available materials.

FUEL CELLS

Chemical energy has long been considered as a direct source of electricity, and of course common batteries accomplish this particular conversion. As early as 1839, Sir William Grove constructed his "gaseous voltaic battery," which generated an electric current from the combination of hydrogen and oxygen. Toward the end of the last century in England and Germany, further work in converting the chemical energy of the oxygen-hydrogen reaction into electricity was performed, and the term "fuel cell" came into use.

Since 1944, fuel cells have again attracted interest. Today we hear them hopefully suggested as compact, clean, and silent energy systems for everything from automobiles to spaceships, and as a future means of generating low-voltage electric power on a large scale.

The fuel cell is perhaps best visualized as a battery in which hydrogen, or some other fuel, is oxidized at the anode, and oxygen is reduced at the cathode. An electrolyte carries the current internally, and electrons flow from anode to cathode in the external circuit.

A point to be made is that, as a result of recent work, the fuel cell is emerging not as one device, but as a variety of devices with quite different performance characteristics.

I imagine that the fuel cell which we announced in April 1959, is about the simplest form of the device that has been contrived thus far. The driving force back of much fuel cell research is the theoretical efficiency of 100 per cent. The device is not a heat engine and does not suffer the defect of "Carnot inefficiency." The fuel cell models that we have made actually convert the oxidation energy of hydrogen to electricity at 70 per cent efficiency. The active element is an organic membrane, and the disadvantage of this simple system is the requirement of a special fuel–hydrogen.

A bulk electric power producer of the fuel cell type would have to use common gaseous fuels, and probably a variety of them. Thus far such cells have employed molten salt electrolytes at relatively high temperatures, and the problems of durability, efficiency, and reliability have not yet been solved satisfactorily. Typical fuel cell performance can be compared with the performance of conventional converters in the table.

Fuel Cell Performance Data

Type	Watts/lb. (cell only)	Watts/cu.ft.
Organic Membrane (H_2, O_2)	4–20	300–1500
Molten Salt Cell (H_2, CO)	4–10	300–800
Molten Salt Cell (Kerosene Vapor)	2–15	150–400
For Comparison–Heat-Acid Storage Battery (6 Hr. Rate)	1–2	90–180
For Comparison–Steam Turbine Generator	200	600

The fuel cell is almost certain to develop practical usefulness. A 70 to 85 per cent conversion efficiency, even for special fuels, is quite an accomplishment. In addition to portable power sources for a variety of applications, we should be thinking of fuel cells as potential balancing elements in chemical processes. If we had a chemical process which required low-voltage direct current for the energy balance, and made available by-product hydrogen for the chemical balance, this organic membrane fuel cell would be just the ticket.

MAGNETOHYDRODYNAMIC GENERATORS

Another old idea for energy conversion has been showing up recently under a fancy new name–the magnetohydrodynamic generator. I will abbreviate this to MHD, which helps to put the scheme in proper perspective, because basically it employs a very simple idea. In conventional electric generators, a conductor is moved through a magnetic field to generate electricity. In an MHD generator, the moving conductor is an ionized gas column. I think, but cannot prove, that the recent revival of interest in MHD generation has been stimulated by the extensive world-wide research on the fusion process, which has focused new attention on ionized gases. In any event, this old idea is getting a new look, and the preliminary results indicate that MHD deserves a thoughtful modern consideration.

An MHD generator with an output of 1 kilowatt was recently developed at our Missile and Space Vehicle Department in Philadelphia. Needless to say, the power input of this 1-kilowatt generator was substantially greater than 1 kilowatt. Let me be quite clear at this point. The measured overall efficiency of this experimental equipment is very poor. However, by taking account of several obvious defects in the equipment, one may calculate plausible conversion efficiencies of 40 per cent for such devices.

Since this efficiency, if it could be realized in practice, is in the competitive range, some thought is being given to the use of this principle for central station use. Even though much is unknown about this method, and even though the configuration of a large plant which might employ the MHD principle is quite speculative, there is an irresistible temptation to calculate the cost of producing bulk power by this means.

However, we need to know a great deal more about MHD before we are able to undertake meaningful calculation of this kind, and extensive programs are under way to better determine the important parameters. For one thing, MHD generators have to run more than a few seconds at a time if they are going to be useful in central power stations.

A durability of ten seconds might not be a disadvantage in certain applications, however. For example, short bursts of power might serve the special needs of satellite communication equipment.

And the required hot plasma might be obtained from a solid propellant rocket motor.

I am well aware that the performance figures that I have given on the various types of energy converters are less optimistic than those you may have seen quoted or published in the newspapers.

My advice, however, is that one must discriminate carefully between "preliminary calculations" based on an unknown set of assumptions on the one hand and reproducible measured performance on the other. The fine print and the asterisks may, in fact, contain the real truth, and should be carefully read as part of the principal text.

FISSION AND FUSION

Now on this occasion I am not going to say much about fission power generation. The vast, world-wide effort on this subject is, of course, the major modern development of basically new energy sources. But the work is so much in the news, and is so much before the general public and the technical world, that one can hardly escape an appreciation and an understanding of the broad outlines of fission power. So I will hurry on to a newer source of mass energy—fusion power.

Up to this point, I have mentioned fusion power only in passing. Actually, there is very little new to report. The world-wide effort on this outstanding scientific and technological challenge of the age continues, and it is no secret that the problem is at least as difficult and probably more difficult than originally contemplated. Unfortunately, engineering handbooks do not include a section on heating a gas to multiple millions of degrees and holding it in a magnetic bottle. So we have to learn from the ground up, and the people who are striving to do this have not yet got the knack.

For one thing, the bottle leaks a bit, and we cannot stand for any leaks. It is the same problem that Socrates had with his quest for a universal solvent—where do we find a bottle that will hold it? However, I detect no lack of enthusiasm on the part of the people working on this fascinating subject, and it is of course not at all surprising that it is taking some time to solve the complex problems of generating electric power by the cosmic process of fusing light atoms.

It is thrilling to contemplate the multiple avenues which research is now opening up for power conversion in the future. It is clear that scientists in many fields will find genuine excitement in advancing this work. Chemists, in particular, will be in the center of the action, and will play a vital role in the development of many of these processes. The future of the fuel cells, for example, rests largely with the chemists, and the development of future semiconductor compounds for energy converters depends heavily upon their skills.

Whether any of these energy converters will ultimately replace the turbine-generator for bulk electric power generation is an interesting and important question, but for many reasons this question is not urgent. The problem of the various energy conversion schemes is not that of determining the one victor to whom all the spoils will belong; it is, rather, the problem of matching these energy conversion schemes with the various power demand situations—situations, incidentally, whose number and complexity are ever on the rise. It is very likely that every one of the described energy conversion methods will develop a significant area of practical usefulness in the future, and we must learn enough about them to make wise choices.

While we are sorting them out, we might be permitted to dream a bit about the combinations and permutations of energy conversion phenomena that will be available for power engineering in the future. We might think, for example, about systems which degrade reactor heat in cascade through several energy-conversion systems to provide a super topping unit for the turbine-generator, and thus add valuable points of efficiency to an already efficient system. A great many other combinations and permutations are possible.

We may also pause for a moment in appreciation of the space vehicle, for it is incidentally providing an out-of-this-world opportunity to field-test energy conversion methods that cannot yet compete on earth. Later generations may come to regard the space vehicle as a modern Pegasus, a mechanical winged horse symbolic of the freedom from plodding toil that new energy conversion systems, fathered by the special demands and opportunities of space travel, have made possible.

It may be that the major contribution of the space age will be full, economic, and efficient energy conversion sufficient to stimulate and satisfy the power demands of a rapidly expanding civilization and a constantly rising standard of living. If so, the space age will be the prelude to the most ambitious age of all, the age that is the culmination of all others and the fulfillment of an ancient hope—the *age of abundance*.

An Appraisal of Technological Progress

An address at the National Industrial Conference Board Meeting "American Enterprise: The Next Ten Years," New York City, May 19, 1960.

The face of American civilization in our time is distinguished by a number of social and economic features interwoven in a complex growth curve with a definite upward swing. Among these features are the increasing worth of our urban-industrial economy and our rising level of education and of working skills. We could point also to the growing mechanization and efficiency by which we draw energy from our mineral fuels and transform our ores and other primary resources into useful products, and to the expanding and increasingly efficient communication and transportation networks through which we disseminate information and distribute goods. The increasing rate at which we combine research, development, engineering, manufacturing, and marketing to effect technological progress is perhaps the unifying feature of our society. The face of American civilization, seen in these terms, is certainly looking up.

I wish to discuss certain of these matters in some detail, with particular regard to the character and the future of technological growth. To provide a base line for this purpose, I shall cite some statistical information drawn from the varied and voluminous data that indicate the trends of our advancing economy.

THE STATISTICAL BASE OF TECHNOLOGICAL ADVANCE

Consider first, if you will, the starting point of technological progress. In this germinal area, which is the area of scientific research, development, and engineering, we are making increasing effort to foster growth. The curve which describes our support of research, development, and engineering (RD & E) illustrates a typical growth phenomenon, which will be no surprise to those who are associated with this activity (Fig. 1). The rate at which human effort has been drawn to research, development, and engineering corresponds, as we would expect, to the upward trend of support in this area. (A word of explanation of the term RD & E may be in order. The most complete industry information which has been collected purporting to show research and development expenditures is taken for a set of definitions which explicitly include engineering expenditures—hence the term RD & E.)

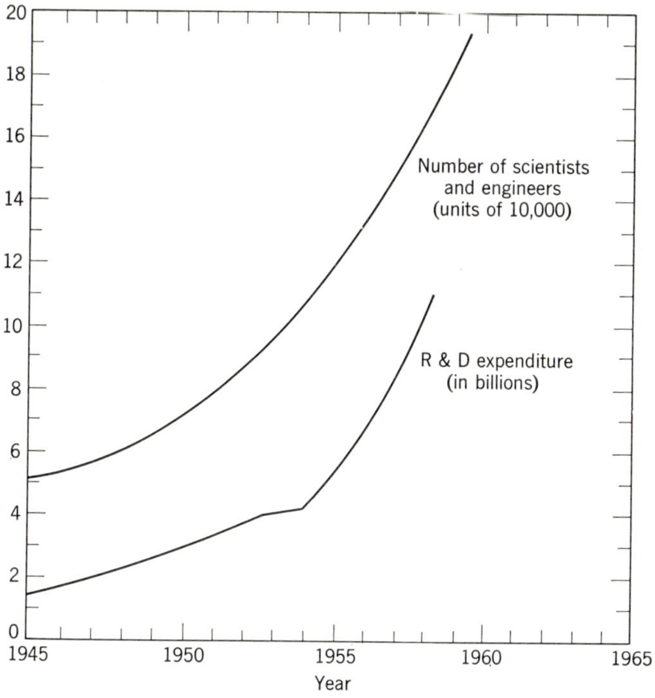

Figure 1. Growth of research, development, and engineering.

To show RD & E growth rates in perspective, we should display these curves in relation to some basic reference points. This is done in Figure 2. It is clear that *expenditures for technology* are growing much more rapidly—25 per cent per year, recently—than is gross national product at 3 per cent per year; and that the *growth rate of technical employment*, at 12 per cent per year, is surpassing that of total employment of 2 per cent per year.

HOW MUCH TECHNOLOGY IS TOO MUCH?

Obviously, these trends cannot continue indefinitely. If the growth rate of the RD & E function could continue to outstrip that of the gross national product, it would eventually constitute the entire economy—an anomaly that even the most partisan supporters of RD & E would view with dismay. It is apparent that, in the present differential between the related growth trends of RD & E and gross national product, we are still appraising and

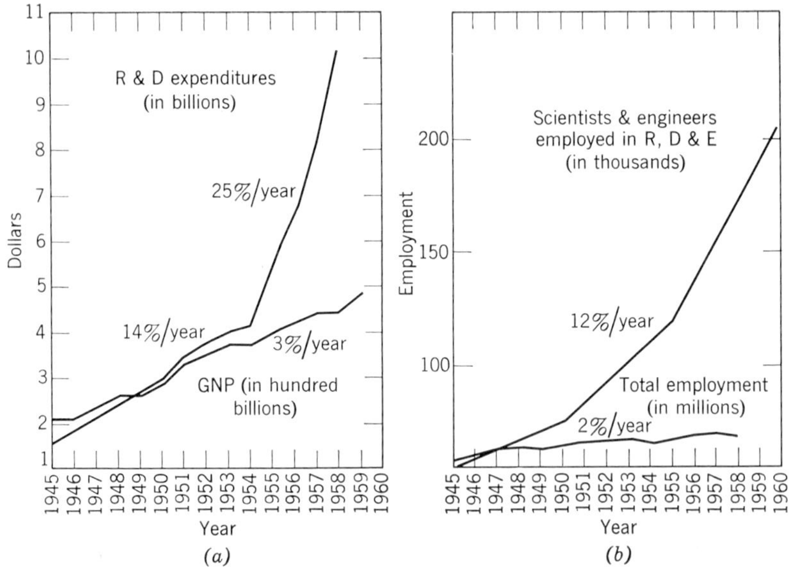

Figure 2. (*a*) Growth of research, development, and engineering expenditures. (*b*) Growth of employment in research, development, and engineering.

exploring the technological component of our economy, and are seeking the proper balance between RD & E and the other industrial functions.

What this balance will be and when it will be reached is a complex question. The problem of determining the balance between the various industrial functions derives from their essential interdependence. If we spend more money for RD & E, we will generate more business enterprise, which will require more RD & E to support its growth, and so forth. RD & E is itself one of the major determinants of the gross national product, as are such other functions as manufacturing, marketing, and manpower. Clearly, the optimum combination of these interdependent activities is not a constant, but a variable that must reflect such changes in the economy as are brought about by its growing complexity.

In spite of this complexity, one is tempted, when examining the abrupt growth curve of RD & E, to ask when this curve will flatten—when the saturation point will be reached. This is a natural question to ask, because saturation effects are a characteristic feature of all growth phenomena. Nothing increases without limit. At some point the rate of increase starts to decrease in response to one or more limiting factors. We propose to look briefly at some of these factors.

THE MANPOWER LIMIT

The availability of technical manpower has been suggested as a factor that will bring about this possible saturation. It has been observed, for example, that if the present trend of technical employment continues, we shall need one scientist or engineer to every 90 persons by 1980, and one to every 40 persons by the year 2000.[1] Only by greatly increasing the rate at which scientists and engineers are presently produced could such a trend be sustained.

However, it is by no means clear that the rate of production of technical manpower will *not* rise in response to need. In addition to increased output, a traditional effect of advancing technology has been an upgrading of working skills. As mechanization has assumed the functions of routine physical labor, the working force has been freed to gain the higher skills needed in the expanding mechanized industries. So also, as automated systems assume routine mental labor, the working force will be freed to fill the demand for still higher skills in the expanding and more complex automated economy.

But this advance from the bench to the lathe, and from the lathe to the computer console, does more than raise the level of working skills. It also enables increasing numbers of capable youth to be diverted into specialized and highly advanced educational programs, upon which progress of the RD & E effort depends.

Among the important benefits of technological progress are increased opportunity and increased reward for the highest development of scientific and technical proficiency. Thus the advance of the RD & E function tends automatically to foster the increase of technical skills upon which its continuing progress depends.

THE CAPITAL INVESTMENT LIMIT

If manpower does *not* limit technological advance, will the availability of capital do so? This is certainly far from the case at present.

[1] H. S. Brown et al., *The Next Hundred Years*, New York, 1957, p. 120ff.

I hesitate to offend my RD & E associates with an unflattering analogy, but one of the classic illustrations of the saturation effect in growth phenomena is that of bacteria in a controlled culture. With the input of nutrient held at a constant rate, growth of the bacteria rises rapidly to a saturation point that represents equilibrium between the consumption of the nutrient and its supply. This is, of course, not only an unflattering analogy; it is also an inexact one, for the RD & E function does not rely passively upon contributed capital for its support. Rather, it adds to capital by creating new markets, new businesses, new jobs, and increased output. Thus also does the RD & E function tend automatically to add to the supply of capital necessary for its continuing advance. Nevertheless, the growth of new technology at present rates *may* strain our future capital resources; and I hope it does, for this is pleasant trouble to contemplate.

THE MANAGEMENT LIMIT

A further limitation on the growth rate of new technology is our ability to manage it, either in an individual industrial unit, or for the country as a whole. The development of atomic energy as a power source is an example of an advance in technology that has strained our aggregate international, national, and industrial management capabilities to the utmost. The technical problems of atomic energy power production have been unusually complex. But in addition it has been necessary to solve or contain equally difficult problems of health and safety, public liability and insurance, public versus private enterprise, as well as the relationships to the military applications of atomic energy. In spite of intensive effort, large question marks remain in many of these areas.

Increasingly, companies in growth industries are hard pressed to acquire or train enough capable managers at all levels who can successfully "get their arms around" a high rate-of-growth technical business. Increasingly, new technology is so new and different from current practice that its application to products and services draws far more heavily upon the rare human resources of imagination and ingenuity, than upon the safer and more dependable refuge of experience. Thus the management of our technological opportunities is becoming ever more challenging, and it is a question whether our skill can and will keep pace with our future bounty of discovery.

The extent of the RD & E function is determined, in the final economic analysis, by its ability to provide return on investment. Indications, based on the past economic contributions of this function, are that it will continue to grow at its present rate for some time. Although a lessening of the growth of RD & E is inevitable at some future point, it will never completely saturate so long as a rising gross national product and the resultant advance of living standards and upgrading of working skills remain primary objectives of our society.

INDUSTRY GROWTH PATTERNS

Let us now consider certain industries in which the RD & E function plays a prominent role. The rapidly growing sales of these industries, as compared with the growth of the gross national product, makes it plausible that the emphasis of RD & E has favored the high growth rate. It can be safely assumed that this particular emphasis of RD & E will continue.

It is no secret that industry growth rates vary widely, and that they are spectacular in some cases, as in the semiconductor products industry, where sales have virtually doubled every year in the past decade. This is an example of a growing number of industries whose life blood is technological innovation, which we shall discuss in more detail shortly.

We may conclude that research, development, and engineering are not equally appropriate—or perhaps we have not yet learned how to make them equally applicable—throughout the various industrial categories. Much of the technical advance, however, that is directly enjoyed by the RD & E-oriented industries passes eventually to those industries in which the technical function is less active and less germane. Thus, for example, the advances in electrical machinery, in data processing and industrial control equipment, in communications systems, and in new and improved materials provide for a general rise in the level of technology throughout all industry.

Man-Made diamonds and silicone products are illustrative of research products whose improved characteristics benefit industry generally. *Note that the supplying of technological innovation to industry is itself a business, and is the means by which all industry partakes of technological progress.*

TECHNOLOGY AND PRODUCTIVITY

One important consequence of the ability of a society to generate and apply new technology is increased productivity, defined as the efficiency with which resources are converted into useful output. When properly calculated, it measures the output of goods and services in relation to the total inputs of management, labor, invested capital represented by plant and equipment, fuel and raw materials. When all of the inputs are considered, we find that productivity in the United States is rising at the modest rate of about 2 per cent per year.

Clearly, technological innovation is a key factor in this rise, for the contributions of the other inputs remain comparatively static. The contributions of raw materials tend actually to decline, as greater value added is achieved; and although we strive mightily to maintain the level of human energy input, it probably decreases also. Only the inputs derived from technological advance—new investment opportunities, steadily improving production techniques, increasing mechanization, more efficient exploitation of energy reserves, and the like—reveal an upward trend sufficient to account for the rise in productivity.

INNOVATION AND OBSOLESCENCE

By far the most glamorous, as well as important, use of new technology is for innovation in products and services, for innovation is a powerful driving force in the industrial economy. New discoveries in the sciences lead to new materials, processes, phenomena, and ideas which provide the opportunity for innovation. The successful application of these results may greatly expand the market for old products, or may generate completely new markets and the business enterprise to serve them.

A corollary of innovation is obsolescence. It is interesting to examine in some detail just how an "innovated" product creates obsolescence, and how the old product and the total market reacts to the pressure of innovation. The fast-moving electronics industry provides many illustrative cases, from which we select one.

The electronics industry, since its origins at the start of this century, has been based on the technology of electron tubes. A tremendous perturbation was created in electronics technology by the invention of the transistor in 1948, providing a basically new technology of semiconduction in solids rather than electron flow through vacuum or rarefied gas. When this discovery was made, it was commonly predicted that electron tubes would experience a rapid decline in favor of the transistor and related solid-state devices. For various sound reasons this did not take place, and on the contrary the electron tube business has been growing ever since.

This is not to say that the transistor and other semiconductor devices did not make good on their promise of success. They have performed handsomely, both electrically and economically, and the semiconductor product business, with a doubling period of one year, is regarded by many as an authentic growth business. Only a small part of this growth, however, has been at the expense of electron tubes that have been made obsolete. The semiconductors created important large *new* markets, most of which could *not be* served effectively by electron tubes. For example, the miniature portable radio first appeared on the market as an electron tube set, but because of marginal performance and extremely short battery life, it never achieved acceptance. The transistor portable radio, with good performance and long battery life, has created a large market for this product and the component semiconductor devices. The same case history, with but few changes in detail, could be related for many other applications of semiconductors, such as the transistor hearing aid, the semiconductor computer, semiconductor communication equipment, and the multiple applications of semiconductors to military technology.

Meanwhile, the technical competition of semiconductor devices has greatly stimulated research in electron tube phenomena. At the time of the discovery of the transistor, large R & D capabilities were in place to provide the technological base for the growth of the electron tube business, and these laboratories rose to the challenge.

As a result, impressive electron tube progress has been made, leading to types which are smaller than transistors, although not lower in power requirements, and with greatly improved signal-to-noise, and very high frequency performance. The progress in

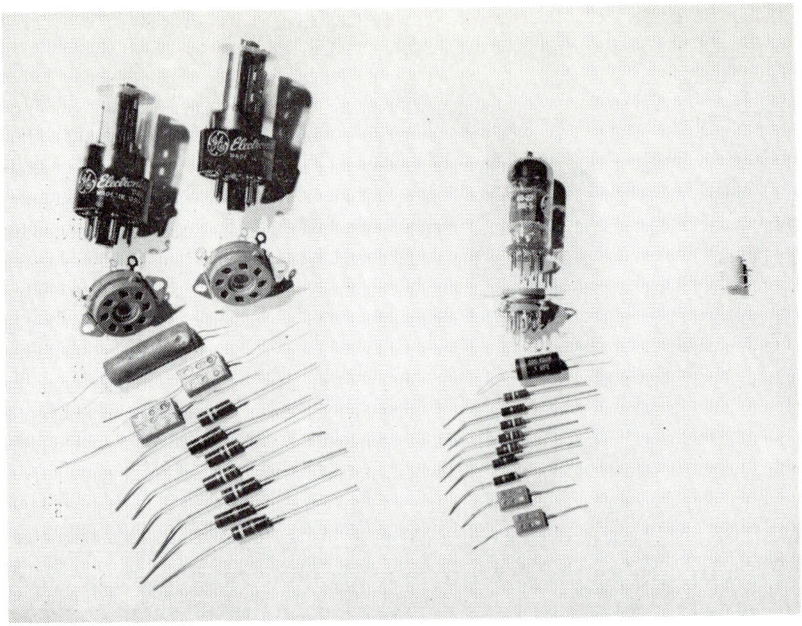

Figure 3. Progress in electron tube and component miniaturization.

electron tubes and circuit elements is even more impressive, as shown in Figure 3. The tiny unit on the extreme right performs the functions which formerly required the handfuls of equipment shown on the left. *It is important to observe that the R & D capability required to support a high rate-of-growth technical business is itself a hedge against obsolescence in that business.*

In the case described, the net result of the technological shuffle has been completely new markets for semiconductor devices, some obsolescence in some types of electron tubes, but with net growth in the electron tube business as a result of technological innovation in this older field. So in this typical example of a technological race *nearly everyone wins.* Most importantly, the consumer wins, for he receives the bounty of improved goods and services.

With the announcement of the tunnel diode by General Electric in 1959, we now contemplate a new contender for honors in the electronics race. In this case the new component is a semiconductor device which will in many areas compete with the now old semiconductor device—the transistor. Although it is too early to say how the race will run, I think it is safe to say that in this case, also, *everyone will win.*

Already the tunnel diode is showing that it has some unique capabilities that are certain to lead to unusual new products. For one thing, the tunnel diode is peculiarly adapted to FM circuitry. For the first time one can make a real Dick Tracy wristwatch radio, and I think it inevitable that we shall use them for hundreds of short-range communication tasks in the home and in industry. In any event, the tunnel diode is certain to open new markets where none exist today.

VARIETIES OF INNOVATION

At this point in the discussion of innovation a distinction should be made between varieties of innovation. When a couturier innovates by rearranging the geometry and trigonometry of feminine apparel, his economic objective is to create obsolescence by styling. The same motive—styling innovation—is applicable to many consumer products. It accounts for a sizeable industrial design activity characterized by a frantic annual arrangement and rearrangement of chrome strips, accent emblems, tail fins, knobs, handles, feathers, and buttons. This large but wattless energy may be regarded as somewhat frivolous, but it *is* solidly based on some *invariable human traits.* Most people agree that the new cars *do* look better than the old ones; and few ladies are willing to retain their old headgear, once the new trend is evident. There can be no doubt of the economic impact of style innovation, and a huge industrial activity lives or dies by the quality of its styling. But technology plays a negligible role in this area. *As a generalization, technological innovation satisfies human needs, while style innovation satisfies human preferences.*

INDUSTRY'S PLACE IN BASIC RESEARCH

I wish to call attention to one final aspect of the relationship between the advance of the RD & E function and the rise of the economy, an aspect that will have profound effect upon future progress. This is the shifting of the "center of gravity" of basic research. This type of research is no longer *exactly* centered in the universities where it has had a traditional home. Prior to World War II very little basic research originated in industrial laboratories, and it was a matter of constant amazement to scientists in the university that any "pure" research was done in *any* industrial laboratory. In some quarters it was believed that preoccupation with the practical usefulness for basic research efforts was essentially sinful. Sad to relate, that archaic idea still persists in some segments of our society.

Since World War II the weight of much evidence has shown that basic research is an eminently practical and economic investment for industrial growth. Constant progress is being made on the achievement of the climate, managerial environment, and facilities for basic research in industry. The outcome has been that the industrial participation in basic research is now widespread, and the results are significant in many fields. Literature studies soon to be published by some of my associates show that the industrially contributed fraction of basic science publication in the field of physics (including all subdivisions) is 19 per cent, in chemistry 22 per cent, and in metallurgy and ceramics (combined) it is 37 per cent.

We observe also that government laboratories have been making impressive additions to the output of basic research. The study I have mentioned reveals the following contributions of government laboratories: physics 22 per cent, chemistry 15 per cent, and metallurgy and ceramics (combined) 20 per cent. University contributions were found to be 56 per cent for physics, 59 per cent for chemistry, and 35 per cent for metallurgy and ceramics (combined).

In many of these areas of basic research, industry participates in a leadership capacity; and the number of such areas is destined to increase in the future. This is a satisfaction to scientists in industry, for *to an increasing extent industry is now "paying its way" in science—by contributing new basic knowledge and understanding of nature to replenish the storehouse from which heavy withdrawals have been made in the past.* Clearly, this contribution will add to the foundations from which dramatic new technological progress of the future will rise.

The Postgraduate Training of Physicists

An address given at the International Conference on Physics Education in Paris on August 3, 1960.

The point of view of industrial physics needs to be expressed at this conference because this is a large and rapidly growing field of opportunity for the physicist. That this is so is shown very clearly by the statistical information published by J. Henry Barton, of the American Institute of Physics, in an issue of *Physics Today*.[1] The data show that in the United States as of 1958, industry was the largest employer of physicists, with 47 per cent, compared with 35 per cent in educational institutions, and 13 per cent in government employment. The trends over a five-year period indicate that the industrial employment of physicists is increasing. The figures express the total employment—as of 1958; the current rate at which physicists are going to industry must be substantially higher. Henry Barton's information indicates also that, judging from the employment of physicists, industry has a substantially greater interest in such fields as solid state, electricity and magnetism, optics and acoustics, while the academic physicists have a relatively greater interest in nuclear physics, theoretical physics, atomic and molecular physics, and a catch-all category of general physics; whether this is good or bad, I do not know, but I will comment on it below. In any event, the percentage of present employment in these fields is not a sure guide to the percentage of educational instruction in the same fields. I will quote one additional number from the Barton paper: of the physicists employed by industry, a large fraction of them—18 per cent—are engaged in the management of research and development activities.

The increasing employment of physicists by American industry is, of course, evidence of the increasing complexity of the industrial technology of materials and processes, especially in the technical industries where most of these physicists are found. It is also evidence of an increasing level of basic research effort in American industry, where the physicist plays an indispensable role. This is true for example in the General Electric Research Laboratory where physicists pursue science-oriented research in a great variety of fields, frequently on an individual basis, not infrequently associated with researchers in other fields. Somewhat more than 70 per cent of the research staff of this laboratory are of the

[1] *Physics Today*, January 1960.

Ph.D. level, while the company as a whole employs about 1000 Ph.D.'s, principally in the physical sciences.

My company's point of view of the graduate training of physicists is, of course, that of an industrial employer of physicists. We take the products of our system of graduate education for the pursuit of scientific research. To be more precise, we regard the Ph.D. physicist not as a product of our educational system alone, but a product of his heredity, his environment, and his graduate training. Perhaps we have been focusing too sharply on his training, and have been overlooking the importance of his natural endowments as a factor in his performance in scientific research. I will try to interpret our observations in terms of the graduate study and research exposure which these young men have experienced; naturally I shall be on much less firm ground here.

By and large we are well pleased with the Ph.D. physicists who come to us. They are generally bright and ambitious and their scholastic preparation is usually excellent. They generally turn out to be able researchers, although it usually takes a few years to determine their particular research abilities and interests.

In spite of this general feeling of satisfaction with our young physicists we are puzzled by a few things. When a young Ph.D. joins our staff we hope he will be able to do original research—the more creatively, the better. Yet we are puzzled to know how to find out from his graduate record whether he is in fact capable of original research. From his formal course mark we can easily determine that he has been a good student, but that is not enough. He did not receive an academic grade for originality, or self-reliance, or energy and drive, or physical intuition. We obtain the impression that, in some exceptional cases, the young man may have gone through graduate school and have received his Ph.D. without ever having discovered whether he can do research. Even in this situation, however, it generally works out well, for in the course of his apprenticeship he and we learn about his abilities, and we can usually accommodate his program to them.

Not infrequently the young physicist comes to us to work in one field, say, semiconduction, having taken his thesis research in quite another field, say, high-energy nuclear physics. The theory of this case is perfectly straightforward, for the dissertation research was supposed to be a trial run and not necessarily a life work, and we rely upon the fact that the student has had a good broad foundation of formal preparation that will, or should, serve him well in almost any branch of physics. Some fields of graduate research tend so to dominate graduate study that by the time the physicist gets his degree he is already highly specialized, rather than generalized. For example, I have the impression that nuclear research, especially high-energy nuclear physics, tends to make for intense specialization, and that solid-state research may encourage more generalization—but that is just an impression, and I would expect to find different points of view on this subject. This particular situation relates to the differences in the fields of physics which are emphasized in industry and in educational institutions. We have an interest in the matter of specialization because flexibility and a broad range of interests are desirable traits in an industrial physicist. We wish to be able to turn to important new fields, as we or others indicate their importance, and this need can be illustrated by a current example. This year a variety of new fields are open to physicists, under the heading of energy conversion fields which, but a few years ago, would have interested few physicists. These subjects include the thermionic converter, the semiconductor thermojunction, the fuel cell, magnetohydrodynamic power generation, and several more.

The case of the young man with the Ph.D. who has hardly explored his research ability might be worth some further discussion. There are plausible reasons for this situation to arise. An important part of the problem is, I think, the fact that scientific knowledge is expanding so rapidly that formal training is crowding research out of graduate training. By the time the student has satisfied the formal course requirements and, in addition, the optional course offerings which have become almost indispensable, there is very little time left to accomplish a respectable experimental dissertation. The life of the young graduate student is complicated by the fact that in four or five years of graduate work he probably has acquired a wife and family. His relatives and friends are beginning to ask his mother when he is going to get out of school and get his degree. This, of course, everyone now assumes is a foregone conclusion; it is just a matter of time. By now his professors are in a somewhat difficult position, too, for the young man has an excellent scholastic record and is very well spoken of by everyone, and to fail to get his degree at this point would be rather harsh treatment. Although there may not have been time for a full thesis research, there usually is some research work which can be assembled and, if the typing is double spaced, it will provide a thesis which will be thick and fairly respectable. Now I am sure that this hypothetical case must be unusual and probably exaggerated, but I am quite sure that one could find such cases. Regardless of the circumstances of individual graduate study, it is a fact that the fund of knowledge with which the modern young physicist must be familiar is increasing at an alarming rate, and although I have not seen statistics to support the view, I think the time required to complete the Ph.D. work is getting longer and longer. Something has to yield, and I suspect it is the quality of graduate research. And I think that is the reason why some of our Ph.D. graduates have not thoroughly explored their research aptitudes by the time they receive their degrees.

This is an interesting but not a serious situation; I present it more as an observation than a complaint. If we, in industrial research, had a choice as between abbreviated formal course work and full research training on the one hand, and full formal course work and a resultant shorter period of research training on the other hand, I think we would choose the latter. The student can, if necessary, serve his research apprenticeship in a large industrial research laboratory. What may happen in the distant, but not too distant, future is interesting to contemplate. Will graduate training in physics then require six or seven years, rather than the present four or five, and former two or three? It is a question of what we will omit from the future course of study for I would view with considerable alarm any further encroachment on the period of productive effort of the research physicist.

A very interesting report on the Postgraduate Training of Physicists in British Universities [2] indicates that much the same problem of scope is being experienced there, although in slightly different form. This report recommends that formal course work should be extended into the postgraduate training period which, in Great Britain, is presently given over almost entirely to research training

[2] *The Postgraduate Training of Physicists in British Universities.* The report of a joint committee appointed by The Institute of Physics and The Physical Society, March 1960.

and apprenticeship. This is certainly sound, and is a necessary consequence of the increasing extent of knowledge. The report notes also that British industrial laboratories would prefer, under present circumstances, to take the physicist at the Bachelor level and provide his research apprenticeship in the industrial laboratory. From conversations I have had with research directors in Great Britain I know that this is true. If formal course work were extended into the graduate period, as recommended in the report, it would probably change this situation substantially.

In the laboratory with which I am associated we very seldom employ a Bachelor in science.[3] For laboratory assistants we prefer graduates of two-year technical colleges, and for research scientists, Ph.D. graduates. There are exceptions, of course, and in some fields we prefer a research engineer. He is generally a Bachelor in engineering who has had a heavy exposure to science and technological courses, frequently into the graduate level. Such specialists are becoming very valuable men in industry, and we need more of them. Because of their emphasis on science and technology, they have had to forgo some of the old standbys of the engineering curriculum, such as surveying and machine design; but on the other hand they know something about nuclear resonance and Fermi levels. If they cannot survey, this is of no consequence to us.

I intentionally have not referred to certain classical problems of graduate training in the United States, because they have been much discussed and much deplored and I have no new suggestions to make. They include the use of the graduate student as a member of a very large research team—where his opportunity for individual development is correspondingly circumscribed. They also include the problem of the graduate student who "inherits" from a graduating student a complete set of experimental equipment and an established procedure for performing one additional research in an extended series of investigations of particular interest to his professor. This student's opportunity for individual development is also circumscribed. Similarly, the employment of graduate students to accomplish relatively menial but necessary technical tasks such as those required in the construction of a particle accelerator, or in the satisfaction of the requirements of a government

[3] It is perhaps necessary to point out the different standards of the Bachelor's degree in Europe and the United States.

or industrial research contract, may not provide an optimum type of activity for the training of graduate students. I hope these practices are uncommon, and that the practical necessities of graduate research can be managed without seriously compromising the student's opportunity for exploring and developing his aptitude for individual creative scientific research.

I must comment on originality and creativity in scientific research. No graduate training which has been devised thus far will ensure the development and flowering of these admirable traits of a researcher. Yet they are of superlative importance to the future progress of civilization. The creative spark in science nucleates a vast resultant and dependent activity that may have immense material and practical consequences; but I do not need to develop this theme in an audience of physicists.

Why do we not do something about creativity during the period of research training? I think there is little merit in the idea of "developing" originality, but that there are abundant opportunities to recognize creative traits and aptitudes, and to encourage and further them. If creative talents have not been in evidence prior to graduate study, I do not think the student is likely to discover them suddenly during this period. In fact, creative aptitudes must appear very early in life, and at secondary-school level they should perhaps be perceptible. In the belief that this may be so, Union College at Schenectady, New York, and the General Electric Research Laboratory are embarking upon an interesting experiment, which has not yet been announced because we are still working out details of the plan. The main outlines of the plan, however, are these. We will bring a group of secondary-school science students to the college for summer study and tutelage. These students will not be selected only for their scholarship, but an effort will be made to appraise exceptional interests and aptitudes in science. The most unusual feature of the plan is the following: each student will undertake a little research project under the individual tutelage of a member of the Research Laboratory staff. We hope that in this way we may discover natural endowments and aptitudes, and encourage them to mature for the benefit of science. Whether we can do anything to recognize and enhance creativity at the graduate level, I do not know. It is so important that it would be worth exploring all possibilities. We now appear to teach problem-solving, and not how to find problems and to appraise them; we teach how to understand, but not how to discover. In effect, we teach the logistics and tactics without teaching the strategics of scientific research. This is a great area of challenge in the teaching of physics at graduate level, and we must make a start at progress toward this important objective.

I would like to return to a statistic from the Barton paper concerning the employment of physicists. Recall that 18 per cent of the physicists employed by U. S. industry are engaged in the management of research and development and that even in educational institutions 4 per cent of physicists are concerned with such work. I consider that this is a very significant statistic, and that it points to an increasing area of opportunity, and obligation, for physicists in particular, and scientists in general. I have not seen the corresponding statistic relating to industrially employed chemists, metallurgists, or engineers. But one might accept that a similar situation exists there. Philosophically, it seems to me that with an ever-increasing impact of science on human welfare, on economic well-being, and on the affairs of nations, it is inevitable that scientists should interest themselves in the management of technical work. I refer to it as an obligation. It is more than that—it is a necessity. Either scientists must concern themselves with the management of their kind of work, or their work will be managed by nontechnical people who are not in a position to support their decisions with necessary technical understanding. In my experience physicists, chemists, metallurgists, and engineers generally make very good managers of research and development. I think that physicists must become increasingly aware of the opportunity and obligation in this area. The day has long since passed when the physicist can retire to his ivory tower in isolation from the world. He must, it seems to me, increasingly and appropriately concern himself with the interpretation, guidance, and management of the affairs which have been created as a result of his work. This fact should perhaps be recognized and provided for in the undergraduate and graduate training of physicists. Perhaps we must insist that the already overloaded curriculum include an irreducible minimum exposure to history, economics, and sociology as partial preparation for life in a world of scientific affairs.

I am sure that I have presented more problems than I have solved. They include:

(*a*) The expanding scope of graduate training;
(*b*) The problem of recognizing individual research ability, and providing an opportunity for its development;
(*c*) The preparation for creative research, rather than problem solving;
(*d*) The recognition of the role of the physicist in the management of scientific affairs;
(*e*) The problem of overspecialization in graduate physics.

I recommend these problems for the most serious consideration.

The Synthesis of Diamond—
A Case History
in Modern Science

Presented before the American Chemical Society, Rochester, New York, November 3, 1960.

The history of attempts at diamond synthesis probably started in 1797, almost at the moment diamond was first shown to be a form of carbon. Since that time, the efforts of a great many people have focused attention on the attractive idea of converting ordinary carbon to the diamond form. The tremendous scientific challenge of making the world's hardest substance, the diamond crystal, has caused many eminent scientists to give serious effort to this problem. The tantalizing prospect of converting an abundant and inexpensive substance, carbon, to a comparatively precious material, diamond, has also proven to be an irresistible temptation not only to scientists but also to numerous pseudoscientists and pseudoinventors. In aggregate, the history of diamond-making has been one of such frustration and failure as to repel all but the most courageous investigators, even in the light of modern scientific techniques.

The decision to mount a major new assault on this old and bloody battlefield was made in 1951, and was carried to a successful conclusion in the next few years at the General Electric Research Laboratory at Schenectady. This work [1,2] has provided a new milestone in man's conquest of matter, and a crowning chapter to the epic history of diamond-making. But it also provides an interesting case study of modern scientific research which demonstrates some of the strengths as well as difficulties of a team approach to the solution of a problem.

In particular, this adventure into the unknown illustrates research teamwork to an unusual degree; I find it difficult to imagine that a single investigator could have carried the work through to a successful conclusion. There were too many blind alleys and booby traps that had to be probed in detail before the path was clear. There were too many specialized subinvestigations and exploratory sorties—some mechanical, some chemical, some physical, and many metallurgical in character—that had to be accomplished before the foundation stones were laid for the final solid progress. In an investigation of this kind, the knowledge that road A is blocked by an immovable obstruction is almost as valuable as the knowledge that road B is clear and passable. Negative results contribute heavily to the final positive success. Finally, in a project which pushes the natural limits of materials and technology literally to the breaking point, the mutual stimulation of joint effort is an almost indispensable ingredient of morale and courage.

These advantages of mutual support and stimulation carry some penalties: credit for progress is shared, and individual identification is partly subordinate to group achievement. But if the aggregate achievement is great enough, as in this case, there is credit enough for everyone.

At this point, I should make it clear that, beyond the scientific challenge, the synthesis of diamonds has a great deal of relevance to the General Electric Company. The Metallurgical Products Department (formerly the Carboloy Department) is one of the largest consumers of industrial diamond in this country, for the use of diamond is essential in the processing of cemented carbide materials. Furthermore, the cemented carbides made by this department compete with other hard substances, including diamond, in many cutting, grinding, and polishing applications. Hence, from the standpoints of both markets and manufacturing processes, industrial diamond is a vital material in this area of General Electric business. Equally significant is the fact that the future of many products of the company is vitally linked to developments in hard and strong materials. The route to diamonds, we presumed, might lead to other new materials of high strength.

EARLIER INVESTIGATIONS

The earlier claims of diamond-making have been examined in critical detail by many people, and I would not attempt to cover them comprehensively. The review paper [3] by Dr. Percy Bridgman in the November 1955 issue of the *Scientific American* is a particularly good summary of this history. Dr. Bridgman has become a symbol of high-pressure research, and is in an exceptional position to appraise the significance of work in this field. The overwhelming fact which summarizes the results of all prior efforts to make diamond is very simple and eloquent: no diamonds made by a synthetic process had ever appeared in world markets.[4] The ghosts of Moissan,[5] Hannay,[6] and Sir Charles Parsons [7,8] loom large in any discussion of diamond-making, but a host of lesser lights such as Despretz,[9] Ruff,[10] Guenther,[11] Geselle, Rebentisch, and many others has also been identified with this subject.[12] There has been, hovering near the periphery of serious diamond research, a still larger host of crackpot inventors, charlatans, and plain swindlers with secret processes which could be bought, preferably sight unseen, for a slight down payment equivalent to a king's ransom.

In the course of describing the work which led to a practical production process, I will have occasion to refer to many of these earlier processes which were claimed to make diamonds, for some of them were considered seriously in the course of our work. Moissan's process of cooling a molten mass of iron with carbon in solution, for example, has long been popularly regarded as an authentic diamond-making reaction, and it is so described in numerous reference sources. Some residual doubt was once admissible because of uncertainties in the boundaries of the pressure-temperature domain for the graphite-to-diamond transition, but these uncertainties no longer exist. It can now be said with positive assurance that the required pressure could be neither developed nor contained in an iron reaction vessel. Moissan certainly did not make diamonds. A possible explanation for Moissan's belief that he made diamonds has been given by Bridgman,[3] who wrote: "In 1924 Parsons informed me that Moissan's widow had told him that Moissan had been the victim of fraud by one of his assistants, who had introduced diamond fragments into the cookery, in order to please the old man and to avoid the tedium of the long digestions."

It has been characteristic of many of the processes which we have investigated that tiny transparent hard crystals are formed. These crystals may be clear spinels, alumina, chromite, silica, or other substances which scratch glass and sparkle brightly and suggestively. Not unnaturally, when these shiny crystallites are first revealed in a new experiment a scientist's pulse may quicken and his hopes may soar. But the thrill is transient and the denouement is complete when X-ray tests show a complete absence of diamond diffraction lines. The propensity of high-pressure reactions to yield diamondlike crystals of other substances has been the origin of many false hopes and much self-deception in the discouraging history of diamond-making.

A case in point is described in the notebook record of one member of the General Electric team of investigators, who wrote:

I've been looking in the microscope all day at the material from runs made yesterday and today. I think I have made diamonds in these runs! In each sample I found many absolutely perfect octahedrons! Some octahedrons scratch glass. There are thousands of triangles. Most of them have pits and steps (as do also the octahedrons). They certainly look like the real thing.

After critical examination during the following week, a brief and unenthusiastic notebook entry states:

Examination of [new] run. Octahedra present here, [but] since there was no carbon present [in this run], the octahedra cannot be diamond.

In this particular case it was determined that the attractive crystals were chromite, originating in the chromium of the stainless steel reaction chamber.

Although conclusions concerning Moissan's experiments are unequivocal, it should be mentioned that there remains some mystery and appreciable controversy concerning what was once thought to be a related phenomenon: the diamonds found in meteorites.[13,14,15] These diamonds are sometimes found in an iron matrix, and sometimes in a conglomerate matrix. In either case, the matrix may have originated at great depth in a planet, which would provide a plausible explanation for the presence of the diamonds. Or, as in the case of the Canyon Diablo meteorite, it may have been subjected to high transient temperature and pressure as a result of impact. Both cases leave room for some conjecture, however.

One of the most comprehensive modern investigations of diamond-making was undertaken by Dr. Percy Bridgman, starting in 1941 and extending into the war period. Largely through the initiative and interest of Dr. Zay Jeffries, joint support for a five-year program was provided by the Carborundum Company, the Norton Company, and General Electric. A 1000-ton press was obtained, and experiments were planned and partly carried through with graphite at pressures up to 30,000 atmospheres (440,000 lb/in^2) and at temperatures up to 3000°C. The war interrupted this work, and only two years of the contemplated five-year program were completed.[16] Following the war, the Norton Company became interested in a continuation of the work. They acquired the press from Dr. Bridgman and undertook a program under the direction of Dr. S. S. Kistler. Late in 1950 Dr. Kistler wrote to me that some encouraging results had been obtained, but that they were unable to support the scale of effort and wished to propose a joint project with General Electric. This proposal was discussed subsequently with Dr. Kistler and other representatives of the Norton Company, who seemed to be convinced that they had some evidence for the synthesis of diamonds in their experiments. In order to have a free hand, it was stipulated in all of our discussions that no process information would be released to the General Electric Company; hence we did not learn about the details of this work, and, so far as I know, it has not been published subsequently.

THE DECISION TO GO AHEAD

This, then, was the vantage point from which a new appraisal of diamond-making was initiated in the early 1950's. In summary, we contemplated the long history of prior investigations in this field. We had jointly supported the war-interrupted effort of Dr. Bridgman, and were familiar with the many pioneering contributions he had made to high-pressure research.[17] We had been informed that Dr. Kistler had continued some of this work at the Norton Company. We had restudied General Electric's interest in industrial diamonds in relation to its business in cemented carbides and broad goals in high-strength materials technology. Most importantly, the subject had stimulated tremendous enthusiasm on the part of some very able members of the research staff, and a host of new instrumental and theoretical approaches to the problem were being proposed. These young men were not in the least intimidated by the history of failure of diamond-making; on the contrary, they were anxious to flex their muscles in this old battleground. From a philosophical as well as a practical standpoint, we noted that every decade brings a host of new materials, processes, knowledge, and understanding which may provide new keys to long-locked doors. What was impossible yesterday may be feasible today. All these reasons encouraged us to believe that it might be a strategic time to mount a new assault on diamond synthesis.

Once the decision to go ahead was made, exploratory work was started in several directions. A. J. Nerad and two associates, Drs. Francis Bundy and Herbert Strong, comprised the starting group. Mr. Nerad's senior mechanical engineering experience in many major undertakings was especially pertinent to the task. He had pioneered in mercury boiler and turbine practice and was a principal contributor to development of the combustion chamber which had given Whittle's gas turbine a vitally needed impetus on its path to the successful jet engine of today. Francis Bundy was trained as a physicist, with heavy leanings in the direction of mechanics, optics, and heat transfer phenomena. Herbert Strong was also a physicist with interests in optics and thermal technology. None of these men had prior experience in very high-pressure practice, which in a way was an advantage, since they had no convictions about what could not be done. As a practical matter, very few experts in this field existed at that time.

Although the high-pressure, high-temperature approach seemed most promising many alternatives were considered. For example, work on crystal nucleation theory by Drs. John Fisher and David Turnbull of the laboratory raised a very real hope that diamond crystals might be grown at low pressures. I will describe the high-pressure work first, and later, somewhat out of chronological order, the low-pressure experiments.

A study of phase transition data for graphite and diamond was the initial step in high-pressure work. For temperatures up to about 750°C and pressures up to 50,000 atmospheres, the thermodynamic calculations of Rossini,[18] Jessup,[18] and Simons [19,20] were available. Although thermodynamic data provided essential guidance, the range of uncertainty in calculated pressures and temperatures was quite large. Furthermore, stability data help very little on the all-important question of reaction rates.

The reaction rate for the diamond-to-graphite transition is evidently negligible at room temperatures; at least one hears of very few diamond gem stones which spontaneously revert to graphite. To cause this reaction to go at an appreciable rate at atmospheric pressure, it is necessary to heat diamond to temperatures of around 1200°C. Conversely, Bridgman had frequently taken graphite far into the diamond-stable pressure range,[21] but at low temperatures, without obtaining any conversion to diamond. It was therefore presumed that a high temperature would, in any event, be required to cause the graphite-to-diamond transition to take place, even at diamond-stable pressures. From such considerations, it was speculated initially that pressures in the range 50,000 to 100,000 atmospheres and temperatures above 1700°C would be required to accomplish a diamond synthesis. How long these temperatures and pressures would have to be sustained was hardly a matter of speculation; it was a complete mystery.

Although Bridgman had produced pressures much higher than 100,000 atmospheres in very small samples, the development of high-pressure chambers which would also permit sustained high temperatures and larger samples was an unsolved problem. This problem was extremely formidable and was the principal bastion to be conquered if the pressure synthesis of diamond was to become feasible.

Initially, no one knew how to tackle this problem head-on, so some exploratory sorties had to be made. The starting point was the anvil apparatus of Bridgman,[21] which pressed a tiny specimen in the form of a thin disk between flat anvils. If the specimen is electrically conducting, it may be heated by current between the anvils as electrodes, provided insulation is introduced at some point in the press. A "retaining ring" of pipestone is used to surround the edge of a graphite disk. This naturally occurring stone has some very useful properties. It is electrically and thermally insulating, and flows appreciably at high pressure while retaining high strength in thin sections. As a high-pressure gasket in the anvil chamber, it permits the movement of the pistons required to develop high pressure, yet restricts flow sufficiently to retain the sample.

INITIAL EXPERIMENTS

The first attempts, however, to develop high temperature in this apparatus were highly discouraging. Bundy's notebook record of November 23, 1951, records an experiment with Carboloy flat anvils:

> The test was run between 9:30 and 11:00 A.M.—The sample was loaded to 150,000 atmospheres and 20 bursts of heating current were passed through it. The calculated temperature was 1400°C.
>
> Then the load on the press was gradually decreased. The Carboloy anvils snapped and spalled as the load was diminished. When the upper anvil was lifted clear, the pipestone was scattered around with many flaky fragments of Carboloy, but the graphite sample remained complete as a nice one cm. dia. disk on the bottom anvil. The top surface of the anvil fell away completely.

This inauspicious start, however, did lay foundations for pressure chamber progress. Although the properties of pipestone as a slightly plastic stone gasket were fascinating, they continued to be highly variable and a source of erratic performance. It was noted wistfully that Bridgman had dealt with this problem by cultivating the acquaintance of an individual Indian, who picked out the best specimens of pipestone from a natural source in Minnesota. We had apparently neglected to develop an equally effective local acquaintanceship. After a particularly discouraging series of experiments in which eleven pipestone gaskets had blown out successively, doubts were expressed as to whether the project could succeed without a friend among the red men.

At this critical junction, the Research Laboratory's Dr. Louis Navias, long experienced in ceramic science and technology, was brought into the problem. He suggested the use of pyrophyllite (also known as wonderstone), a hydrous aluminum silicate, as an alternative high-pressure gasket material. This proved to be a more uniform, more available, and less expensive material which essentially solved the high-pressure "gasket" problem. Thus, an expedition to an Indian reservation in Minnesota was made unnecessary.

It was hoped that initial experiments with the flat anvils and electrically heated sample would yield a few points on the graphite-diamond equilibrium line, and thus better define future pressure chamber requirements. Numerous attempts were made to establish points on the graphite-diamond equilibrium curve, but the net result was to impress a full realization of the experimental difficulties on the group.

The pressure experiments during this period were carried out on an extremely ancient 400-ton single-acting press (believed to be the original press used by Dr. Coolidge in ductile tungsten work), which was the only equipment available pending delivery of a 1000-ton double-acting press. Nothing on this old press was very tight, or true, and the ram guides were worn so much that the plunger moved partly in an axial direction and partly in unpredictable radial directions. Measurements of pressure were uncertain at best, and breakage of chamber parts was common. However, pressure chamber development had to be carried out on this dilapidated equipment in the early stages of the work.

The first important step in pressure chamber progress was a modification of the Bridgman design due to Bundy.[22] Bundy's improved anvil chamber, called the "saucer" from the shape of the anvils, provided ceramic inserts in the anvils for thermal insulation. The plastic flow of a pyrophyllite retaining ring allowed a small reduction of volume in the specimen, and hence the increase in pressure. The graphite specimen, by virtue of its smaller cross section, was heated much hotter than the large graphite end plates. The measurement of temperature and pressure in this cell and in all subsequent cells turned out to be a continuing source of difficulty and uncertainty. Phase transitions involving sharp changes in electrical resistance of cesium, bismuth, thallium, and barium, previously studied by Bridgman, were used for pressure calibrations, and additional points were determined, but mechanical hysteresis phenomena imposed troublesome uncertainties. Temperature measurements were made principally by thermocouples; single temperature calibration points were obtained from the melting points of pure metals, and some measurements were made with temperature-sensitive color materials.

During this period, two additional scientists joined the team, Drs. H. T. Hall and R. H. Wentorf, Jr. Both are physical chemists and both were neophytes in superpressure research, but not for long. They both were soon contributing heavily to the growing progress in techniques and instrumentation.

After initial difficulties and subsequent refinement, the "saucer" chamber became a very useful apparatus. It was the workhorse for the great range of experimental work up to pressures of about 35,000 atmospheres (515,000 lb/in^2) and temperatures limited by the melting point of the insulating inserts–approximately 2700°C. The bismuth transition at 25,000 atmospheres was definitely attainable, and pressures probably greater than 40,000 atmospheres were reached.

A large number of graphitization tests were carried out with this chamber during 1952, tests which it was hoped would establish a more accurate graphite-diamond equilibrium curve, as well as giving a reasonably accurate value for the activation energy required for the transition. However, uncertainties in temperature and pressure measurements continued to be a serious problem and, unexpectedly, determination of the onset of graphitization of diamond proved to be difficult. Different natural diamonds behaved differently. Some showed surface blackening, some became "smoky," and some remained clear and resisted graphitization even at relatively high temperatures. Experiments indicated that activation energy increased rapidly with pressure, leading to doubts as to the feasibility of direct conversion of graphite to diamond by pressure and temperature. It was believed for some time that a vapor phase reaction might be required to accomplish the transition.

NEW PRESSURE CHAMBER DESIGNS

For a while, pressure chamber design was in need of some new ideas, and many were tried. In a saucer chamber with Carboloy anvils, the maximum pressure was set by the length of stroke allowed by deformation of the pyrophyllite gasket. This gasket would not hold pressure if its initial thickness was greater than 0.030 inch, and under pressure it would extrude down to an edge thickness of about 0.003 inch. The resulting decrease in active volume due to a stroke of 0.027 inch thus determined the maximum pressure. The geometry and materials of the portion of the chamber in the immediate region of the active volume were most critical to chamber performance, for it is here that the greatest stresses occur. The outer structure of pressure vessels almost invariably employs concentric ring assemblies with tapered fits. These rings subject the innermost chamber to an extremely high initial compressive stress. While the design of these parts and the choice and processing of the materials are by no means routine, they are a manageable problem. Multistaging, that is, placing one pressure chamber within another, is a standard method of reaching higher pressures, but this approach has some inherent difficulties for the present purpose. The first penalty of multistaging is a considerable decrease in the active volume available for the higher pressure. A more important problem is the difficulty of leading the electric current required for heating into the inner chamber. For both of these reasons it was decided to concentrate on a single-stage design.

The year 1953 was characterized by intensive effort and final success in high-temperature, high-pressure chamber design. A great many modifications of the saucer chamber were attempted with the objective of increasing its stroke, but these attempts failed to extend the top pressure appreciably above the 35,000 to 40,000 atmosphere limit. Basically, a piston and cylinder will provide any desired stroke, and Strong, Wentorf, and Hall built and tested a series of piston and cylinder designs during this period. Strong carried his design of a supported-piston chamber to 58,000 atmospheres and, later in the year, to possibly 70,000 atmospheres, and he and others used this equipment for a great deal of work during the next two years.

Strong's chamber [23] permitted work in the diamond range of temperature and pressure, and it gave the group one of its two major types of pressure chambers for experimental use.

The trouble with the piston and cylinder geometry is basically the lack of lateral support for the piston—it crushes at extreme pressure. The trouble with the saucer-anvil chamber is the lack of adequate stroke. Hall pursued an alternative in this dilemma, which led him to a geometry midway between the piston-cylinder and the saucer. He reasoned that, if flow of the stone gasket was limiting the stroke in the saucer anvil, a conical gasket would provide an increase in stroke for the same decrease in gasket thickness. This resulted in the "semi-piston and cylinder." [22] Tests of this chamber were very encouraging; it was repeatedly used up to 40,000 atmospheres, and at temperatures limited only by the melting point of the inserts, as in the saucer design. Hall next applied the same concept to a symmetrical structure which was, in effect, two "semi-pistons" working into one cylinder. This chamber is known as the "belt." [22]

By employing two pistons, the net stroke was doubled, and subsequent improvements and refinement in geometry added additional volume and stroke to the belt chamber, which has become one of the most useful of the high-pressure, high-temperature chambers.

The first model of the "belt" chamber turned out by the shops performed remarkably well and easily produced the bismuth transition at about 25,000 atmospheres. A few weeks later the same unit demonstrated its potential for reaching very high pressures. The notebook of J. E. Cheney, who was assisting Hall at the time, records this run as follows:

> Helped H. T. Hall make a run in his belt anvils. We got the bismuth transition. From this data, it is not surprising that we broke an anvil on the first test, as it must have been about 100,000 atmospheres.

Cheney was evidently overenthusiastic, for this chamber employed steel parts which could not have reached much more than 60,000 atmospheres.

The single gasket initially employed in several of the experimental pressure chambers, including the saucer, Strong's "stubby piston," and the "belt" was eventually replaced by a sandwich gasket, in which two or more layers of stone were separated by metal segments. In an assembly consisting of two stone gaskets and one metal separator, it was found that each of the two stone gaskets can be 0.030 inch thick at the start of the stroke. The net result of the metal layer is thus a desirable doubling of the active volume of the chamber.

As these pressure chamber developments were being made in 1953, there were increasing opportunities to do high-temperature, high-pressure experiments, and some interesting and tantalizing preliminary results began to appear. By midyear, Hall and Wentorf were giving most of their time to high-pressure reactions. Strong's chamber was working well, and assisted by H. P. Bovenkerk, he spent increasing time on high-pressure chemistry and physics. Bundy continued to devote his major effort to adapting the best current experience to the design of chambers for the 1000-ton press which was expected in 1954. During this period, Nerad, as the senior mechanical advisor and mentor, contributed heavily from his experience to many essential details of mechanical design. Most importantly, he contributed an unvarying, firm determination to reach the project goal.

THE "SMELL OF DIAMONDS"

Wentorf undertook the preparation of sodium carbide (Na_2C_2), and he subjected this material to pressure and temperature in the presence of mercury. The result was carbon and sodium amalgam, but no diamond. He then prepared lithium carbide (Li_2C_2), and decomposed it at high temperature and pressure. In this experiment, a few tiny crystals were identified and an X-ray diffraction analysis showed a few weak diamond lines. This was the first evidence that diamonds were produced in the project. Strong's notebook record of this result says:

> We obtained X-ray photos taken on Bob Wentorf's sample of lithium carbide. These photos showed faint diamond lines. From many tests on scratching glass with his samples, and this X-ray [evidence], we now feel quite certain that diamonds were made.

Quite naturally, the excitement at this point was high, and Wentorf pursued a great variety of experiments with lithium carbide decomposition, in which he varied chemical environment, temperature, and pressure, as well as their rates of change. No improvement in the yield, or in the strength of the X-ray evidence, was produced from this work, and late in the year the line of approach was essentially abandoned. But the "smell of diamonds," as Nerad put it, left a deep impression on the group, and sustained enthusiasm over a somewhat discouraging period directly ahead.

Hall studied a lithium-lithium carbonate system next, and also found some tantalizing but incomplete evidence for diamonds. He identified tiny crystals which would scratch boron carbide and which showed faint diamond lines in X-ray diffraction photos, but as in the case of Wentorf's experiment, the identification was not positive, and further experiments with the same system failed to improve the result.

Later, Strong developed evidence for diamond growth on a seed crystal. The seed had been prepared by etching in an oxidizing medium, and its surface was covered with etch pits. Strong's notebook record states:

> I just had my first look at the diamond we heated under pressure. All traces of the pits are gone. The edges were rounded and the faces had the appearance of a contour map as though rapid growth had occurred.

The questions of solvent action and catalysis had been under consideration for a long time, and it was now becoming a very active avenue of attack. Strong's notebook discusses the solvent objective:

> The central idea in this effort over the past several months has been to provide sufficient pressure so that a molten metal can be used as a carbon solvent for growing diamond, the molten metal to act as a medium of transport from a source of carbon.

The diamond project at this time—mid-1953—had been en route for two years and was about midway on its exciting journey. We now know that we were closely approaching the tools and techniques required for final success. There is little doubt that diamond crystal had been grown in numerous experiments, but no ponderable quantity was available for comprehensive tests. The enthusiasm of the investigators was high, they were confident of success, and they were excited about the prospective results of forthcoming experiments. It is also clear in retrospect that there was no logical way to choose a unique right course to bulk diamond production at that time. There were too many unknowns and, in spite of a great accumulation of knowledge since then, there are still baffling aspects to the diamond growth process. A great deal of exploratory work had to be done to discover the best process, and it was to take about a year more to do it.

Several processes were under active investigation. In one, carbon was to be released from a carbon compound, followed by its movement through a "mineralizer" or mineral solvent, and condensation as diamond. In another, a metallic solvent for carbon was to serve as a "catalyst" for the graphite-to-carbon transformation.

During 1954, the primary activity consisted of a continuation of the study of these high-temperature, high-pressure reactions. There were some refinements in equipment—Hall obtained Carboloy anvils and cylinders for his belt which permitted operation at about 100,000 atmospheres. However, because of short chamber life at the higher pressures, most of his experiments were at pressures below 70,000 atmospheres. At this time also the new 1000-ton press was put in service. Clearly the stage was now set for some exciting work.

Strong restudied the iron-carbon system and made many runs at pressures up to 60,000 atmospheres and 1200°C. These experiments produced a thin yield of the now familiar tiny crystals which gave weak X-ray evidence for diamond.

Hall made a major effort to convert graphite directly to diamond, using the maximum pressure and temperatures available in the Carboloy belt at that time—about 100,000 atmospheres and 4000°C. In one experiment he made a short excursion up to an estimated temperature of 7000°C—and ruptured the gasket system in the belt. A striking feature of the techniques which had been developed was the extended time for which an experimental charge could be held at high temperature and pressure. For example, Hall maintained 100,000 atmospheres and 2000°C for over an hour, and could have held it indefinitely. However, no evidence of diamonds was detected in any of these direct conversion experiments.

Wentorf took a detour, to make garnet,[24] the starting material being hornblende (an iron-aluminum-magnesium silicate) from near the Gore Mountain garnet mines in north central New York.

This conversion takes place at about 30,000 atmospheres and 1200°C, and is of interest because of the occurrence of garnet in diamond-bearing strata. The economic value in the abrasive field of this conversion, however, is not impressive, and has not in the least discouraged natural garnet-mining operations.

A SUCCESSFUL PROCESS

Strong and his co-workers next embarked upon a series of experiments which were to lead to the first *ponderable* diamond to come from the project. In these experiments, a seed diamond, carefully weighed before being placed in the pressure chamber, was wrapped in iron foil and immersed in a proprietary carburizing agent containing sodium and barium carbonates. This charge was held at 54,000 atmospheres and 1260°C for sixteen hours and produced several ponderable crystals—the largest of which was 0.9 x 0.2 x 0.4 mm in size—and a net volume of diamond much greater than the seed. After many months of faint diffraction lines and microscopically detectable crystals, the appearance of a thoroughly *visible* crystal was the occasion for much jubilation. X-ray examination left no doubt about the identity of these crystals; they were diamonds, and the fact that they could be seen with the naked eye was an intoxicating stimulant to the researchers who had been waiting so long for this moment. However, attempts to duplicate this initially happy result at similar pressures and temperatures were disappointing. After much study, it was concluded that the complexity of the environment and the apparently critical dependence on temperature were probably responsible for this failure.

But this experiment by Strong and co-workers gave a valuable clue, and a feasible process was discovered only a few days later. A charge of seed diamond immersed in iron sulfide in a graphite tube closed with tantalum end disks was brought to an estimated 95,000 atmospheres and 1600°C for several minutes. When the chamber was opened, the sample from this experiment was examined with some excitement; Hall's notebook record of this event states:

> Near the tantalum disk at the top end are dozens of hard (makes terrific scratches in glass), transparent, isotropic, triangular faced crystals! The refractive index is high, luster is adamantine, and there are triangular growth pits. The material is insoluble in boiling concentrated HCl or aqua regia. It looks more like diamond than anything I've seen yet!

Hall quickly repeated the experiment, this time omitting the diamond seed, and for some reason unknown at the time, he drew a blank—there were no crystals in the chamber. Then he repeated the experiment with the natural diamond seed, and observed the original result—many shiny crystals with triangular facets. In the next few days, he scraped enough crystals off tantalum end caps to perform all of the now routine tests for diamond. The material scratched silicon carbide and tungsten carbide readily. It burned in oxygen without leaving a residue. It had a density greater than 3.325. The intensity of the work during this period is illustrated by a notebook entry by Hall several days later: "Well, I've worked over 12 hours today. Time to go home."

A few days later, the X-ray evidence was in and there was no doubt that the crystals were diamond. Subsequent experiments by Hall, and especially those of Wentorf, have bracketed the diamond growth environment and explained many of its features, although some mysteries remain. It is now known that no seed is required for diamond growth—it will take place spontaneously and quickly when the right conditions are reached. The pressure should lie in the range between 55,000 and 130,000 atmospheres for temperatures between 1400° and 2350°C. Most important is the presence of the metal interface which functions as a catalyst-solvent. The first successful material for this purpose was iron, but since then a long list of metals have been successfully used, including chromium, manganese, tantalum, cobalt, nickel, ruthenium, rhodium, and platinum. The temperature must be high enough to melt the metal saturated with carbon, in each case. With a great range of experimental conditions available through choice of temperature, pressure, catalyst-solvent, and time, it is not surprising that the size, crystal habit, color, clarity, and other properties of Man-Made diamonds are subject to a significant degree of control. Diamonds can be grown for specific industrial applications; for example, phenolic bonded cutting wheels, which are substantially better than natural diamonds for the same purpose.[25, 26, 27] Needless to say, by now, after nearly four years of manufacturing diamonds, a great deal of very detailed technology has been developed for the diamond growth process.

In relating this adventure in science, I have, in the interest of brevity, omitted much detail, some of which must now be mentioned for the sake of perspective. I have referred to blind alleys and detours. Some of them took a great deal of effort, and in the aggregate they absorbed almost as much human energy as the successful portion of the project. For example, the possibility of a low-pressure synthesis [28] persisted throughout the whole project history, and, in a somewhat emaciated form, still does. It *ought* to be possible to deposit carbon atoms on a diamond lattice in the form of diamond; for example, on a hot diamond surface in vacuo. If the diamond substrate is hot enough to ensure surface mobility of the carbon atoms, and if the nucleation of graphite can be prevented, and if the right degree of carbon vapor supersaturation is provided, the diamond lattice should grow.

Alternatively, it ought to be possible to build carbon on diamond by electrodeposition. Diamond becomes a sufficiently good electrical conductor at high temperatures, and it is known that, at low temperatures, carbon can be deposited electrolytically upon a metal from a carbonate bath. Or, it ought to be possible to decompose a suitable carbon compound and cause the carbon to condense to the diamond lattice. All of these stone walls, and many more, were beaten and battered in the course of this work, but they yielded not at all. Mother Nature has thus far insisted on high pressure and temperature, and we have had to acquiesce to her will.

It was pointed out earlier that measurements of temperature and pressure are difficult in superpressure work. There are several sources of error. Accurate measurement of temperature is hindered by the unknown effects of pressure upon temperature-sensing devices such as thermocouples.[29, 30] Accurate measurement of pressure is difficult because pressure distribution on tapered pistons is complex and does not permit a simple calculation of pressure as the ratio of force to area. Instead, pressure calibrations are generally made by plotting press force against pressures inside the chamber as measured by phase changes in certain metals. These calibrations must be performed at room temperature, but the apparatus is used at high temperatures, thereby introducing some uncertainty. The phase changes provide a reproducible basis for calibration, but uncertainty as to the exact pressures at which they occur has recently arisen. These points were originally determined by P. W. Bridgman, but it now appears that some of the pressure values should be corrected downward. Recent work by F. R. Boyd,[31] G. C. Kennedy,[32] H. G. Drickamer[33] and George Jura[34] gives lower values of pressure for phase changes of the metals, values that more nearly correspond to those obtained by Bridgman using a volumetric method[35] instead of an electrical method.

Up to this point, the pressures I have given do not take into account a correction which has since been made by Kennedy and co-workers at UCLA. It is probable that pressures formerly thought to be 100,000 atmospheres will be corrected to a new value of about 75,000 atmospheres, and that there will be smaller corrections down to the level of 24,800 atmospheres where the bismuth transition remains a firm calibration point. From here on I will be speaking of revised pressure values on the approximate basis which has been indicated.

The ultra-high-pressure, high-temperature process for forming diamonds, which at first strained existing high-pressure technique almost to its limits, has in only a few years become a routine production process which has turned out millions of carats of diamonds. As a result, this country has been relieved of a critical dependence on industrial diamonds from the Congo.

But regardless of this large measure of success, there is still a great deal to be learned about the details of diamond-forming reactions. Fortunately, tools for the study are largely developed, and the growing scientific effort in superpressure research, largely as a result of successful diamond-making, is certain to yield a great deal of new knowledge in the future.

OTHER SUPERPRESSURE RESEARCH

There is now available a vast new domain of high temperature and pressure to be explored, and the research opportunities are manifold. It is now possible in single stage apparatus employing boron nitride insulation to reach pressures of 110,000 atmospheres (1,600,000 lb/in^2) at temperatures of 3500°C and hold these conditions indefinitely.

Under transient conditions temperature excursions to much higher levels can be accomplished at this same pressure level. In two-stage apparatus, which I have not described, it is now possible to reach 160,000 atmospheres (2,400,000 lb/in^2), and the same temperatures. This environment corresponds to the temperatures and pressures found at a depth of about 250 miles in the earth. For comparison, the recently accomplished deep ocean descent was to a depth of seven miles.

As Wentorf has shown in his brief study of garnet,[24] the application of this new technology to geological research can yield valuable new information concerning the earth's structure and formation. Detailed study of the effects of experimental environment on the diamond crystal habit has been shown by Wentorf [36] to be useful in attempts to determine the origin of both terrestrial and meteoric diamonds.

Strong has measured the melting point of iron [37] up to about 100,000 atmospheres because of the importance of these data to earth scientists. For similar reasons Wentorf [38] has studied the forsterite-spinel transformation. This mineral is believed to be an abundant component of the earth's mantle at depths of 50 to 1000 kilometers. Strong, Bundy, and Wentorf have studied the fusion points and high-pressure modifications of a long list of metals and inorganic compounds up to 100,000 atmospheres.[39, 40, 41, 42] Most interesting is the transformation of boron nitride into the cubic crystal form accomplished by Wentorf.[43] The common hexagonal form of this compound is a white slippery substance sometimes known as white graphite because of its lubricating properties. These properties result from a layered structure remarkably similar to that of graphite, in which alternate boron and nitrogen atoms occur in six-membered rings in each layer. Because of these similarities Wentorf reasoned that it might be possible to condense these layers by heat and pressure to produce the cubic form, analogous to diamond. This transformation does take place in the diamond-forming environment, and the resulting cubic boron nitride, called borazon, is a very interesting substance. It has a density of 3.45 and a hardness equal to diamond. It is stable at 2000°C in vacuo and oxidizes only slowly in air.

As indicated at the outset, there were a number of reasons for wishing to collect the detailed history of this project. First and foremost was the objective of preserving the historical record. Traditionally, the history of political events and warfare between nations is recorded in minute detail, and rightly so because of the tremendous impact of these forces on the course of civilization. But we have not yet established a tradition of recording in historical detail the events of science and technology, which are increasingly playing a decisive role in human welfare and the affairs of nations. In many recent and important scientific and technological events, we have found it remarkably difficult subsequently to reconstruct the chronology and especially the relationships among multiple participants. The purely scientific publications provide certain indispensable chronological ideas, but they generally fall far short of a satisfactory historical record. I am indebted to Richard T. Miller of the General Electric Research Laboratory for his painstaking study of all of the original records of diamond-making in the Research Laboratory which has provided the factual background information for this presentation.

A second reason for presenting this case history of the diamond project is that it provides an admirable example of scientific teamwork. Since the realities of teamwork effort tend to obscure individual recognition, we wished to show the truly great individual contributions to diamond-making in appropriately sharp focus. Although I have stressed the essential team effort in this fascinating project, I have very incompletely credited many coaches, mentors, and counselors whose advice and guidance were essential. Dr. A. L. Marshall was the senior chemist whose experience was invaluable through the entire project. Drs. Herbert Hollomon, John Fisher, Richard Oriani and many of their metallurgical associates played indispensable parts in the program. And the work could not have succeeded without the skill and perseverance of some of the most able people in our shops. It is clear, I think, that a structure of accomplishment of this kind is built one section at a time. The top of the edifice is supported by, and could not be reached without, the solid foundation elements. Each member of the team contributed to the structure, as did many prior workers in superpressure research.

REFERENCES

1. Bundy, F. P., H. T. Hall, H. M. Strong, and R. H. Wentorf, Jr., *Nature,* **176,** 51 (1955).
2. Bovenkerk, H. P., F. P. Bundy, H. T. Hall, H. M. Strong, and R. H. Wentorf, Jr., *Nature,* **184,** 1094 (1959).
3. Bridgman, P. W., *Scientific American,* November, 1955.
4. Eyring, H. and F. W. Cagle, Jr., *Z. Elektrochem.,* **56,** 480 (1952).
5. Moissan, H., *C. R. Acad. Sci.,* Paris, **118,** 320 (1894) and **123,** 206, 210 (1896).

6. Hannay, J. B., *Proc. Roy. Soc.*, **30**, 450 (1880) and *Nature*, **22**, 255 (1880).
7. Parsons, C., *Proc. Roy. Soc.*, **79**, 532 (1907); *J. Inst. Metals*, **20**, 5 (1918); *Phil. Trans. A.*, **220**, 67 (1920).
8. Desch, C. H., *Nature*, **121**, 799 (1928).
9. Despretz, C., *Bibl. Univ. Geneve*, **24**, 281 (1853); *Compt. Rend.*, **28**, 755 (1849), **29**, 48, 545, 709 (1849), **30**, 367 (1850), **37**, 369, 443 (1853).
10. Ruff, O., *Z. anorg. allgem. Chem.*, **99**, 73 (1917).
11. Guenther, P. L., P. Geselle, and W. Rebentisch, *Z. anorg. allgem. Chem.*, **250**, 357 (1943).
12. A lengthy list of references, dating back to 1828, is given in S. W. Mellor's *Comprehensive Treatise on Inorganic and Theoretical Chemistry*, Vol. 5 (Longmans, Green & Co., London, 1924).
13. Moissan, H., *Compt. Rend.*, **118**, 320 (1894), **139**, 773 (1904), *The Electric Furnace* (Edward Arnold, N. Y., 1904).
14. Nininger, H. H., *Out of the Sky* (Dover, 1952); *Arizona's Meteorite Crater* (American Meteorite Museum, Sedona, Ariz., 1956).
15. Urey, H. C., *Astrophysical Journ.*, **124**, 623 (1956).
16. Bridgman, P. W., *J. Chem. Phys.*, **15**, 92 (1947).
17. Bridgman, P. W., *The Physics of High Pressure* (G. Bell & Sons, London, 1952).
18. Rossini, F. D. and R. S. Jessup, *J. Res. Nat. Bur. Stand.*, **21**, 491 (1938).
19. Simon, F., *Handbuch der Physik*, Vol. **X**, 350 (Berlin, 1926).
20. Berman, R. and F. Simon, *Z. Electrochem.*, **59**, 333 (1955).
21. Bridgman, P. W., *J. Appl. Phys.*, **12**, 461 (1941).
22. Hall, H. T., *Rev. Sci. Instr.*, **31**, 125 (1960).
23. Wentorf, R. H., Jr., *Modern Very High Pressure Techniques*, p. 14 (Butterworths, Washington, D. C., 1962).
24. Wentorf, R. H., Jr., *Am. J. Sci.*, **254**, 413 (1956).
25. Kennedy, J. D., *American Machinist*, Mar. 23, 1959.
26. Mueller, J. and J. Ripple, *American Machinist*, Mar. 23, 1959.
27. Nisula, C. A., *Grinding and Finishing*, Feb. 1959.
28. Oriani, R. A. and W. A. Rocco (unpublished work).
29. Hall, H. T., *J. Phys. Chem.*, **59**, 1144 (1955).
30. Strong, H. M., *American Scientist*, **48**, 58 (1960).
31. Boyd, F. R. and J. L. England, *J. Geophys. Res.*, **65**, 741 (1960).
32. Kennedy, G. C. and P. N. La Mori, *Progress in Very High Pressure Research*, F. P. Bundy, W. R. Hibbard, Jr., and H. M. Strong, eds., John Wiley & Sons, Inc., New York (1961).
33. Balchan, A. S., and H. G. Drickamer, *Rev. Sci. Instr.*, **32**, 308 (1961).
34. Jura, G., private communication (to F. P. Bundy, 1960).
35. Bridgman, P. W., *Proc. Amer. Acad. Arts and Sci.*, **74**, 425 (1942) and **76**, 55 (1948).
36. Wentorf, R. H., Jr., *J. Phys. Chem.*, **63**, 1934 (1959).
37. Strong, H. M., *Nature*, **183**, 1381 (1959).
38. Wentorf, R. H., Jr., *Nature*, **183**, 1617 (1959).
39. Strong, H. M. and F. P. Bundy, *Phys. Rev.*, **115**, 278 (1959).
40. Bundy, F. P., *Phys. Rev.*, **115**, 274 (1959).
41. Bundy, F. P., *Phys. Rev.*, **110**, 314 (1958).
42. Mackenzie, J. D., W. L. Roth, and R. H. Wentorf, Jr., *Acta Cryst.*, **12**, 79 (1959).
43. Wentorf, R. H., Jr., *J. Chem. Phys.*, **26**, 956 (1957).

Progress and Prosperity

An address at the Ninth Annual Management Conference, University of Chicago Graduate School of Business, March 1, 1961.

I shall begin this discussion by qualifying the rather general title that I have selected, for I wish to be as explicit as possible in considering certain complex economic questions that are of major concern today. By progress I mean the technological progress that extensive scientific research and development make possible; and by prosperity I mean the economic growth that technological progress, in turn, largely stimulates and supports. The operation of these related factors—research, technology, and economic growth—will be my central theme.

Some of you, moreover, may have guessed that I selected such an optimistic title as a play upon the rather negative "Progress and Poverty" that Henry George chose for his historically interesting treatise of an earlier day. It is significant that it was not until long after George's time that large-scale and systematic exploration of science became a powerful force for economic growth, and not until even later that this force began to be recognized.

I have no intention of arguing the pros and cons of George's thesis which has suffered some obsolescence. Simply stated, George's belief was that land tended to usurp income that properly belonged to labor and capital; and that the benefits of progress were reaped by a few, at the expense of many. The main point is that when George wrote, eighty-odd years ago—and even up to much more recent times—economic growth was viewed only in terms of land, capital, and labor. Perhaps, with the emergence of the new economic force of planned and purposeful scientific research and development, some optimism in present analyses of economic growth may be justified, and the concept of progress and prosperity brought closer to realization.

ECONOMIC GROWTH

However it may have been viewed in the past, economic growth has always been an important subject, but it is of particular importance at the present time. For the last five years the United States economy has been faltering and hesitant, while vigorous economic growth has been taking place at numerous points overseas. In the modern world our economic vigor is not only a matter of domestic concern; we are becoming increasingly aware of the international political impact of domestic economic strength or weakness. I certainly shall not attempt an appraisal of this economic cosmos in its political complexity, but for the purpose of perspective it will be necessary to discuss some of the broad industrial economic framework within which technological innovation operates as an economic force. My point of focus will be technological innovation, and I shall approach this sufficiently complex subject by examining certain cases drawn from actual experience, in which it is convenient to observe the relationships between technological innovation and economic growth. Let me emphasize that in the time available I only hope to introduce the subject, and to discuss it to some extent, but any conclusions expressed will be quite tentative.

But let us first return to the broad framework of economic growth. I have already mentioned the challenge to United States industry that springs from rapid industrial growth elsewhere—especially in Western Europe, Japan, and the U.S.S.R. It is important to remember that, outside the Iron Curtain, this competition is not only for sales but also for capital, and that we are witnessing at present some flight of United States capital to the strong magnets of rapid growth rates abroad.

We must also remind ourselves that these external considerations, involving such vital matters as world confidence in the democratic as opposed to the totalitarian economic system, are closely related to an internal problem that is generic to industrial growth: that of utilization of the working force. The bumper crop of war and postwar babies will enter the working force in the next five years, and will increase our working force by 1.1 million each year. It is obvious that economic growth must at least keep pace with this rise in the working force if the United States standard of living is to maintain its present level.

In view of popular concern over so-called automation and its impact upon employment, it is well to remember that the historic trend of increasing efficiency in the production of goods and services has always been associated with more and better jobs. Automation is merely the latest phase of this long-term progress in industrial productivity. In the past, as mechanization has assumed the functions of routine physical labor, the production of goods at lower unit costs has expanded the market and created new job opportunities to more than balance the lower labor content per unit of production. With automation this trend continues, and ever more complex production functions are being accomplished by machine and control equipment. It is inevitable that this overall progress in production efficiency will be accompanied, as in the past, by the inconvenience of changes in requirements for worker skills and the consequent necessity for worker retraining, and by shifts in plant locations and other disruptions.

In an economic context, however, the drive for production efficiency is a basic necessity for competitive survival. Worldwide industrial developments of the last five years make it crystal clear that the competitive arena for much of American industry is not only the domestic market, but also the world. In many product areas the improved production efficiency available through automation is the only countermeasure available to combat drastically lower labor costs overseas. It is also worth noting that automation has been in operation much longer than most people realize: witness, for example, telephone services and the process industries, particularly in chemical and petroleum production, where automated operations are already far advanced.

Let us now take a closer look at the economic force exerted by living standards. It is a well-established American tradition that standards of living continue to rise, and we can expect that this tradition will in itself exert force upon the economy. Perhaps the easiest measure of living standard is the wage scale of the total working force, expressed in constant dollars. For a long period of years we have experienced a fairly general rise in this wage scale made possible by the growth trend of the overall economy. However, in a situation such as the present one where the economy is on a plateau, pressure for increasing wages continues. If this cannot be resisted, obviously, something will have to give, and the casualty will be the value of the dollar, or the return upon investment, or both. Of these two unpalatable eventualities, perhaps the latter—return upon investment—may suffer most. This is because campaigns to educate the working force with regard to the fallacy of inflation have succeeded to some degree in convincing most workers that more—but inflated—dollars in the pay envelope are an illusion, and that the real measure of a rising pay scale is a rising purchasing power.

Consequently, if a rise in the real purchasing power of wages were forced upon a static economy, return upon investment would necessarily decline. The result would contribute to the flight of American capital. This is admittedly an oversimplified picture of the operation of the economic complex, but it still leaves little room for doubt that we require sound economic growth to support a genuine rise in the standard of living.

SCIENTIFIC RESEARCH AND DEVELOPMENT AS AN ECONOMIC FORCE

In the early history of the country, the emergence of economic commodities was basically related to the expansion of our frontiers. As the land frontier moved westward, the resultant new farming areas, new transportation and communication facilities, and new grazing, mineral, and forest resources provided the driving force for the generation of new wealth and new employment. In a later stage, as manufacturing and marketing concepts were developing and increased efficiencies were being achieved in the production and distribution of goods, increased yields for invested capital and labor provided increased wealth and a growth in employment. Our present economic niveau is characterized by the emergence of a new economic force provided by the large-scale exploration of science, which yields a source of new technology for the creation of new and improved products and services—both for the civilian economy and the defense establishment.

The significance of this development has not gone unnoticed, as I indicated earlier. Current economic theory has not only added scientific research and development to the classic economic forces of land, labor, and capital, but also it is now beginning to assign to scientific research and development the major role among the forces which will in the future account for the overall growth of the economy.

This latest trend in economic thinking has been noted with interest by industrial scientists also, because the abundance, promise, and fascination of scientific exploration have long provided motivation and challenge for their efforts. They see striking current examples—and the promise of many more in the future—of new and large businesses and even industries, such as the polymer industry, the many-faceted electronics industry, and the semiconductor industry, arising from new technology based on new scientific knowledge.

To be sure, these modern technical industries exploit native raw materials as did primitive industries; but in many cases the similarity is superficial. The ratio between the value of a technical product and the cost of the native raw materials—for example, raw materials like coal, hydrogen, oxygen, germanium, and silicon—may be large, and in some cases may be very large indeed. In other words, the *value added by technology* is predominant in the products of many new technical industries, although, of course, the native raw materials—silicon or carbon, for example—are still essential raw materials for the technical product. In some technical products the expenditure for research and development far outstrips raw material cost as the most significant element of product cost. On balance, Research and Development is thus the principal raw material from which many new business enterprises are being created.

Let us now turn to some specific case histories that bring out the actual relationships between technological innovation and economic growth in modern technological business ventures. The relationships involving human effort should be of primary interest, so let us examine this aspect first. Figure 1 shows typical employment for research, development, and engineering, and for the resultant manufacturing and marketing over a period of time. This typical employment picture is based upon four projects that were undertaken at the General Electric Research Laboratory, and were carried to manufacturing activity by the company.

We see how a relatively small research effort provided the basis for a second, somewhat larger development and engineering effort which, in turn, accomplished the transition to a manufacturing and marketing effort employing a still larger and growing group of workers. The role of industrial science in nucleating economic growth is here shown in an actual relative magnitude for currently important new technical business enterprises.

This illustration does not show, however, the important relationship of the financial factors concerned with these four projects. These relationships in terms of *net cash flow*, for the same four ventures, are shown in Figure 2.

We separately identify research expenditures in this figure, but because of the problem of accounting definitions, we do not attempt

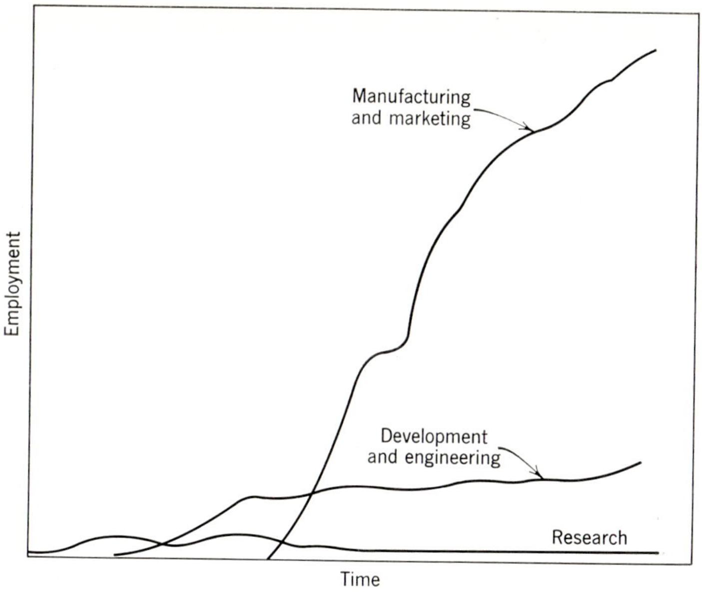

Figure 1. Typical employment in research-originated enterprise (based on four case histories).

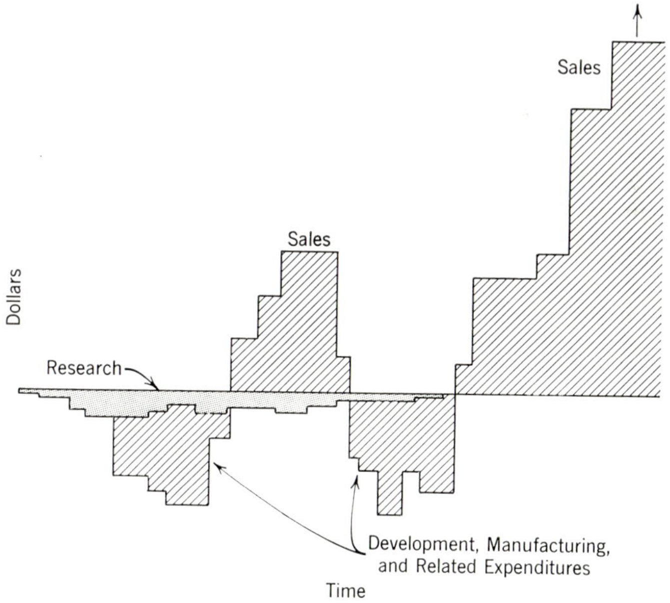

Figure 2. Typical cash flow in research-originated enterprise (based on four case histories).

to separate completely the other contributions to cash flow, but lump them together. The lumped quantity includes, as negative elements, the expenditure for development and engineering, for manufacturing facilities, for market development, and for related expenses. The sale of products generates positive cash flow, which at some point leads to a net operating profit position for the enterprises. It is only much later that a net positive cash flow for the entire group of ventures is achieved.

EFFECTIVE UTILIZATION OF SCIENTIFIC RESEARCH AND DEVELOPMENT

If growth by technology is to be a pattern for economic progress—and I agree with the economists who see this as an increasingly important factor—then we must increasingly learn how to understand and manage our technical resources upon which this progress is based. I shall, therefore, consider a few of the interesting questions, including unsolved problems, related to the scientific and technological factor and its potential for industrial and economic growth.

The requirements of a manufacturing business for raw materials, manufacturing facilities, process knowledge, engineering designs, marketing plans, and the organizations to manage them are generally explicit considerations in the planning of a business. However, only in recent years have we approached the corresponding degree of specific planning for the integration and organization of the technical resources of research and development required for the growth of new businesses. It would be fair to say that numerous unsolved problems in this area continue to present challenges and opportunities for industrial management. Although the broad relations between industrial scientific research and resultant business enterprise are quite clear, the specifics, as they apply to a given area of business, are frequently quite unclear. We are, of course, surrounded by evidence that new scientific knowledge has frequently led to new technology, which has generated completely new areas of business opportunity or even new industries. This information provides historical statistical evidence for the general thesis of economic growth through technology, but it does not lead automatically to safe and reliable guide lines for use in the planning and organization of future progress on this road. This is because we are still in a learning phase in technological business growth, and there is a great deal that we still need to learn.

The discussion up to this point has emphasized the place of technological innovation in the growth of entirely new businesses. It would be a mistake, however, to overlook the less spectacular, but equally impressive and important, rejuvenation of old products and businesses by new science-originated technology. An example is the electron tube industry which has refused to "fold up" in the face of the versatile competition of semiconductor technology, but has on the contrary extended its own technology into new fields.

As a rough generalization, to which there appear to be many exceptions, the technology of new *materials* and *processes* frequently rejuvenates existing products and businesses, while the technology of new *phenomena* frequently stimulates completely new business enterprise.

At this point there is an interesting question of semantics. One cannot be sure in some cases whether one is dealing with a new business, or a revitalized old one. Take for example the business of Man-Made industrial diamonds. This venture as a company activity has all of the earmarks of a great growth business. On the other hand, industrial diamonds—the natural rather than the man-made variety—have long been an established product business. From this viewpoint the new growth business rests on a *process* development—a new process for making an old product.

In spite of all of the evidence of the economic impact of industrial scientific research, it must be observed, at least in passing, that the management of research remains somewhat mysterious and metaphysical. "Playing by ear," rather than from a score, is commonly practiced, and many believe that the intrusion of management practices in the research area can result only in the withering of the delicate flower of creative thought and output. In truth, the application—to research—of management practices that have been developed primarily for production operations will have exactly this distressing effect, and many examples of this misdirection could be cited. However, the aesthetics of science and the details of science management are extensive subjects in themselves, and they cannot adequately be developed here. The complex, point-and-counterpoint relationship of industrial science, technological innovation, and opportunities for business enterprise is sufficient to engage us in this discussion, so we shall pass on to another aspect of this relationship—that of the systematic planning of business growth via technological innovation.

INDUSTRIAL SCIENTIFIC RESEARCH IN RELATION TO OTHER INDUSTRIAL FUNCTIONS

The systematic planning of industrial scientific research, as the origin of opportunities for business growth, is essentially a problem of *balance* and *scope*. The *balance* involves the allocation of available resources to the many competing industrial functions, all of which are required for the performance of a successful business venture. The *scope* concerns the identification, analysis, and planning of the industrial research work in relation to the state of the technology and the resultant technological needs of the business, and both in relation to the state of world science. The planning for technological business growth must extend well beyond the boundaries of the research and development functions, however. This is because the transition from a science-originated opportunity to an economic development is frequently extremely demanding of all industrial functions and resources.

For the purpose of discussion let us consider some idealized relationships between research, development, and engineering, a product business, growth and decline of sales, and net profit. (See Figure 3.)

Preceding the sale of products, expenditures for research, development, engineering, and manufacturing facilities generate a negative cash flow which is a measure of the initial investment in a business. At some following period, expenditures for marketing take place, which further contribute to the negative cash flow. Finally product sales are achieved.

It is evident that the scope of the development and engineering effort, and the manufacturing activity, need to be appropriate in amount, in kind, and in timing to achieve a planned level of output. Moreover, a suitable organizational framework must be provided within which these functions can operate effectively.

The sales curve shown, commonly called the S curve of business growth, may rise steeply in the initial period of sales and require that the business not only reinvest all of the retained earnings but also that it obtain additional financing for necessary expansion.

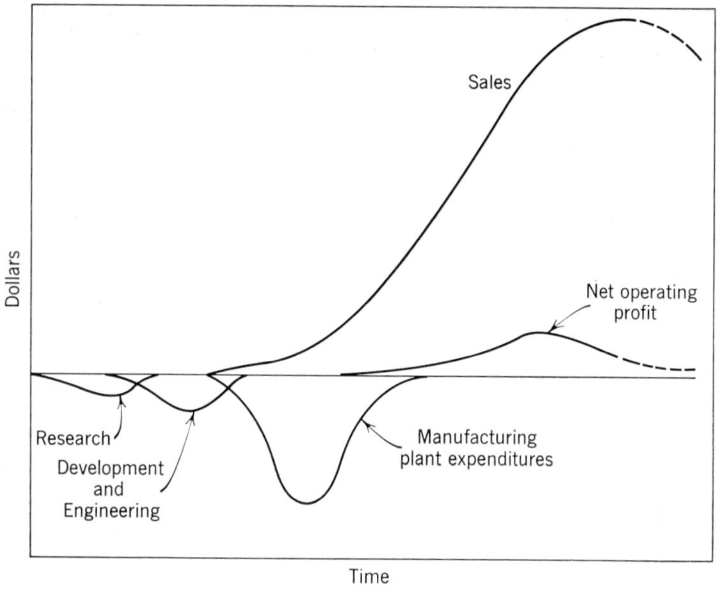

Figure 3. Idealized relationships among various industrial functions.

The time scale associated with these theoretical quantities may be very great, as in the case of the development of a major power system such as a reactor power plant. On the other hand, both quantities may be quite small, as in the case of consumer products with a high obsolescence rate. Whatever the period required to reach the position of an operating profit, a still longer time is required to achieve a *net positive cash flow* for the entire product or business development investment.

Although all of the expenditures for Research, Development, and Engineering (RD & E) and for manufacturing facilities are "investments in development of the business," they are taxed quite differently. Corporation tax laws permit RD & E expenditures to be treated as current expenses of the business, which in truth they are, while expenditures for facilities are amortized over a period of years on a formula basis. If a corporation is in the 52 per cent income tax bracket, which with minor deductions applies to all businesses with annual net profits in excess of $25,000, the tax treatment of RD & E expenditures thus appropriately acts as a stimulus to economic growth.

THE DIVERSIFIED VERSUS THE SPECIALIZED ENTERPRISE

There are some interesting comparisons between the way in which a new business grows as a separate enterprise on the one hand, and how it grows within the structure of a diversified corporation on the other. A diversified corporation, in the first place, must invest in new businesses, preferably with the promise of high growth rates, to balance the fact that its older business operations may be high on the S curve, or even declining. In passing, however, this frequently-referred-to S curve of business growth (see Figure 3) is in need of some analytical consideration. Any theoretical S curve of business growth should, of course, show a declining region and should finally come to zero.

Some hesitation at this point, indicated by the dotted portion of the curve in Figure 3, is justified by the fact that some businesses in some industries have never actually experienced this decline during the history of the industry. In some cases this is clearly due to the fact that the constant infusion of new technology progressively modernizes the business and counteracts obsolescence. In other cases, competing technology fails to develop effectively to challenge the business. For example, in the electrical industry, one of the oldest product businesses is that concerned with electric motors. Actually, we identify many different motor businesses according to motor size, technology, and the markets. All of these businesses have been growing since the birth of the industry, and they continue to show great promise for the future.

For the moment, however, let us overlook this fact of "durable" businesses which fail to go over the hill and concentrate on mature declining businesses, examples of which certainly can be identified. These "mature" businesses may be operating in an area of diminishing returns with respect to product, process, or marketing innovation, as a result of which they grow at a reduced rate, or decline. Because the investment requirements are also diminishing, the mature business may be in a period of relatively high profits, and in fact *must be* in a relatively profitable phase to balance the long, early period of negative cash flow.

Thus a diversified company which is financing growth from retained earnings can "afford" new rapid-growth businesses by virtue of the mature profitability of its older businesses. Furthermore, because these older businesses will eventually go downhill and decline both theoretically and practically, the company must invest in new growth ventures as a simple matter of long term survival.

The choice of investment in a "mature" business versus investment in a new business, is a question of unusual complexity which faces not only the amateur and professional investor. This question also presents a constant dilemma for the industrial manager. If one could be sure that a business or a product is mature in terms of future growth potential, life in industry would be much simpler. It is, however, surprisingly difficult to determine when a business is "mature" in the intended sense.

Not all cases are as clear, for example, as the product business in horse-drawn vehicles. Here, because of a source of obsolescence which is extremely evident on every highway, the market appears to be in a definite decline. Another case in which it is possible to make an almost certain determination of product obsolescence is that of feminine wearing apparel. No matter when these perishable products are introduced or sold, it can be said unequivocally that severe obsolescence is about to set in, or more likely, has already taken place.

Except for these crystal clear cases–and a few others could be mentioned–there is a vast range of industrial products and services where a determination of a locus of the business, with respect to a theoretical S curve, is a highly uncertain and speculative matter. Innumerable examples could be cited of apparently old businesses which have suddenly taken off for a new, higher level of activity. In many of these cases new technology, or the rejuvenation of old technology by new materials or processes, has provided the stimulant for new growth. Hence it is quite difficult to marshal facts to permit a calculated choice between extending investments in an old business or investing in a new business believed to have high growth potential.

There are, of course, spectacular cases of completely new growth businesses, such as semiconductor products where, for the first ten years of the industry, sales nearly doubled every year. There are also, as indicated, multiple examples of very old product businesses, such as the fractional horsepower motor business, where significant innovation and vigorous growth is still taking place.

Up to this point, I have been discussing these matters in the frame of reference of growth within a large and diversified company. A striking feature of the present industrial scene is the rapid emergence and growth of new technological business through private ventures. This is sometimes in addition to, and sometimes alternative to, growth of a similar business within existing corporate enterprises. Some interesting differences in these alternatives can be seen, and are worth discussing in any consideration of economic growth.

It might be that the financing would present differences. However, the traditional problem of privately financing a new technical business venture can be largely discounted in the present space-age climate. It is my impression that there is plenty of risk capital available, and that if a business proposal fails to obtain financial support, its qualifications are extremely nebulous or its sponsor is extremely inarticulate, or both.

On the other hand, the same technical business proposal which has to make its way within a framework of diversified corporate resources and opportunities has to meet a much more experienced and hence critical audience. Competing internal investment opportunities are weighed in the balance, and necessarily a high "internal discount rate" is applied to proposals. This usually leads to some impatience on the part of the project sponsors. The impression is created that it takes too long for the large company to get under way with its growth ventures, and that the private entrepreneur has the advantage of mobility and sharp focus in developing a growth business. Of course, it is entirely possible that, from his point of view, the private entrepreneur might completely reverse the advantage I have expressed.

Wherever the advantage of optimum mobility and business judgment may lie, the success ratio for corporate growth ventures which are finally undertaken appears to be quite high. I cannot prove it, but I have the impression that the success ratio for private growth business ventures may be quite low, contrary to a general impression in the financial community. This impression may exist because, when misfortune befalls an embryo business venture, the interment service is usually a quiet family affair, attended only by the next-of-kin. While the occasional spectacular successes tend to make very good news and headlines, the obituaries tend to become newsworthy only as statistics.

The industrial manager is, in a true sense, a financial investor. For some of the same reasons that are important to investors, he must view alternative opportunities for the direction of his effort and resultant expenditure and investment in perspective. This perspective includes the fact that not all proposed growth businesses experience high growth rates. Many of them show a high mortality rate as well, and this probability must be appraised and evaluated for balance.

Of course, the safest way to evaluate the potential of new technology for business growth or development is in miniature—in laboratory scale if at all possible. In some areas of technology—for example, in chemical process and material development—it is frequently quite feasible to explore the proposed business "in miniature scale" and to arrive at conclusions which are applicable to a substantially scaled-up operation. This is ideal and economical. If a project interment is finally necessary, it can be accomplished at much lower cost at test tube, rather than pilot plant or factory, scale.

INDUSTRIAL RESEARCH AND NATIONAL DEFENSE

I shall conclude these remarks with a brief discussion of the relationship between industrial research and defense capabilities, and of the impact of defense RD & E on the economy.

There is no doubt that the scientific sophistication and the overall capacity of our industrial complex play a decisive role in the strength of our modern defense establishment. Yet a popular misconception has somehow come into being that defense-oriented research and technology determine, or at least are primary factors in, the advance of nondefense industry. If this were so, the tail would certainly be wagging the dog.

It is true that defense RD & E expenditures are large in the national total and that a large portion of these expenditures are made in industrial establishments. However, we must bear in mind that the primary object of these expenditures is the development of defense capabilities and hardware. It is also a fact that there is an extensive and mutually beneficial interaction between defense technology and nondefense technology; in some fields the two are so interwoven as to defy separate identification. Illustrative cases are electron tube and semiconductor technology. The field of computer development might fit into this category also.

However, I have the impression that, in view of the remarkably large magnitude of defense RD & E, there has been a surprisingly small amount of technological by-product which has provided nondefense consumer and industrial products and services. There *is* a general feeling that the defense contractor by virtue of his defense work acquires access to a rich opportunity for civilian products and markets. But this feeling does not appear to be justified by the facts. Perhaps it is engendered by the relatively small number of spectacular exceptions. By and large, however, the defense contractor acquires access to the technology of defense capabilities and hardware, and he must make his living in the defense market.

This is not to minimize the importance and economic impact of the business of defense materiel and services. The defense business includes not only the building of defense hardware, but also the large and growing business of defense RD & E, which now furnishes employment to a very large fraction of our technical manpower. All of this is a primary factor in the economy. But my point is that its by-products do *not* provide a *primary* source of technology for our nondefense industry.

To be somewhat more specific in this matter, let us consider the nondefense products or businesses which have grown from the "civilian" application of defense-originated technology. We can start the list with the civilian air transportation business, which is certainly a prime example. It is true that the aircraft manufacturing industry also depends heavily upon technology developed for the defense establishment, but this industry relies upon defense business rather than "civilian" business, for its *primary* sustenance.

By-products of World War II include microwave communication and radar. Both have led to nonmilitary businesses, but these have been and continue to be small in comparison to the defense businesses in the same or closely-related products. The computer business has had a heavy infusion of technology developed from the application of computers to weapon systems, and is perhaps a good example of the kind we are seeking. In the field of electronic devices and components, particularly electron tubes, the situation is somewhat involved. Here the technological origins preceded World War II, and the early developments were unrelated to the military. The wartime electron tube developments were extensive and significant, but their present applications are principally in defense systems and equipment.

The case of synthetic rubber may also come up in any discussion of this subject. I believe, however, that this is an example of industrial pump-priming under the pressure of wartime needs, and is not a technological by-product. Atomic power originated in the wartime atomic bomb development, but it has by no means been a simple by-product of the latter. To beat this particular sword into a plowshare, it has been necessary to throw away almost all of the sword, and start with new raw materials. We may have used the small atomic escutcheon on the hilt of the sword, but that's about all.

This impression—that defense technology is specialized in its application, and is limited principally to defense—is sustained by looking around one's home, one's community, the farm, and the nation's highways. We see that the satisfaction of the needs for food, shelter, clothing, and local transportation looms large on the national economic balance sheet. Yet in these important segments of human activity defense technology plays a very small role.

Defense requirements are no less specialized today than they have always been, and these requirements are primarily—and necessarily—sharply focused on defense. It is, for this reason, perhaps *not* surprising that the applications to nondefense industry have not been greater and this situation may be expected to continue as long as defense technology accomplishes its vital, but highly specialized, purposes.

Finally, in relation to defense let us remember that, under the rules of the cold war, our security is not determined by military materiel and manpower alone. The growth and vitality of our domestic economy, as we have seen, is also a matter of major international political impact. Now that industrial scientific research and development has emerged as a major factor in industrial and economic growth, we must increase our efforts to determine how this force can best be nurtured, and how it can most effectively be employed.

The New Engineer and His Scientific Resources

At the Eta Kappa Nu Society on the Occasion of the 25th Anniversary of the "Outstanding Young Electrical Engineer" Award, New York City, New York, January 29, 1962.

The engineer has long been characterized as the "middleman" between science and society. He is uniquely able to translate new knowledge and understanding—which is the output of science—into design capability for our modern civilization. *The dominant characteristic of this modern civilization is rapid change.* Inevitably, therefore, the engineer is in the swiftest current of the stream of events which is creating rapid change in the world, and it is not surprising that he, himself, should be caught in some of the turbulent eddies of this stream.

ACCELERATED CHANGE

Modern civilization is characterized not only by *change*, but also especially by *the rate of change of change*. In an earlier historical period, rapid changes in the world were created largely by the impact of wars and pestilence; between these twin scourges the status of the status quo varied hardly at all. The farmer who learned to farm could farm productively throughout his lifetime. The artisan, the tradesman, the shopkeeper, the baker, and the candlestick maker could pursue their livelihood in a relatively static social and economic environment; certainly their world was not dominated by change. But in the modern world change has become the order of the day. This is expressed simply and effectively by some lines attributed to Ogden Nash, which I quote in part:

When Dickens wrote his Christmas Carol
Space suits were unforeseen apparel.
Christmas has changed since the days of Dickens;
The plot, when revised by Science, thickens.

As civilization has moved from the time of Dickens to the present, man has become remarkably conditioned to change, particularly if it does not seem to affect him personally, and at the moment. In fact he appears to have become somewhat blasé about innovation, and tends to be impatient if the pace seems to falter somewhat. Many events have set the stage for the contemporary acceptance of change, and most of them have had their roots in science and technology. As evidence of our conditioning to a world of change, consider the dramatic advent of the atomic age. The shock of this event, and the appraisal and appreciation of its impact and consequences, took, one might estimate, about ten years. Since then an event of possibly equal significance has burst upon the world, and now monopolizes public attention: the space age. In this case, I judge, it has taken only two or three years for people and nations to accommodate their thinking to an equally dramatic historical and technological milestone. Even space travel, as a logical consequence and concomitant of the space age, has been accepted as though it were now a part of our way of life. It seems that only a few details remain to be worked out by the engineers.

INNOVATION IN INDUSTRY

The acceptance of change as a way of life is nowhere so complete as in industry, where innovation is the watchword of the day. As evidence of this, it may be noted that some companies even identify progress itself as being more important than specific products or services, while other companies talk about better things for better living through chemical innovation, and still others claim that research is the key to tomorrow.

As suggested, the rapid reduction of new knowledge to products and practices by engineering has created a dilemma for the engineer himself. The question has been raised as to whether it is realistic to believe that an engineer's stock in trade—acquired during the years of formal education—will suffice for a professional career. This has interested educators for some time, and is a matter of particular concern in the engineering colleges of the country.

At this point I should like to examine in some detail the idea of obsolescence, in engineering training and practice, and, to put the matter in focus, consider two extreme and opposite points of view. One extreme is that obsolescence of engineering and engineers is a problem of crisis proportions that threatens to demoralize the profession. This point is usually proven with the aid of what *Fortune Magazine* might well characterize as "unknowable statistics."

The statistics show, it is said, that since the "doubling time for technical information is ten years or less," in a professional career of three or four decades the original investment in formal professional training will become "watered down" or diluted to almost nothing. A moment's thought, however, will show that there may be some defects in this argument.

Let us consider the other extreme point of view. This one is embodied by the mature engineer, thirty or more years out of college, who is completely effective and happy in his work. Moreover, his employer and the organization with which he is associated are also happy about his continuing contributions and output, and his prospects for further progress. This engineer might very well say to himself in a contemplative moment, "If I'm obsolete, why do I feel so good?"

Having thus posed the problem in somewhat theoretical terms, let us consider the circumstances that confront the engineer in real life and in actual engineering practice. In spite of some reservations concerning the statistics, we accept as a completely valid concept the proposition that new engineering knowledge and practice *is* being generated at a high rate from new scientific and technological sources. Let us consider some specifics bearing on this point.

A 25-YEAR VIEW

Since this is a Twenty-fifth Anniversary occasion, it seems appropriate to look back over the past twenty-five-year period for examples of how rapidly the sophisticated subjects of science *have* made the transition to practicing engineering disciplines.

As a reference point, let us take a look at the scientific world of just twenty-five years ago: 1937. It makes fascinating reading. Among other things, we learned in our historical review that during 1937 a bill was introduced in Congress which would have prohibited the granting of any patent on a labor-saving device. Apparently this proposal failed in committee.

In 1937 nuclear physics was a fashionable subject for scientists; but certainly no engineering professor in the land at that time was so visionary and lacking in judgment as to be considering the future teaching of nuclear engineering, which has since become a large field of opportunity. The chemical engineer of 1937 was principally concerned with oil refining and the manufacture of inorganic heavy chemicals. Today, there is superimposed upon these earlier fields a series of chemical industries that are concerned with the manufacture of synthetic organic materials, replete with engineering opportunities in the production of antibiotics, synthetic hormones, insecticides, plant growth regulators, detergents, synthetic elastomers, fibers, and structural polymers. Just to remind ourselves how long ago 1937 was, consider that in 1937 the announcement of nylon was still two years in the future. Although polymers were in production prior to nylon, the modern polymer industry was ushered in by the nylon stocking, which everyone will agree was a very important and eye-opening event.

Although the idea of computing machines is ancient, the first public demonstration of an electronic computer took place in 1945. Subsequently, the impact of computers upon industry as a whole and upon engineering practice in particular has become one of major proportions in multiple areas.

A microwave engineer would have been lonely, miscast, and unemployed in 1937, for the principal hardware in his trade was produced during World War II, and has only since then proliferated in engineering practice.

Even later, solid-state and semiconductor engineering have arisen and grown to large new fields that only dreamers could have believed plausible in 1937.

There is only time to mention in passing the large fields of engineering that have arisen since 1937 in aircraft jet propulsion and in rocket and space propulsion, in plant hybridization, in high vacuum practice, in information and control systems, and in many other areas.

A TENFOLD INCREASE

A comparison of the technical program of the AIEE Winter Convention of 1937 and the meeting this week adds substance to the point being made. In 1937 the program consisted of thirteen technical sessions, and the program chairman was distressed by the necessity for concurrent technical sessions. With but few exceptions all of the technical papers were concerned with these five subjects: power transmission, rotating electrical machines, protective devices, communications, instruments and measurements.

In striking contrast, the meeting this week includes 128 technical sessions, up by nearly tenfold, and on the average there are fourteen parallel sessions in progress at one time.

Although the subjects discussed in 1937 were all represented by papers presented on this week's programs, there is a heavy flavor of atomic and space-age technology. Witness the titles: *Digital Computer Workshop*, *Instrumentation for Survival*, *Millimeter Wave Generation*, *Computer Systems and Devices*, *Solid-State*

Devices, Space Communications, Adaptive Control Systems, Biomedical Engineering.

Surely the point may be sustained—*that vast additions and changes in engineering practice take place in a period which is less than the span of a professional engineering career.* The question of just how this affects an individual engineer remains to be answered. How, for example, does he manage the flood of technological progress? How does he selectively acquire the knowledge essential in his particular field of engineering? In a few words: how does the modern engineer keep up-to-date?

Fortunately, this problem is much more manageable for an actual *individual* engineer than that theoretical entity, the *statistically-average* engineer, who has become engulfed in the flood of new knowledge and practice. To focus on the problem of the individual engineer, let us, for example, reconsider the happy fellow we talked about earlier—the engineer who was productively employed to the satisfaction of his employer and associates, although the statistics might indicate that his training is now obsolete.

THREE GENERAL OBSERVATIONS

Perhaps this individual is a designer of motors—a subject that has appeared in AIEE meetings this week, in 1937, and probably in every year since the Institute was founded in 1884. It is noteworthy that not only motors, but also each of the technical subjects on the AIEE program of 1937 is of current interest and importance today, and this fact has a significant bearing on our discussion of the impact of science and technology on the engineer. Scientific progress has not made motors obsolete, or is it likely to in the future. Electromagnetic induction and the galaxy of products and services based on this fundamental phenomenon of nature have not been superseded by atomic energy or space science. Transformers, circuit protective devices, transmission line studies, and many other traditional and fundamental technical subjects of electrical engineering will appear on AIEE programs for the indefinite future. Let me generalize, with full knowledge of the risks, and state three observations:

1. Although new *science-based technology occasionally renders an existing field obsolete, more frequently it creates a new technical discipline, and completely new additional opportunities for products or services.*

A careful study of the 1937 and 1962 AIEE technical program certainly shows that, in electrical engineering at least, this is a valid conclusion, for we still have all of the old, and have added much that is entirely new.

Thus one may identify "replacement technology" which makes obsolete an old field, and "addition technology," which adds to our engineering capability; and my point is that the latter appears to be far more common.

2. As a second observation, it appears that *the impact of new science on mature and long established technical fields is principally in new materials and new process technology.*

Thus in the cases under discussion—for example, motors—there have been great changes in rotating electrical machinery due to the development of improved materials of construction, especially insulation; significant changes in design capability due to computers; and in manufacturing processes due both to design and new materials. But the basic electrical and magnetic phenomena are still dominant.

3. As a third observation, *new or "addition technology" is frequently brought into engineering practice by new practitioners entering the field.*

This has an important bearing on the basic discussion, for the influx of new engineering talent is taking place at the same time, and probably at about the same rate that new engineering technology is coming into use.

This is of course no coincidence, for the most recent engineering graduates will tend to be the most knowledgeable in the most recently established technical subjects, and may thereby be conditioned to enter the related industrial fields. This is of course what my associate was referring to when he said, "if he's outstanding, he has to be young." Perhaps he should have said that "if he knows all about solid state, atomics, cryogenics, space and what not, he must be a recent graduate."

THE "SCIENTIST-ENGINEER"

Among the "new practitioners" entering engineering, in addition to the young graduate, we might include the "scientist-turned-engineer." Irving Langmuir, in a talk presented during that year we were discussing, 1937, made some observations pertinent to this latter phenomenon. He said, "Let us suppose that through the discovery of a new scientific principle or fact the possibility of some new application is opened up. The men trained in pure science are usually not the men to make most rapid progress in the applications; on the other hand it is not possible to turn the work over immediately to a separate engineering research laboratory. The growing idea, like a child, must not be weaned from its mother too soon.

"Before the continued development of the idea can be assured in the hands of an engineering staff, it is necessary for a relatively large amount of engineering research to be carried out by the originators of the idea, or those closely associated with them, for only these have the necessary familiarity with the subject and the deep personal interest required for success."

Some geniuses, and Langmuir was an authentic example, have been able to lead their discoveries along the engineering paths toward application and then, when they see their ideas fully communicated to others, have been able to shift gears and return to completely fundamental research in a new area. Langmuir did this literally dozens of times, as has been emphasized to us recently during the compiling of his scientific papers into a single collection, a set of some twelve large volumes. The diversity of his scientific interests was truly astounding; but no small part of his genius was his ability to communicate what he learned. One of my associates, who is an electronic engineer, commented after reading through some of Langmuir's more than 200 papers on highly technical subjects: "Why, these are written so an engineer *wants* to read them."

Other scientists, once they have become enmeshed in applying a particular result of their research work, will quite properly become full-time engineers and move from the research bench to the model shop and—in some spectacular cases—to the factory, then to the sales office, then the manager's desk, and, eventually, find themselves with all the worries of a full-time Wall Street financier.

It is, of course, a matter of individual aptitude and choice—and no doubt there is a good deal of chance and fortuity involved along the way. In any event, the "scientist-turned-engineer" is and will continue to be not uncommon in our fast-moving technological society, as a direct consequence of the rapid transition of today's science to tomorrow's engineering.

OPPORTUNITIES VERSUS OBSOLESCENCE

From this discussion of the impact on engineering of the rapid growth of scientific knowledge, I hope some highlights of the picture will emerge. The great growth of engineering knowledge during the period of an engineering career does not invariably fall with a loud statistical crash on the shoulders of every engineer, for no individual practices in all of these fields of engineering. Obsolescence of engineering does take place, but a more characteristic consequence of our flood of new science-based knowledge is the creation of completely new opportunities for the engineering of new products and services.

An engineer who works in the established field must, of course, be increasingly aware of new developments and must be able to appraise their application to his work, and significance in his area of specialization. This is particularly true in the new materials and new process areas. Certainly the technical societies, with their sponsorship of major technical meetings such as the one we will be attending this week, play an important role in the engineer's objective of keeping up-to-date.

However, in pointing out the factors that somewhat relieve the impact of technological change on the members of the engineering profession, it would be a serious error to give the impression that no problem exists. Evidence to the contrary is shown in a particularly disturbing form in employment and compensation data. For example, an engineer who is ten years out of school, looking through the employment advertisements in New York newspapers, could easily come to the conclusion that the principal engineering employment opportunities lie in fields for which he is not trained.

Similarly, the fact that salary survey information generally shows that starting rates for graduating engineers are rising faster than "going" rates, and that salary/age plots tend to flatten out, or even turn downward for mature members of the profession, is far from comforting. This "salary compression," or the excessive maturity of "maturity salary plots," should not be ascribed solely to engineering obsolescence, but the latter must contribute a significant component to the trend. For these and other reasons, the concept of engineering obsolescence has reality and must be taken seriously.

EDUCATORS' PROBLEMS

The role of formal education in this picture needs some discussion. Although the circumstances described lessen the impact of new technology on the practicing engineer, they do not provide the same buffer for the engineering colleges. The whole statistical weight of the change in the content of engineering practice must somehow be encompassed by the curriculum of the college, and it is widely recognized that this presents a large and continuing problem to engineering educators.

A friend of mine, who likes to reduce such dilemmas to quatrains, put the educator's problem this way, under the title of "Lament of the Dean of Engineering."

We've contrived a curriculum that will surely create
The best engineer alive—
But the day before he's to graduate
He'll retire at sixty-five.

Although the technical societies, like the AIEE, present opportunities to their members of keeping abreast in their field of science or technology, the feasibility and possibility of actually doing so depends upon many practical factors. For example, the spare time which is left over after the demands of full-time professional employment may be completely inadequate for this purpose. Obviously the engineer who, by virtue of his environment, is able to do original engineering rather than routine or repetitive engineering is in a favorable position for keeping up-to-date. In general the "learning situation" at the advancing edge of engineering experience—where original concepts are being born—presents an exceptional opportunity to keep up-to-date.

The industrial employer of engineers can, of course, do much to stimulate and encourage the latest practices in engineering. The R & D programs, including the advanced product engineering and exploratory product and process development, are basic sources of origination and dissemination of new engineering principles and knowledge.

ON-THE-SPOT EDUCATION

When completely new engineering technology is thus being generated, almost invariably the necessity arises to provide on-the-spot training programs to provide leverage for the new knowledge, and in the technical companies multiple instruction activity of this kind is constantly in progress.

For example, Dr. Charles W. Merriam of the General Electric Research Laboratory recently developed some new general mathematical concepts for optimizing systems design, and these methods have excited great interest in many parts of this company. In response to this interest, Dr. Merriam and his associates established a lecture-study course in the new methodology, and last Friday thirty engineers "graduated" after three weeks of intensive study of this subject. Our company's General Engineering Laboratory has in progress a somewhat similar program directed at the specialized engineering of controlled solid-state rectifiers, presently of great interest to many engineers of the company. This work will generate the knowledge needed to provide the new chapters for future engineering handbooks on this subject.

The mimeographed notes of such germinal engineering activities, much refined by later progress and additional contribution, not infrequently are reduced to book form and thus become the first published reference material in the new field of engineering.

The large number of engineers who do not have the advantage of a "learning situation," and for whom extensive self-education is impractical, must seek other solutions. As indicated earlier, and for the reasons stated, this problem does not seem to me to present crisis proportions at this time; rather it is a significant growing need which deserves an orderly and thoughtful solution. With this problem in mind consideration is being given to engineering refresher courses by some educators and some educational institutions. Rensselaer Polytechnic Institute, a nearby friend of ours, is planning such a course, and other courses must be under consideration by other institutions. There appears to be a real need here, and it seems inevitable that some type of formal, rather than informal spare-time, re-education programs will become established in the curricula of our engineering colleges.

REFRESHER RE-EDUCATION

To explore joint industrial-educational approaches to broad-front engineering refresher re-education, the University of California at Los Angeles has recently developed a six-weeks' course taught by members of its own faculty as well as by people from other campuses and from industry. A version of this course is about to be presented by UCLA at our company's Crotonville facility, with an enrollment of thirty senior engineers. These are initial steps, and from them we will hope to learn how to proceed on a continuing basis in the future.

Irving Langmuir, in the 1937 address from which I quoted earlier, also was concerned with the rapidity of change in science. "Judging from the volume of scientific publications in the last thirty years," he said then, "there seems to be an increase of about

6 per cent per year in scientific activity, or a doubling every eleven years. Such a change," he added, "has not taken place without bringing many new and serious problems."

Today, as is well known, the rate of increase of scientific research is nearly three times what it was when Langmuir spoke those words in 1937. It seems likely, therefore, that our technical world will change as much in the next ten years as it has in the past twenty-five years. But this is no time to be frightened by the prospect of new problems that face the new engineer as he seeks to use his scientific resources. What Langmuir said then we must ourselves say today: It *is* a time to be challenged by the new opportunities to apply new science for the benefit of mankind throughout the world.

Selectivity and Timing in Research

An address at the Industrial Research Institute, Colorado Springs, Colorado, May 15, 1962.

Participating in and encouraging the progress of a new concept from its research origins to the beneficial service of mankind is one of the great satisfactions and rewards of our industrialized society. The lighting of the home and the highway, the extension of man's arm by labor-saving machines, the lengthening of his stride by modern transportation, the bounty of food from modern agricultural production, the alleviation of pain and pestilence by modern drugs—all of these and many more milestones in civilized progress are marked by spectacular examples of bending nature to serve man's will.

These *bending* activities are a major preoccupation of the technical segments of industry, where it is hoped that the creation of worthwhile services and products will produce income commensurate with the expense, investment, and risks associated with such developments; and this hope is not infrequently realized. However, enthusiasm for the industrial triumphs of science and technology need to be tempered by the knowledge that even the most carefully planned technological innovation strategies sometimes backfire in a spectacular manner, thus creating distress rather than profit for the hopeful and courageous entrepreneur.

A current case in point is the transistor business, which but a short time ago was a bright shining star in the galaxy of growth industries. For an amazing ten-year period of soaring industry growth, sales doubled every year. Then, about a year or so ago, the roof fell in, and great wails of financial anguish are now heard from voices that, but a short time ago, were making only sweet talk about growth and profits.

What happened to our innovation prodigy? The diagnosis is still in progress, but I judge that it *is* a diagnosis and not an autopsy. In fact, I think one can without too much difficulty detect a not unfamiliar pattern here, a pattern that is difficult to avoid in very high-rate-of-growth innovation businesses. The electron tube business, the home radio industry, and many other illustrative examples could be cited from industrial history.

The pattern is something like this: a new business, or industry, is incubated by a technological development. Rapid growth takes place, in response to demand for a new product or service. Many new companies enter the business, and each grows at a high rate. *Each* company forecasts future growth as a continuation or acceleration of current growth trends, and each conceives that in consideration of its own inner resources it is destined to provide manufacturing capacity not only for the growth market, but sufficient to allow an increased participation in the future growth market. Blending equal parts of growth, optimism, and extrapolation makes an intoxicating brew. Result: great overcapacity for production. This overcapacity eventually outruns the actual demand, and prices collapse. Quite a few companies go out of business, which ends the abnormal growth cycle.

When the growth pains of the transistor industry, which seems to be experiencing a pattern of this type, have been analyzed and some overoptimism has been tempered by realism, I am sure that the industry will emerge from its present crisis with firmer foundations than before, and with excellent future potentials. For the time being, however, the patient is still under anesthetic.

I would like now to consider some management aspects of technological innovation, particularly from the R & D point of view. But, I will not attempt to cover the many additional considerations—marketability, manufacturability, serviceability, and the like—that would necessarily enter into any decision to go to market with a new product or service.

It is no secret now that an investment in new technology as a source of business growth is a speculative venture, and that great care and some luck are required to make it pay off in the intended manner. One of the most difficult aspects of technological innovation from a management standpoint is the general problem of *timing*, for the day on which a required critical new discovery will be made, or a new technological development will come to fruition, tends to resist precise scheduling. With respect to the economic consequences, it is almost *equally* distressing to have the new fact appear too early as too late. In one case, the flower of creative effort withers on the vine because the market, or the customer, or the related and supporting technologies are not yet ready. In the other equally distressing case, the inspiration may arrive too late—when the market opportunity has passed, or when the development has been overtaken, superseded, or made obsolete by other competing developments.

For various reasons, which have developed logically as a result of experience with the management of technological innovation activities, the matter of timing is a far more serious and crucial element of business growth strategy today than formerly. At an earlier era of industrial history, the planning of a product or business venture on the basis of scientific or technological innovation started, logically enough, *after* the scientific discovery or an invention had been made. This was the case because not uncommonly scientific discoveries were completely unexpected and unanticipated by the business community, and hence they were not preceded by any business planning or strategy activities.

Many illustrative examples could be quoted from industrial history; I have chosen one from the company with which I am most familiar, and I hope you will regard this reference as an example rather than a "commercial." Many years ago Irving Langmuir discovered that very high heat transfer rates could be accomplished at normal atmospheric pressure by the dissociation and recombination of hydrogen. At this point Langmuir reverted to a habit which I always found completely charming and delightful—he wondered whether this interesting scientific observation might have some practical use. He speculated that the new effect might provide the basis for a new type of welding, and his suggestion led to the initiation of a specific project to investigate this possibility, which, in turn, led to the development of the well-known atomic hydrogen-welding process. It is important to note the chronology in this early example: Langmuir's study of hydrogen dissociation was not preceded by a Task Force Study of the business opportunities in the field of metallic welding, or was it motivated by a market survey of metal fabrication practices and needs.

On the contrary, he was engaged in a scientific voyage of discovery, and his alert sponsor simply took advantage of his perception of an opportunity, when and as it became available.

A gradual evolution has been taking place in industry with respect to the strategy of employing technical resources and capabilities for business growth. It cannot be said any more that *all* of the discoveries of science are completely unanticipated. The same is true, and to an even *greater* degree, with technological developments. Nowadays we are not inclined to wait and be completely surprised if they turn out well, as we had hoped they would. The reason for this is the fact that the cost of setting the stage for discovery and invention has become so high, at least in industry, that we *must* make every effort to anticipate new scientific and technical progress wherever and whenever possible, in order to take maximum advantage of the business opportunities that may thereby become available. Thus, we initiate a certain amount of planning as if for an expected happy event, although we do this with some apology to ourselves and our associates because, of course, we do not know for certain whether matters will turn out as we hope. Certainly, we are unable to judge the gestation period with anything like the accuracy which enables young husbands to plan with a reasonable precision when the nursery decorations will have to be completed.

In a given case, the type of consideration described might very well lead an industrial sponsor to the conclusion that, if the anticipated degree of progress is actually achieved in the R & D venture, the new discovery might create quite a turmoil in business A, in which he is now engaged, or, might make it possible for him to enter new business B, in which he is not now engaged.

So, to be sure that no time would be lost if this turn of events should take place, which of course is still speculative, he may set quite a few wheels in "anticipatory" motion under the general heading of planning and strategy. In this context, studies of markets, organization, staffing, and related matters may acquire a considerable momentum and may unintentionally create the impression that the new discovery is practically in hand, or if not in hand, is at least *assured* of attainment. It is evident that delay or dalliance on the part of our scientists at this point in time would be most unbecoming and might create embarrassment in many quarters.

I hope it is evident from this not atypical set of circumstances that the *timing* of industrial scientific progress cannot, to the same degree as formerly, be viewed with relaxed detachment. So much may hinge on the results of research that we cannot avoid a certain degree of breathless anticipation and resultant activity. It is evident that in these circumstances the research director must steer a very carefully calculated course between two extremes. If he is too conservative in his appraisal of the prospects of success in a research project, appropriate components of his company may be completely unprepared to take advantage of the new result, if it *should* be achieved. Conversely, too much optimism on his part may generate a level of anticipation, planning, and expenditures that will be difficult to explain in the event the project is not successful. And it is clear that errors in either direction are undesirable, and could conceivably lead to cross words from his president.

Many case histories could be cited to illustrate the complexity of timing on blue-chip developments, but a particularly good case has recently reached the product stage, and I believe it is worth a detailed examination for its several interesting and illustrative features.

The example to which I am referring is the high-power vacuum circuit interrupter and I must again apologize for presenting a General Electric case history, and I hope you will believe me when I say that my purpose is not to sell a product but to make a point. However if anyone should want to order a vacuum circuit breaker I believe Mr. Phillippe could arrange to have the order accepted.

To set the stage for a discussion of vacuum high-power circuit interruption, some background information should be related. The interruption of power circuits by switchgear is a basic requirement of electric power systems, and over a long period of years highly refined methods of accomplishing this result have been developed. The basic technology relates to the fact that the separation of electric contacts in high-power circuits produces a gaseous discharge with a very high temperature and with consequent capability for what can be politely termed "destructive thermal effects." The metallic contact material is vaporized at a high rate, and the housing and related structure must be designed to accommodate the high-temperature plasma, provided it is of short duration.

The interruption of the power circuit is actually accomplished by reducing the arc temperature and hence its ability to conduct current, at the earliest possible "current-zero," which is the preferred time between the cycles of alternating current.

Presently in use are a variety of methods of reducing the ionization of gas between the separating contacts at the current-zero, including the movement of the arc through a heat-transfer structure called an arc chute, or the use of an air blast, an oil blast, or similar methods. There has not been any universal best method of interrupting power arcs, and the choice of a particular type of switchgear is determined by the voltage and power level, by the electrical characteristics of the circuit and the load, and by many operating requirements. So much for circuit interruption background.

An old dream among scientists and engineers in the field of circuit interruption is the idea of separating contacts in high vacuum. In a high-vacuum environment the vaporized metal from separating contacts, and the inevitable high-temperature ionization products, would condense at very high rates and, in the absence of "permanent" gas, would provide practically instantaneous interruption following any current-zero. It is as though one applied a super vacuum cleaner to the arc products, and removed them without fuss or bother.

This concept was first proposed and explored by Professors Sorenson and Millikan at the California Institute of Technology in 1923. The studies which they published confirmed many attractive features of current interruption in high vacuum, and created an interest in the practical application of this technology which has persisted ever since. Their early switch was operated in a vacuum system that was continuously pumped, and no attempt was made at that time to solve the problems of a sealed-off vacuum switch, which is a necessity for any practical vacuum-switchgear application. There were a number of other vexing problems, some nearly as serious as the maintenance of vacuum, but all of these problems were subordinate to the basic question of vacuum integrity; without a continuous high vacuum, the whole concept was unworkable.

General Electric acquired rights under a patent which resulted from the Caltech work and initiated a feasibility study under Prince and Wellman in 1927. Very early in this work the serious dimensions of the vacuum-maintenance problem emerged. What was meant by "high-vacuum" at that time were the pressures of about 10^{-4} millimeters of mercury characteristic of incandescent lamps and the slightly better 10^{-5} millimeters of mercury in electron tubes. The reliable vacuum switch, however, requires a pressure one-hundred times lower (that is, two orders of magnitude better vacuum) which must be maintained for a period of at least twenty years under conditions that may include both frequent interruptions, or practically no interruptions at all.

This requirement brings into focus the factors which will slowly destroy a sealed-off vacuum system: the gas which comes right out of the volume and surface of metals within the system, and the gas which leaks, by slow diffusion, through the glass, ceramic, and metal walls of the container.

Eliminating gas from the metals themselves is a much more serious problem in a vacuum switch than in lamps or electron tubes because a different metal, copper, must be used. We long ago learned how to "outgas" refractory metals, such as tungsten, for use in lamps and tubes, but outgassing the copper contacts in the switches turned out to be quite a different matter.

Marshall and Norton tried melting copper in a vacuum furnace in 1930, and although much gas was evolved, the metal was unsuitable for sealed-off switches. Kelly and Charlton in the same laboratory later tried heating the molten copper up to 2000°C in a high vacuum.

The results were somewhat better, but the copper still was not sufficiently gas-free for vacuum switches. This problem resisted valiant efforts in this laboratory, and presumably many others, for over three decades before it was finally solved by virtue of a development in a completely unrelated field. Learning how to make super-pure semiconductor materials for transistors and other new solid-state devices led to the development of new techniques which finally solved the problem of making the pure, gas-free copper that had been sought for so long.

Solving the "diffusion and leak" problem was aided by refinement of methods for measuring vacuums, which methods now extend down to 10^{-16} millimeters of mercury, the latter being essentially what is—or, if you prefer, is *not*—found in outer space. Another way of expressing 10^{-16} millimeters of Hg is this: it is about one molecule of gas per quart. Meanwhile, the sealing of metal to glass, which was a poorly understood art in 1930, and was limited to small parts and conditions of low mechanical stress, made a transition to a science which made possible coping with the very serious structural problems of a massive vacuum envelope for power switchgear.

Any one of these critical problem areas—gas in metal, production and measurement of super-high vacuum, and vacuum structures—could alone have blocked the early vacuum-switch development. In aggregate, they provided for many decades an absolutely impregnable defense of nature, which you will recall, abhors a vacuum.

Now I want to come quickly to the end of the research and development narrative, so that we may examine the whole case study and hopefully draw some useful conclusions from the example. Through the efforts of one of our research scientists, James D. Cobine, a number of his associates, and an intensive development, engineering and manufacturing program, the story was brought to a happy ending.

The basic problems of high-power vacuum switchgear were solved in a most satisfying manner. The dream of nearly four decades has been achieved, and a basically new and very powerful switchgear technology has come into use.

If I continued along this line, you might question my promise not to deliver a commercial, which I *still* assert is *not* my objective. Rather, I would like to learn what we can—with respect to timing and other factors—from an examination of decisions, strategy, and resultant experience in this interesting case, all details of which are fully available for detailed retrospection.

Some points which now appear quite clearly in this development are the following:

First: A key technical problem was identified early in the project.

Success or failure turned principally on the question of maintenance of vacuum. Unless a positive solution of this problem was found, the concept was impossible of realization. There were many other important and interesting problems, but the temptation to solve the peripheral problems was resisted in favor of concentration on the pivotal problem. This key problem was attacked repeatedly as new approaches were devised, or became available. Thus, *if there is a pivotal problem, and if you can identify it, economy of effort favors concentration on the solution of this essential problem first.*

The moral seems to be: Before you cook a rabbit stew—catch the rabbit.

Second: A massive project attack too early may be inadvisable, and very wasteful.

A decision to mount an "allout" attack on a broad project basis, including research, development, and engineering aspects of the task, would have been excessively costly and—in the 1930 period—would have been doomed to failure. It is inconceivable that any reasonable massive effort could have led to a successful vacuum switch in this early period. There were too many gaps in knowledge of the basic problem of gas-in-metals, and there were grave deficiencies in vacuum technology and practices.

Moral: *Do not beat your head against a stone wall, unless you can find a few loose stones.*

Third: If you cannot do it today, and the problem is still important, do not forget the problem.

The final solution of the gas-in-metal problem came from a development in the completely unrelated field of semiconductors. The development of the zone-refining process originally made possible the parts-per-billion impurity level for the significant impurities in semiconductor materials. An investigation of this process for high-vapor-pressure metals, including copper, revealed the fact that a reduction of gaseous impurities to parts-per-billion could be achieved for this case also. This discovery solved the pivotal problem.

Moral: If you feel sad, do not bury your head in the sand. *Be alert to progress in other fields. World scientific effort is now so massive that there is an ever increasing possibility of a completely new solution for your old problems.*

We have thus far mentioned two examples of new products resulting from scientific discoveries in the laboratory. In the case of Langmuir's atomic-hydrogen welding, the discovery came first and the product development followed. There were no very serious timing problems. In the example of the new circuit interrupter, scientific knowledge of what "should happen" in a vacuum switch—if it could be achieved—coupled with close familiarity with customer needs, provided incentive to stay at the research job. In the latter case, it was necessary to appraise constantly the chances of success, and adjust the amount of effort, planning, and strategy in accordance with new scientific developments in many fields.

There is yet another new-products situation in which the three "morals" I have just cited can and must be applied with prudence by those responsible for research and development. I refer to the situation, becoming more common in these days of massive market surveys and psychological studies of consumer wants, in which the research director is told: "People want this, discover it."

One could, of course, compile a very long list of things which are worth discovering. As an example, let us take something with which everyone is familiar: the automobile. It takes no marketing wizard or motivational psychiatrist or soothsayer to determine that people would *like* to have their cars run by an electric power source about the size of the present ignition battery, and one which would be quiet, nonsmog-producing, and inexpensive. This super-battery should, of course, provide enough power for the turnpike speed *and* the jack-rabbit starts to which we have become accustomed, and should not require replacement or re-charging any more often than it is now necessary to fill the gas tank.

Not incidentally, of course, our super-battery would have some other applications such as powering boats and aircraft, providing convenient power in our homes, and making life more comfortable on the moon until the first turbine-generator can be shipped there.

It is easy to be facetious about this, but the fact remains that there *is* a tremendous need for a new, more compact, less expensive way to store electrical energy. In the interests of making best use of our fuel resources, our future electrical generation will undoubtedly become more and more centralized in large and efficient plants, whether powered by fossil fuels or atomic energy in the form of today's fission or tomorrow's fusion. Progress will surely be made in more efficient transmission of this power, but there remains a fantastic bonanza for the inventor who can put electricity from these central stations into extremely powerful, portable packages. And if any of my associates in the laboratory comes up with a suitable answer, I am even prepared to place four gold stars in front of his name in the Research Laboratory Organization Directory.

In view of the tremendous need in this area, it would seem obvious that any research director in the electrical business should have hundreds, or even thousands, of scientists working on nothing else.

However, with the perspective gained from our exercise in retrospection, we might be forgiven for not mounting an allout attack on this problem *right now*.

First, for example, has a key technical problem been identified early in the project? It has, indeed. The problem is that no one has indicated he has the vaguest idea of how drastically and reversibly to rearrange the molecular or atomic structure of matter in a way that will substantially improve upon known energy storage processes. The presently known electrochemical reactions accomplish only minor rearrangement of the outer electrons in a few convenient molecules; and for that reason little energy change is produced; what is needed is a means of storing energy by separating major parts of the atomic nucleus—reversibly. The requirement is for something that might be described as a reversible atomic fission process. This is a key technical problem.

Let us examine the second point: A massive project attack too early may be inadvisable, and very wasteful. This strategy is completely applicable to the case under discussion, where there is no evidence of a tangible scientific opportunity on the basis of today's knowledge.

There is little or no promise in assembling an army of scientists for a massive assault on a wall that does not appear to have even a tiny foothold, or the smallest of cracks in its surface.

But recall the third "moral": If you cannot do it today, and the problem is still important, do not forget it. Here is the hope, of course. Perhaps research in fusion, or superconductivity, or solid-state phenomena, or electrochemical reactions, or some other research field will someday provide the chink in the wall which will yield a promising basis for a massive attack. I trust we will recognize it promptly. But meanwhile, our best approach to the problem is not to assign scientists to the job of discovering "a new battery for electric automobiles" but rather to provide them with opportunities to extend the frontiers of knowledge in the study of fusion, and superconductivity, and solid-state phenomena, and electrochemical reactions, and in the many other scientific avenues that challenge human skill and intellect. If we are *lucky*, and *alert*, we may find a new approach to energy storage.

Scientific Competition

An address at the Engineering Society of Detroit, Detroit, Michigan, January 16, 1963.

A few weeks ago, the *New York Times*, under the headline "2 Concerns Make Same Discovery," wrote: "In a striking coincidence, two companies announced independently yesterday the same scientific achievement, which, they said, promises to open the door to communication by light waves. Both General Electric and IBM announced a laser amplification device that could be operated by a simple electrical current rather than by an inefficient beam of intense light. Compounding the coincidence was the fact that a third group of researchers, at M.I.T.'s Lincoln Laboratories, had reached similar success a week or less after one company."

Farther down the column was the comment, "The competitive aspect of the development was apparent in a simultaneous announcement by General Electric that it would market experimental laser devices of this type at $1,300 each."

A popular subject during almost any forum in technical circles these days is "How to bridge the gap between new knowledge and its commercial application." Concern is frequently expressed over the difficulty encountered in getting acceptance of new technological ideas and processes.

I have just cited a case, however, in which a new product was announced on the same day that a scientific journal (*Physical Review Letters*) published the very first, highly technical "letter to the editor" describing the new scientific phenomenon on which the product is based. This may be something of a record, and I am not suggesting that it is typical. But I am suggesting that as new technology is becoming an increasingly important factor at the market place, *competition* is bringing new ideas into practical use earlier.

The case at hand bears out the point, which is not too surprising, that competition forces technological change. This basic economic competition is being reflected and intensified in the source of new technology—in scientific research competition between laboratories, between research teams, between individual scientists, and finally between national science efforts. Competition in science, like competition in anything else—politics, the sports arena, the market place—makes things happen.

In our own laboratory in Schenectady, the competitive atmosphere of this latest laser race has resulted in a large group of highly enthusiastic young scientists literally working around-the-clock, seven days a week, on the new discovery. And engineers, manufacturing, and marketing people entered the fray with equal dedication, adapting the new discovery to a commercial product in minimum time.

The individual scientist, the scientific teams, and the laboratories thus compete in the professional sphere. The product-people compete at the market place in order to earn a profit to sustain and grow the enterprise. Finally, it is certain that substantial sums of government funds will be spent on exploiting this new discovery for defense and space purposes in a still broader arena of competition.

Does it take too long to gain acceptance and practical use of new discoveries? Sometimes, perhaps, when the economic aspects are particularly stubborn. But when the economic factors *are* favorable, the powerful forces of competition prevail, and progress can be remarkably rapid. Next month we will celebrate the eighth anniversary of General Electric's announcement of Man-Made diamonds. In February of 1955 I must confess that an ill-timed sneeze in the wrong place would have wiped out the entire world supply of Man-Made diamonds.

But the element of competition was certainly present in this instance. At that time, the diamond business was the world's most complete monopoly, and our nation was completely dependent on imports for this critical industrial material. Now, only eight short years later, Detroit has become an important source of industrial diamonds for world markets. Not long ago, the manager of the General Electric Diamond Product Section here assured a Senate Subcommittee, which was investigating the national stockpile, that "the ability to produce Man-Made diamond in America eliminates forever the need to depend upon any foreign supplier for any diamond bort."

Let me cite another recent example—the transistor portable radio—where but a short time ago it appeared that this market would be lost completely to foreign imports. Now, however, American design and manufacturing engineers, spurred by this competition from abroad, have captured the initiative in the manufacture and sale of this important electronic product, which for a time had been lost to the Japanese. Seldom has the economic scene witnessed a business develop as rapidly, or as competitively, as this and many other areas related to the semiconductor business.

Not long after the Bell Telephone Laboratories announced the discovery of the transistor (and it seems almost incredible that this took place less than fifteen years ago), it became popular to suppose that the old-fashioned vacuum tube now was, or shortly would become, obsolete. In research laboratories this idea is certainly sustained by the enthusiasm of the many solid-state scientists working on new kinds of transistors and related semiconductor devices, including that new type of laser we have already mentioned.

The transistor competition, however, was not only at the market place. In the same wing of our laboratory where the solid-state people are covering the blackboards with band-structure theory, there is another large group of scientists working on those "obsolete" old vacuum tubes. One of them has a large sign on his laboratory wall: "Our Job–Stamp Out Transistors." Well, this is not about to happen. But, the spirit of competition certainly prevails, and the obsolete vacuum tube is showing remarkable vitality–for a dead duck. Under the motivation of scientific competition, electron tube researchers have in effect "re-invented" electron emission. Let me explain. The emission of electrons from a hot surface is a basic parameter of all electron tube devices. New materials, new process technology, and new knowledge of high vacuum have now made available unbelievable new high levels of electron emission and correspondingly improved performance for electron tubes. As a result, some of the new types are as small as transistors; others will operate in the presence of nuclear radiation and at temperatures corresponding to a red heat, where even the best semiconductor devices break down. Moreover, modern electron tubes will operate at fantastically high frequencies, well beyond anything the transistor can accomplish, and at power levels greatly above former practice. Over the past seven years, for example, the frequency range for triode-type vacuum tubes has been doubled, and the microwave power level has been increased by a factor of 100.

Once again, I am not suggesting that these improvements might not have come along if there had not been competition from the transistor; but the competition certainly did not hurt—either from the scientists' standpoint or from the standpoint of the businessmen who are now making and selling these new kinds of metal-ceramic tubes in substantial quantities.

One of the most intriguing competitive situations to be found in the world's industrial scientific laboratories these days involves the exploration of new methods for generating electric energy. It must be remembered that some 99.9999 per cent of the electrical energy we now use still comes to us through the courtesy of Michael Faraday, by virtue of his discovery of electromagnetic induction in 1831. (The number of battery-operated toys found under Christmas trees this year has cut into Faraday's percentage only very slightly.)

In spite of the established eminence of electromagnetic induction, there are a large number of basically different ways to convert other kinds of energy into electrical energy. All have been tried at one time or another in the history of technology, but for producing bulk electrical energy in large stationary plants, electromagnetic induction has always won out. It is safe to say that bulk electric energy will continue to be produced by rotating electromagnetic generators for a long time in the future, and they will be driven by steam turbines, or occasionally by water wheels in hydro plants. But, in our competitive scientific and technical world, new materials, processes, and concepts, and especially new knowledge are being produced in a flood tide. As a result, some of these energy converters are being reappraised.

In fact, some of these methods have recently been attracting considerable attention in research laboratories. In particular, these are: fuel cells, thermionic converters, magnetohydrodynamic generators, thermoelectric generators, and photovoltaic cells (or, if you prefer, solar energy converters).

Surprisingly enough, recent progress in all five of these areas has been based largely on discoveries made by researchers working in some other areas of science. The most practical fuel cells yet devised have at their heart a membrane whose discovery was closely related to work on new polymer materials intended for insulation and structural purposes. The thermionic converter is essentially a structure similar to an electron tube in which heat boils electrons out of a cathode and onto a cooler anode, creating an electric current; improvements in the high-temperature vacuum tubes which I have already mentioned are frequently applicable to the thermionic converter, and vice versa. The future of magnetohydrodynamics will depend largely on progress in the science of ceramics and metallurgy plus basic studies of plasma phenomena. Both photovoltaic and thermoelectric generation became the subject of new scientific attention after semiconductor research, inspired by the discovery of the transistor, gave great new impetus to the study of the electrical properties of solids.

How is the scientific competition between these unique energy-conversion methods progressing? Well, one must be brave—or a little foolhardy—to assume the role of "handicapper," not only in horse races but also in scientific races. But a little speculation may be in order.

First of all, our friend Michael Faraday is not yet being seriously challenged in the field of bulk electric energy generation. But there are multiple important applications for power conversion equipment which turbine generators, or hydro generators, will not serve, particularly portable sources for industrial, military, and space applications. In these and related fields, electrical energy will achieve greater versatility and usefulness by virtue of the new energy converters.

It would be a mistake to underestimate the technological significance and economic impact of these new *portable* sources of electrical energy. For example, unique power sources are vital components of our space programs. Solar cells have played a key role in the success of many of our early satellites. A fuel cell will provide the power for the two-man Gemini spacecraft; the two-kilowatt fuel cell power plant will weigh only about as much as an astronaut, whereas batteries required to provide equal power for as long a time would weigh nearly a ton. Applications of thermionic converters, MHD, and thermoelectric generators to future spacecraft are being actively studied in many laboratories.

The fuel cell is clearly leading the "portable power" parade, particularly with the growing evidence that it will be possible to use inexpensive hydrocarbon fuels such as coal, natural gas, propane, and gasoline in such devices. (An important step in this latter direction was revealed in the announcement of a high-temperature hydrocarbon fuel cell on December 26, 1962.)

Basic to the importance of the fuel cell is its high *theoretical* and *actual* conversion efficiency. The 100 per cent theoretical efficiency, as popularly defined, is realizable to the extent of 50 per cent to 80 per cent in some operating cells.

In the automotive hub of the world—Detroit—it is probably unnecessary to point out that if a low cost hydrocarbon fuel cell becomes feasible in the future, it can hardly fail to receive careful attention by the automotive industry. By this route one could expect to gain, among other things, much better gasoline mileage and a final solution of the smog problem. The seriousness and growing magnitude of the smog problem may itself force this solution. In my crystal ball, which is somewhat cracked and battered from being kicked around, I think I see a fuel cell in the future of the automotive industry.

Rated in second place in portable power is the thermionic converter, which is sharing in the progress resulting from continuing research in electron tubes. Actually, in *portability*, the thermionic converter is excellent, but it suffers from the requirement that it must operate at very high temperatures, and its conversion efficiency so far is about 15 per cent. However, this converter may be used as a thermal "topping device" in atomic power plants, where the increased overall efficiency would be quite significant.

The remaining three converters—MHD, thermoelectric junctions, and solar cells—are not doing too well in this race at present. MHD entails very difficult materials problems, and after extensive research on this method, my associates and I have become unenthusiastic about MHD for all but the most specialized applications. Both thermoelectric junctions and photovoltaic cells are substantially less efficient than any of the other converters now in the running.

None of these contenders in the energy converter race, however, can be ruled out. I am only reporting my impression of the progress to date, and a new discovery tomorrow could change the odds dramatically.

A fascinating arena of scientific competition is the world of physical extremes—on the one hand, the absolute zero, and the millions of degrees of temperature—and on the other, the near zero pressures of the modern vacuum, and the superpressures that create Man-Made diamonds. For many years the motivation here primarily has been curiosity—in the words of the mountain-climber, "because it's there,"—but now there are some technological connotations that add competitive zest to research at physical extremes. At absolute zero, or near it, high-field superconductors are creating a tremendous scientific stir. At the other extreme, scientists working with high-temperature plasmas, with thermonuclear fusion in mind, now talk in terms of the "tens of millions of degrees" already achieved, and they look, with considerable confidence, to attaining the temperatures of hundreds of millions of degrees required for fusion-energy generation.

I wonder how many people living in Detroit know that, as I indicated earlier, one of the world's largest diamond "mines" is located here? This has come about through superpressure research which makes possible the attainment, on the *surface* of the earth, of the millions of pounds of pressure and the high temperatures which nature used to create natural diamonds at great depths in the earth.

At the other end of the pressure scale we find scientists devising new kinds of mass spectrometers that will measure pressures of true outer space: 10^{-16} mm of mercury, which is about the pressure to be found halfway between here and the moon. This pressure, or lack of it, is equivalent to three or four atoms per cubic centimeter, as compared with the several quintillion atoms per cc in the air around us.

This vacuum-research race toward more and more of less and less has had some interesting recent by-products. One such was the development of a new vacuum circuit interrupter, which we believe to be the most important advance in the technology of electrical switchgear in several decades.

In another field, it would seem at first that super-vacuum would have little or nothing to do with lubrication. On the contrary, super-vacuum has been a magic key to exciting new knowledge and practice in lubrication, and I am glad to be able to announce some of this work for the first time tonight.

At the Gordon Conference in 1959, Dr. Leon E. St. Pierre of the General Electric Research Laboratory discussed the idea of studying the lubrication of metal surfaces by sorting out the surface-chemical reactions with super-vacuum. It is hardly necessary for me to remind a group of engineers—in Detroit, or elsewhere—the extent to which modern mechanical technology depends on the free ability of one metal surface to move over another with low friction. Rubbing friction inevitably results in "wear," and on a microscopic scale the wear process results in the exposure of *fresh metal* surfaces. The basic chemistry of the fresh microsurface has been extremely difficult to study because, under normal atmospheric conditions, it remains "fresh" for only about a billionth of a second.

The ability to achieve extreme vacuum in the laboratory has recently given our chemists a means of looking at these surface chemical reactions in "slow motion." At 10^{-10} mm of mercury, for example (which is a pressure only about one ten-thousandth that found in ordinary electron "vacuum" tubes), fresh metal surfaces can exist for more than 24 hours before they are contaminated to any marked degree. By means of this technique, St. Pierre and his associates, Robert V. Klint and Robert S. Owens, have been able to study the reactivity of a variety of metals to a variety of organic molecules, with some surprising results.

To make things exciting, they took one of the worst possible metals from a lubrication standpoint—aluminum—and studied it. The end result of this study is that it is now possible to lubricate aluminum and to produce thereby the low friction of the best babbitt bearings. Several classes of organic chemical compounds have been identified which have the required properties and they may be employed either in pure form or as compounded additives. These materials form a chemical attachment to the aluminum surface and form a wear-resistant film. Furthermore, they coat oxide particles and eliminate their abrasive properties.

Now, some of these chemical names are difficult to pronounce, and I hope you will forgive me if I do not even try. No one, nowadays, likes to be caught without a secret ingredient, so for the time being at least, we have a family of secret additives for aluminum lubrication.

Preliminary studies of bearing friction using these new aluminum lubricants have shown some remarkable results. Over a wide range of speeds, the coefficient of friction of aluminum lubricated with the new materials, as I have indicated, matches that of the best babbitt bearings. The coefficient of friction for aluminum lubricated with the new materials is about one-fifth to one-sixth that of aluminum lubricated with conventional oils.

Broad applications are anticipated not only in bearing lubrication but also in aluminum fabrication methods such as die forming, roll forming, rolling, machining, extrusion, wire drawing, polishing, spinning, die casting, and so forth. The advantages are illustrated, for example, in the case of aluminum extruded with the new lubricant, which comes out of the die with a mirror finish. In addition, early studies have indicated that the family of chemical materials effective for aluminum lubrication also has significant lubricating potential for other metals and alloys, both ferrous and nonferrous.

So here, in the early stages of development, is a new competitor in lubrication technology, and we believe that as it progresses it will have an impact on many important industrial practices and processes. At this time, a number of interested components of our own company are already examining the best way to advance the applications.

You will note that *scientific curiosity* was the motivation for this new knowledge. From this point on, we may expect that *competitive factors* will speed the application of the resultant new lubrication capability toward the solution of many practical problems.

Steel—and Its Opportunities

The Charles M. Schwab Memorial Lecture delivered at the 71st general meeting of the American Iron and Steel Institute, Waldorf Astoria, New York, New York, May 22, 1963.

In gathering my thoughts for this occasion, I was aware that I present to you a point of view that has both some advantages and some disadvantages. I should be quite frank at the outset and acknowledge that I know little or nothing about the steel industry. I was first going to put this on the "disadvantage" list, but then I noticed that this list was becoming quite long, so I moved it to the "doubtful" category.

Although a knowledge of the past history of the industry is, of course, a vital source of guidance for the future, we do live in a changing and increasingly competitive world, and past patterns may not be an entirely reliable guide for the future. For example, your industry has had a tradition of process-research and development, but research in the physical sciences has not been a preoccupation in the past. Meanwhile, research done outside your industry, on competitive materials, is now posing appreciable competitive problems.

It is my intention today to make clear a firm belief that research must play an increasingly important role in the future of the iron and steel industry. That I should have this view will perhaps surprise no one, but in addition to a "science-and-research" flavor, you may also detect in my remarks a certain customer flavor. The electrical industry is a large user of iron and steel, and some of my associates in our company have been able to think of a few improvements they would like to see in certain of your products. Also, I intend to describe several new developments, some of them from the company with which I am associated, which I think will be interesting, and which will, we hope, contribute to the further development of industry technology.

With that introduction, let me first say that in order to satisfy the customers of the electrical industry, at least for heavy apparatus products, we are constantly faced with the necessity of using our steel components closer and closer to their limit of performance, be it high-temperature strength, magnetic flux density, fatigue strength, or some other property. Heavy machinery continually grows larger because of the unit economy of large size; thus the ability to make large castings and forgings is a critical matter in many types of equipment. It appears that in the past a distressingly large number of important process and product innovations in this area have originated in Europe, and we do not see why this dependence should continue.

It is also a fact that most of the basic and applied research on metals has been done by consumers of these materials rather than by their producers, and we doubt if this is a safe pattern for the future health and progress of your industry. One of our most extensive facilities at the General Electric Research Laboratory in Schenectady is devoted to metals and ceramics research, and although much of the research in this portion of the laboratory is devoted to materials unique to our industry—superconducting wire, ceramics for electronic applications, and the like—you will note from some examples that I will cite later the extent to which we have become involved in basic understanding, processing, and application of your favorite product.

There is, however, some evidence that the research situation in the iron and steel industry is changing appreciably—that the industry is now much more active in technological developments and is starting to participate in scientific investigations to an increasing degree. Some of the laboratories which have been built in recent years provide an impressive beginning on the facility requirements for such work.

I am sure you have heard of the "revolution in materials" which has been reverberating through our industrialized society for these many years. This is the peaceful revolution which has come from the great flood of new materials based on new scientific and technological developments. It should be appraised and evaluated for its impact on—and opportunities for—the steel industry. In particular, does the materials revolution pose a threat to steel's future competitive position, or to its opportunities for future growth and diversification?

Steel is unquestionably one of the most important bulwarks of modern civilization, and it has occupied a position of unassailable eminence since the Middle Ages. No one questions that mankind, having progressed through the Stone Age and the Bronze Age, now lives in the Steel Age.

Is it conceivable, then, that steel will some day be displaced, to a significant extent, by some newer material or combination of materials? I am sure that few people who have grown up in the industry could be persuaded of that possibility, and even to an outsider it would seem safe to continue running the mills for a few weeks longer—maybe even a few decades.

But just to put matters in perspective, let us take a quick look at another basic industry in our modern society—coal—and note some startling competitive factors at work. One of the largest markets for coal is electric power generation, which accounts for nearly 50 per cent of our present coal usage. Then came the atomic age and uranium. Ten years ago, and even as recently as five years ago, the forecasts of costs of electric power generation by uranium-fueled reactors placed the attainment of a significant segment of the electric power market well into the 1970's. However, reactor technology and economics have advanced much faster than even the optimistic forecasts, so that the age of competitive atomic power is already here.

Recent studies show that, based on 1961 fossil fuel prices, atomic power plants of present design capability are competitive today in about 50 per cent of the total United States energy-production market.

The coal industry is reacting competitively to this threat, and by the "unitrain" and other concepts there is the possibility of significantly reducing transportation costs in selected areas. However, even if coal transportation costs could be reduced broadly to half of the present level, about 30 per cent of the United States energy-production market can still be served competitively by uranium reactors employing present technology. One must bear in mind that today's reactors operate at relatively low temperatures and pressures, do not breed new fuel, and are very conservatively designed, and that their future is certain to be marked by additional economic progress. It seems evident that uranium will take over a significant portion of the electric utility coal market within the decade ahead, although the growing chemical and petrochemical markets will partly offset the impact of this development on the coal industry.

My detour to a discussion of coal, to note the interplay of competitive factors in another major basic industry, has been merely a "for instance." I know of no comparable threat to the iron and steel industry. However, it would be a grave error not to note a number of competitive factors, even in the steel industry. To an increasing extent steel is not only competing with other steel, but also with other metallic materials, particularly aluminum, and especially with polymers and composite structures. The present tonnage use of all polymer materials is relatively small in relation to steel, but the growth rate is striking, and worth noting. And, as in the case of uranium competing with coal, the newer technology seems to present larger potentials for further development.

Some of these magnitudes and trends are evident in some charts to which I would like to refer. Figure 1 shows the relative growth of steel usage versus aluminum, copper, and polymers. Steel retains its leadership by an order of magnitude, but the differences in growth rates are striking.

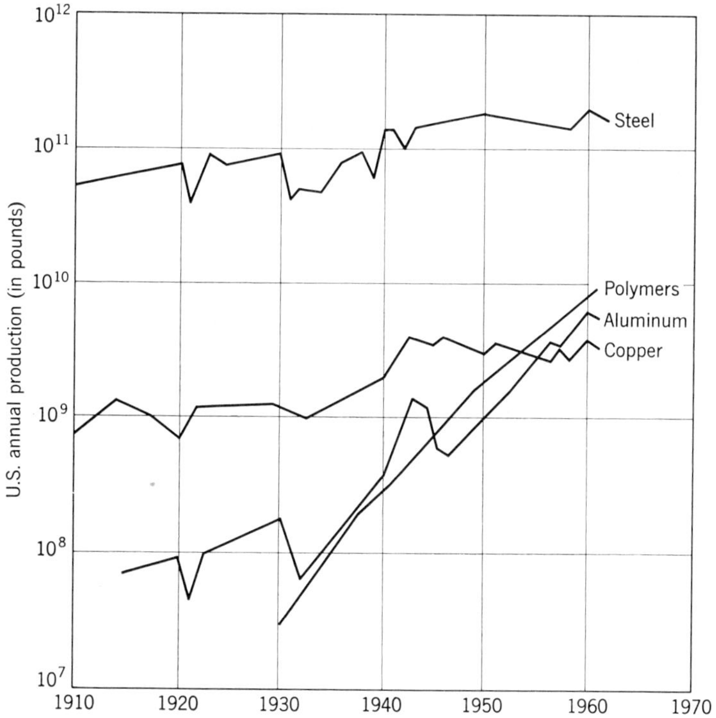

Figure 1. How polymer use has grown.

And on a basis of physical volume—as contrasted with weight—there is an additional factor of six to eight times in favor of polymers.

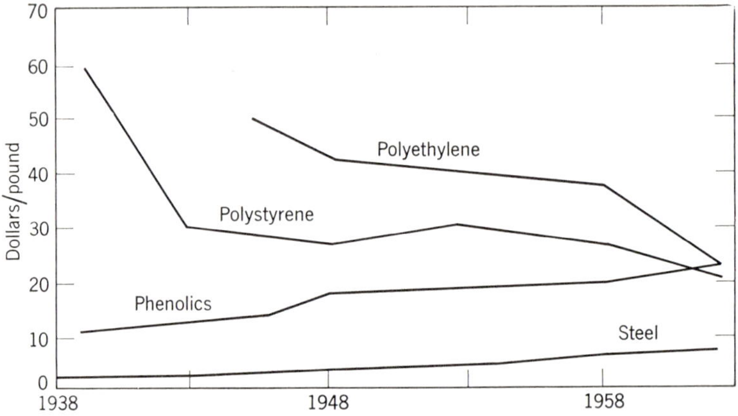

Figure 2. Material prices.

Figure 2 shows the development of cost factors, comparing polymers with steel. Substantial further reductions should be available in some of the newer materials, as the chart suggests.

Figure 3. Graphite lattice (top) and diamond lattice.

But even more striking than sales and cost figures is the continuing improvement in the strength of polymeric materials, even at moderately high temperatures. There is no reason to look askance at the chemists' favorite carbon-carbon bond, which provides the basic framework for polymer structures.

The carbon-carbon bond is, in fact, the fundamental source of strength in diamond; tying carbon atoms together in the compact structure shown in the lower model in Figure 3 provides the difference between diamond, the hardest known material, and soft slippery graphite, depicted by the model on top. The carbon-carbon bond has some distinct, and not yet fully developed, possibilities in polymers.

An example of this possibility is to be found in polymer impact resistance, which measures the ability of a material to absorb a large amount of energy in a very short time. Some of the newest polymers are approaching the metals, including steel, in impact strength. Steel, however, covers a range of values, the top edge of which is well above polymers.

Although polymers in bulk usually exhibit *tensile* strengths considerably below those of metals, polymers in the form of fibers—where structure is optimized and defects are minimized—exhibit strengths comparable to those of metals.

As to the temperature limitations of polymers—which we have generally considered to be useful only near room temperatures—the recent American Chemical Society meeting in Los Angeles heard about a commercially available polymer which is flexible at

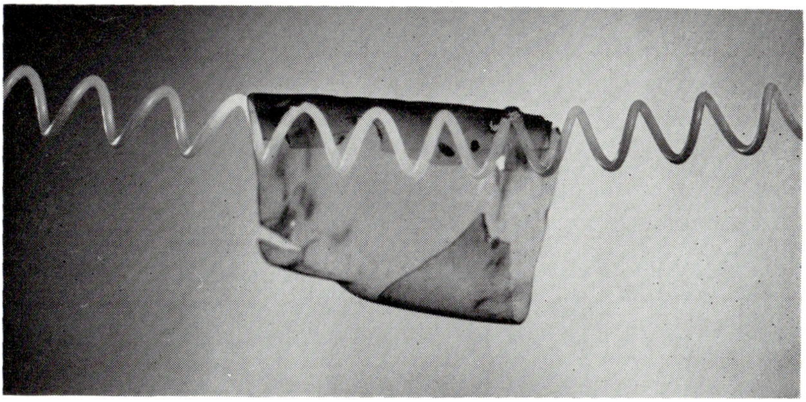

Figure 4. Silicone on a hot filament.

$-450°F$, and which has a tensile strength of 5,000 pounds per square inch even at temperatures above 900°F.

Some experimental polymers, not yet out of the laboratory, have significant strengths even at a dull red heat. The unique new kind of silicone resting on the hot filament in Figure 4 remained intact for a substantial period of time while it was having its picture taken—somewhat longer than I would like to sit for a portrait.

Polymers have thus long since made the transition from novelty decorative materials to structural materials of real consequence, and there is no doubt that additional great future advances are going to be made in polymer properties. Although polymers have, I judge, not yet taken over major structural parts of automobiles, their use is growing rapidly in that industry, and meanwhile they have gone a long way down this road in the case of home appliances.

Foamed-in-place insulation, for example, permits remarkably strong composite structures that are already on the market as refrigerator and freezer components. Here is a case, of which there are many examples, where a higher cost-per-pound may be tolerated because of reduced specific gravity, reduced fabrication and finishing costs, and for some improved properties. Also, the strength requirements are nominal, and the steel formerly used for some of these parts was very much stronger than actually required for structural strength.

More important than strength, in many such applications, are low cost and formability. We might make some steel parts for appliances thinner, and still have sufficient strength, but at the sacrifice of rigidity. Ideally, in using a high strength material like steel in such a structure, we would prefer it to have a lower effective density.

This would give greater rigidity—stiffness—which varies as the cube of the thickness. This, in turn, suggests a steel foam, or some kind of a molecular network. Actually this latter structure is essentially the description of a polymer.

Before focusing on steel, however, there are two other aspects of the polymer-competition that need to be examined thoughtfully: (1) research, and (2) application engineering. In both of these areas, the polymer industry is doing an outstanding job, and this view can be supported with some specifics.

For example, one measure of "who does how much research"

nowadays can be gained by counting scientific publications. In the statistically significant sample that was counted, the chemical and petroleum companies accounted for 83 per cent of the research on polymers. Here the suppliers of the materials are clearly taking the lead in research.

I do not have the corresponding statistics for iron and steel research alone, but the figures representing metals research as a whole show that the electrical and communication companies account for 68 per cent of the industrial scientific research in metals, while the metals producers themselves do only about 10 per cent.

We have no similar handy index for measuring effort in application-engineering, but it is a fact well known in the trade that in many branches of the polymer industry such work is a major area of technical activity. The polymer manufacturer frequently will take the initiative in showing a prospective customer—by sample moulding, extrusion, and test results—how his polymer will better perform in the customer's product, often suggesting displacement of what he refers to as "costly metals." I believe this is known in the trade as "creative selling."

Here is a large area of opportunity for your industry: undertaking exploratory application work with the objective of displacing competitive materials, especially those "messy plastics."

There is evidence that this idea is getting more and more attention. Some of my associates have recently been exposed to examples of very impressive creative selling by members of the steel industry. In this case, they were shown not only a fancy and informative brochure entitled "New Ways to Use Steel in Refrigerators" but also a sample steel refrigerator door that included some new fabrication ideas. The result has been that a plan to start manufacturing this door out of "messy plastics" is now being restudied.

Two observations seem pertinent: first, the polymer "menace" has risen and grown with massive industrial research support, while the metals industries have played a very much smaller role in the industrial research support of their own products and processes, at least in the scientific sector; and second, polymer application-engineering is a highly developed industrial function, while this work appears to be at best in initial phases in parts of the steel industry.

Now, let me deal more directly with the subject of steel. The weight of my remarks will be to the effect that your favorite metal still has great and largely unexplored potentials, that you should make sure that it becomes part of—and not a casualty from—the materials revolution, and that substantial research and development effort in this area should pay off handsomely.

After what may have seemed to be critical remarks on R & D in the steel industry, I will make amends by noting that the industry does rest on process innovations, and that at the present time a great many important and exciting process innovations are in progress. My associates and I have the impression that the overall level of technological development in the industry is now at an all-time high.

Historically, a major milestone in the industry's progress was the Bessemer process, which dates from about 1850. The open-hearth furnace came along slightly later. For a long period, the two processes were in fierce competition, but the open-hearth won out and became the principal reliance of the industry.

Electric arc furnaces and vacuum induction furnaces came along later, and more recently the consumable-electrode arc furnaces have come into significant use. And during the past few years, of course, the improved Bessemer process has been making significant gains.

The development of methods of forming iron has proceeded in parallel with processes for its extraction from ore, and, as I will indicate later, there are still some research opportunities down this road. The rolling mill has come in for major attention, and ever since 1850 there has been a continuing parade of significant incremental improvements in this basic device.

Recently, for example, a continuous finishing mill has been operated by a completely automatic computer control with noteworthy economic improvement in sheet quality. This is an outstanding example of innovation in the industry.

Certainly a major steel-making innovation of recent years has been the basic oxygen process. The original idea is quite old, but the modern development is identified with early work in Germany, Belgium, Switzerland, and especially in Austria, somewhat prior to 1950. It appears to be rapidly growing in tonnage capacity in the United States at present. The still more recent Kaldo process, whose origins are Swedish, is a further refinement of the oxygen furnace.

In discussing new process developments, it is popular to note that in many cases early production use in Europe preceded United States practice by many years, but the interpretation of this fact requires some care. Many of the new processes—for example, continuous casting—are far more easily adaptable to the small mill operation more common in Europe than to the heroic scale of operation found in many United States mills. Thus, the compatibility of a new process with the scale of operation, including its integrations into preceding and following high-investment operations, is a factor of great weight in determining whether a process is economically feasible in this country.

On the other hand, major process innovations seldom spring into full-scale economic use overnight, and some small-scale beginnings are usually necessary to determine if in fact the new idea is feasible for tonnage operations.

Our example of continuous casting appears to be pertinent, and, as will be seen, the small beginnings here have now reached the point where a single installation can produce a reasonable fraction of the requirements of a large rolling mill. Also, the oxygen converter has moved the economic batch size to smaller tonnages with resultant better match with continuous-casting capabilities.

Looking ahead, it is evident that developments in oxygen converters—and related operations—are setting the stage for increasingly automatic process control by computer techniques. McLouth has recently installed the first completely automated hot-strip finishing mill, to which I referred previously. It is a significant milestone. Further integrated and computerized steel mill operations appear to be an inevitable direction for future progress.

Now let us take a closer look at this old idea of continuously casting metal to approximately finished form. I say "old" because Sir Henry Bessemer undertook experimental work in 1858 on the idea of pouring iron between rolls to produce strip. He obtained a United States patent on the process in 1865. Sir Henry was led to this idea by the successful practice in the glass industry, but his experiments showed one vastly important difference between glass and metal: glass exhibits a long viscous range, while liquid metal, upon cooling, freezes to a solid suddenly. This fact delayed the successful continuous casting of metals about 100 years.

The growth of continuous-casting plants for steel, however, has been steady over the past two decades, and this trend can be expected to increase considerably in the years ahead.

Over and above the steel mill application of chilled-mould methods of continuously casting steel, there has been interesting recent progress in the continuous casting of thin sheets and small diameter rods of various metals. The Properzi Company of Milan, Italy, has developed a process in which a section of aluminum is cast continuously in the form of a five-sided bar between a grooved wheel and an encircling belt. The stock may then be hot-rolled to finished dimension rod. We have learned quite recently that Properzi also has developed a similar process for copper.

My associates in General Electric have been interested in the continuous casting of a number of metals and have made some progress, including a continuous-casting process for copper rod which employs the "dip-forming" principle.

In this process, a small rod—usually three-eighth of an inch in diameter—is continuously drawn at several hundred feet per minute through a pot of molten copper, forming a five-eighth-inch-diameter rod suitable for wire-drawing. The core rod is the heat-sink which permits the added hot metal to freeze.

The copper dip-forming process is now in initial operation phases and promises good quality rod on an economic basis. This process will be described in detail in a forthcoming paper being submitted to the *Journal of Metals* by its author, Dr. Roland P. Carreker, Jr.

After our copper dip-forming project was well under way, we learned that the British Iron and Steel Research Association had independently conceived and were experimenting with an idea for the continuous casting of steel strip.

It also was pointed out to us that a certain Mr. Lash had similar thoughts sufficiently well defined in 1904 to have obtained a United States patent disclosing his idea for continuously casting metals. Evidently Mr. Lash was not able to bring this technique to a successful conclusion at the time, and undoubtedly others have considered this same idea since then.

The BISRA experiments are well documented, but apparently

the work was not promising enough to justify a continuation of the program. Our own attempts to "dip-form" steel strip have been encouraging, but they have not been carried to an operational stage. It would be honest to admit that there remain a few details to be worked out to bring steel dip-forming to an operational stage.

I mentioned earlier that there were some R & D opportunities still remaining in the area of metal forming. As noted, one of the very old methods of metal forming is the rolling mill, which over a period of about 150 years has been the subject of many improvements and refinements in drive mechanisms, control systems, precision, automatic materials-handling, speed of operation, and many related matters.

Specialized mills like the Schloemann and Sendzimer can perform unusual rolling requirements, and, as indicated earlier, the completely computerized automatic mill has now been achieved.

Certainly at this late date in the history of metals, we should hardly expect to discover a basically new principle in rolling mill technology. By now, it would seem, all possible mechanical principles should have been explored, utilized, or discarded, with the related patents long since expired. However, this does not appear to be the case.

Some of my associates have discovered and developed a new and exciting rolling mill principle, which I would like to describe. When the idea was first reported to me, I made a mental note that the patent search would promptly reveal the expired patents. But I was wrong. The idea is novel, even in the patent art. I regard the whole development as a modern-science fairy tale—unbelievable but true.

The new principle did not arise from a study of rolling mills, but came as "fall out" from an experiment in pure science. The example illustrates how relevant basic research in a compatible industrial environment can yield both important new scientific knowledge and new opportunities for the sponsor to usefully apply the knowledge. This is the essence of industrial scientific research.

At this point, it is instructive to follow through the logic which has led us to these new rolling-mill concepts, since the incorporation of plastic bending in a rolling process is by no means an obvious matter. For some years we have carried out in our laboratory a research program on the effects of *cyclic plastic strain* on metals. This activity has produced much fundamental knowledge relating to such important technical questions as the resistance of metals to cyclic thermal stress and to low-cycle fatigue. In turn, this new knowledge has contributed to the design of high-temperature equipment such as pressure vessels, steam turbines, and aircraft gas turbines.

During the course of these investigations, it was noted that the surface of test specimens subjected to cyclic plastic strain was easily indented by very light pressure, such as that produced by dial gauges used to measure the change in dimensions of the specimen. In recognizing this phenomenon, Dr. Louis F. Coffin, Jr., wondered whether a new and possibly useful principle was involved. For example, he asked, could deformation be produced by small steady stresses, in the presence of a superimposed cyclic plastic strain?

Subsequent studies of these phenomena have led to a method of rolling in which the metal is subjected *simultaneously* to: (a) roll force, (b) strip tension, and (c) plastic bending. It can be demonstrated that a substantial reduction in thickness can be achieved by these functions—in combination—under conditions in which roll pressure alone, or strip tension alone, will produce no elongation.

A considerable amount of work has been done with experimental mills employing this principle, and some very interesting performance data have been obtained. The support of the tiny bend roll by the large contact rolls provides unusual rigidity and permits accurate rolling to very thin gauges and with large reductions per pass. It is believed, but not yet proven, that the method will permit the rolling of wider stock than is possible in conventional practice. Performance on difficult-to-roll materials is exceptional, and because the process applies cyclic plastic-bending strains it has been shown to overcome some of the prior work-hardening of the material. In some cases, severely work-hardened materials have been reduced as much as 98 per cent without intermediate annealing.

With our experimental equipment we can handle strip up to twelve inches wide, and we have, for example, been able to take stainless steel from 100 mils to four mils in eight passes, without intermediate anneals. It should be emphasized that this new rolling method requires further work to adapt it to specific rolling applications, but some very worthwhile new metal deformation principles have been convincingly demonstrated, and it appears very likely that they will add usefully to metal-working capabilities.

There is hardly time to more than mention another important segment of technological progress that affects your industry—the segment which includes metal-cutting, machining, and all of the related arts. Electrolytic machining, for example, is a new process by means of which one can perform conventional machining operations, and some unconventional ones, employing, however, electrochemical means. The development, over some period of years, of tungsten-carbide cutting tools in grades suitable for the high speed machining of steel has had a very considerable impact on the economy of steel usage, and there is considerable additional opportunity for further advancement here. We look hopefully to wider application of Man-Made diamond tools for cutting steel; development work now in progress will determine whether this goal can be realized.

I have been describing some technological developments in the field of metals and have included some oblique references to basic research in the field. Actually, the biggest opportunity for new future metals technology is to uncover additional new knowledge upon which it can be built. Metallurgy has been an empirical art for centuries, but now it has become a modern science with multiple opportunities for further research progress.

There is abundant experience to support the view that to participate in technological progress, you must play a *contributing* role. For example, if first knowledge of new progress comes from a newspaper item, it is probably too late to get in the act—to effectively and profitably participate in the development. You have a much better chance to participate if your own knowledgeable scientists can alert you to forthcoming events, and the best chance of all comes when they themselves produce the discoveries on which these events will be based.

Let us look briefly at just one of the metal concepts that has emerged from recent work in science and conjecture about its possible consequences. An exciting chapter in metallurgical science has been enacted in the past ten years by the discovery and understanding of the role of crystalline *defects* in metals. It will hardly be necessary to remind you that all metals are crystalline in basic structure. Until this recent progress, which I refer to, we have had the dilemma of extremely high *theoretical* strength of metal crystals, and the relatively low actual strength of these metals. This paradox has now been explained quite satisfactorily and has led to a fundamental understanding of what makes metals strong—and weak. (To insert a word of interpretation at this point, we now know that the strength of practical metals does not reside in the strength of the metal crystals as such, but rather in the peculiar crystal-structure irregularities which are invariably present.)

One can safely say that the matter of the strength of metals now rests on a firm scientific base. A sequel to this mystery has been the ability to produce theoretically perfect crystals which actually do exhibit the phenomenal theoretical strength of metals.

These so-called "perfect" metal crystals can thus far be produced only in the form of tiny metallic needles, or whiskers, and it appears unlikely that they will be produced in large, massive form. But even in miniature, a perfect metal whisker is a very useful accomplishment.

For one thing, it places an upper limit on the strength of metallic substances, and there is plenty of room above presently used strength levels. I believe that these tiny perfect crystals also point a way toward future steel progress. Let me explain what I have in mind.

Practical metals are, of course, crystalline aggregates. We now know, from the work I have referred to, what the ultimate strength of the individual crystals may be, and we are probably at the point of diminishing returns in the exploitation of such factors as crystal size and shape, grain-boundary phenomena, multiple phases and phase transitions, and the many related complex matters which are optimized in metallic-material design.

However, in our preoccupation with what might be referred to as the "composition-and-processing" approach, we may have been overlooking an old but little exploited avenue of metallurgy: *composite structures.*

Consider the example provided by the material we call Carboloy, which is made up typically of tungsten-carbide crystals embedded in a matrix of cobalt. Tungsten carbide itself is a strong substance, but it is much too brittle for most useful purposes. Cobalt, alone, is neither hard enough nor strong enough to have achieved much individual usefulness. The *composite structure*, Carboloy, is, however, vastly stronger and more durable than either component, and it has many unique properties.

The value of composite structures is a fact of which we have not taken sufficient note in metallurgical research and development, and significant future breakthroughs here could have a great impact on metal industries—including iron and steel.

There are other striking examples of the unique merit of composite structures, including one that may appeal to some in the audience: the glass fishing rod. Here, too, is a composite, in this case made up of fine filaments of great strength but little or no ductility, embedded in a compatible and flexible matrix. I am informed that this structure has revolutionized certain recreational practices, but I am here relying on expert consultative advice rather than personal knowledge.

The very old basic idea underlying the glass fishing rod continues to have some fascinating possibilities, and investigators in Great Britain have succeeded recently in making a composite of silica fibers in an aluminum matrix with a tensile strength of 800,000 pounds per square inch in the direction of the fiber.

We have a few other examples of the potential of this approach. A few years ago, permanent-magnet technology experienced a major innovation with the development of Lodex magnets, in which tiny single-magnetic-domain material is embedded in an inert matrix to achieve unusual permanent-magnet properties.

In addition to these few, but wonderful, examples of the capability of the composite-structure approach to metallic-material design, we have one additional persuasive reason for looking in this direction: we know how to make perfect metal crystals, and these tiny filaments would appear to be perfect superstrength components for future composite structures.

In summary, we have noted that your industry faces significant problems at the market place, especially with the rise of competitive materials. However, industry activities in the area of steel process improvement and production automation are most impressive. Additionally, the application-engineering-marketing strategy is an important development which could exert real leverage in the competitive materials arena, particularly if sustained by scientific research as a continuing source of technological innovation. Since metals science is a relatively recent manifestation, the scientific research opportunities for the industry are worth careful appraisal.

Best of all, the *decisions* to innovate—in research, in process technology, in manufacturing automation, or in marketing strategy—are largely in the hands of you, the leaders in the industry. This, I submit, is in itself a desirable innovation.

The Emergence of Atomic Power

An address before the Engineers' Week Meeting, Chamber of Commerce of Kansas City, and Western Chapter, Missouri Society of Professional Engineers, Kansas City, Missouri, February 19, 1964.

difficulties appear with such regularity in complex research ventures that they should be expected, but because they are not, our immediate timetable for research progress is under constant revision. However, the impact of the completed work, five years or more hence, may easily exceed the blue sky predictions made earlier. We hoped that "so-and-so" would be possible eventually, and eventually it was. However, the work took a longer period of time than we planned, but happily we got quite a few unexpected good "breaks" en route that have greatly extended the new concept and its useful application. We had a great deal of bad luck, too, but we simply discarded these results, and retained the good ones.

Atomic power is a case in point. After fission had been discovered and an actual self-sustaining reaction demonstrated by Enrico Fermi in Chicago three years later, there was no unanimous view among the "experts" as to whether, or when, the atom might become a source of power for peaceful purposes. Actually, of course, the wartime necessity of building a bomb somewhat precluded all but occasional thoughts on power generation.

After the war, we engaged in the typical short-range optimism. There was a rather widespread belief that, with the tremendous wartime accumulation of knowledge of the fission process and related technology, beating this particular sword into a ploughshare would be accomplished in short order. As it finally turned out, it took about a decade and a half to do this. That is hardly "short order."

After the first blush of post-war enthusiasm wore off, there came the realization that wartime research was intensively focused on building the bomb, and although much of the resultant knowledge and information was fundamental to atomic power production, there were huge gaps in atomic knowledge and understanding.

The most significant area of technological uncertainty, and the one which imposed an unavoidable ten-year delay in the program, was the nuclear power reactor concept itself. Given a wide variety of coolants, moderators, fuel configurations, neutron energy spectra, and structural materials, it was inevitable that a field day for reactor inventing should have taken place. The possible permutations and combinations of these elements of reactor systems were almost numberless. For a happy, carefree period, reactor systems were being invented far more rapidly than they could be investigated either scientifically or technologically. During this period, no self-respecting atomic scientist would be caught without a new reactor configuration, and if he had reasonable ingenuity, he could accomplish the required inventing over a long weekend.

In the hands of a persuasive and competent individual, a new reactor concept can look very attractive, particularly since it is not possible to determine its basic merit or lack of it by debate alone, although that was attempted frequently. The most attractive concepts became the subject of serious investigation and major support by the AEC, and after some years many reactor projects progressed through the stage of pilot-plant or semiworks construction. It became clear that unless such proposals were carried to the point of a pilot plant with a significant power level, no meaningful appraisal of the basic merit of the reactor configuration could be made. Even in this case, the scale-up of the pilot plant, even to a small power-producing reactor, involved serious additional design and control problems. It became equally clear at about this time that each exploratory reactor project would involve an expenditure in the range of $50 to $100 million, and after many such projects had been funded, there still remained a long and seemingly inexhaustible list of reactor proposals, each complete with proponents, enthusiasts, and a prospectus that could not be proven wrong for much less than $100 million.

At this point something had to give, and the conclusion was reached in many quarters—perhaps simultaneously—that some of the companies that intended to build reactors would have to be brave and make a choice on the basis of the then available knowledge if *any* reactor system was *ever* to produce power economically. The choice—in this country, at least—was made in favor of a *water moderated and cooled* reactor system. This specification still leaves room for a great deal of choice, as for example between high-pressure water or boiling water, and many other important variables. But even this degree of decision was a godsend, and practical atomic reactor development progressed by leaps and bounds following this decision.

Few decisions are unanimous, nor was this one, but the water reactor has become the focus of such intensive development that there is no doubt that it is and will remain a primary nuclear steam source for a long time in the future.

During the ten-year power-reactor Derby, the betting odds changed dramatically from time to time, as new data became available and as new knowledge of basic reactor processes and phenomena were established. The general public was not completely on the outside of this contest, and to a considerable extent shared in the cycles of optimism and pessimism that accompanied the swings in technological progress. I suspect, however, that the public has become somewhat weary of listening to the pros and cons of the atomic power situation, with an understandable lessening of interest in this two-decade debate. Perhaps for this reason, and because the issues have become increasingly complex and technical, we have somehow missed an opportunity to tell today's most exciting technological news. We have missed the boat in letting people know that, as of right now, one of the most significant landmarks in the history of mankind's power progress has been reached.

Atomic power is here now, and we do not have to wait until the 1970 period for this historical event to take place. Atomic power is competitive today—economically and in every other way—with conventional methods of generating electric power in a substantial fraction of the power-generating areas of the United States, and in well over half of the areas of the free world. In new power plants now going into operation, nuclear reactors are successfully battling—head-on, competitively, today—power generated by conventional fossil fuels. Thus, the nuclear power industry has become an important new member of our private enterprise system.

Actually, the sudden emergence of atomic power has surprised even some of those most closely associated with it. For example, the manager of the General Electric Company's Atomic Power Equipment Department said recently, "We are roughly three years ahead of a self-imposed schedule for development . . . and the nuclear industry generally is perhaps four or five years ahead of the stated national goal."

Although the United States government investment in wartime bomb development provided essential foundation knowledge in the field, and the later AEC demonstration reactor projects contributed essential pump-priming efforts for the new industry, the nuclear power industry today still retains the basic characteristics of a private business. This rapidly growing business is presently supported with huge and increasing amounts of private capital supplied through the courage and foresight of some remarkably venturesome private businessmen in the electric power-generating and electrical manufacturing industries.

This is an extremely competitive business; witness, for example, the recent Herculean struggles between manufacturing companies in the nuclear power business to secure contracts for major power plants. Meanwhile, the reaction from the coal, gas, and oil industries—and, more recently, water-power generation—is providing the kind of competition which can only be a source of benefit and joy to consumers of electric power everywhere.

Why is the sudden emergence of atomic power such a surprise to so many people? To seek an answer to this question, I spent some time recently thumbing through some old magazines, scrapbooks, newspaper clippings, and other memorabilia of this exciting fifteen-year period during which atomic power emerged from the morass of a highly complex technology to achieve the status of a billion-dollar business.

First, I would call to your attention one of the most remarkable pieces of science reporting in history: a story by William L. Laurence in the May 5, 1940, issue of the *New York Times.* The headline said, "Vast Power Source in Atomic Energy Opened by Science . . . Relative of Uranium Found to Yield Force 5 Million Times as Potent as Coal." In his story, Mr. Laurence presented remarkably complete facts about Uranium-235, the concept of the critical mass, and "the tremendous implications this discovery bears on the possible outcome of the European war." The article envisions a water reactor and says prophetically that because water can be used to regulate the speed of neutrons, "the process is admirably suited for power purposes."

Incidentally, I can call your attention to Mr. Laurence's story now, although there was a period during World War II when a lot of people wished the entire May 5, 1940, issue of the *Times* could somehow be magically removed from the face of the earth.

Shortly after the appearance of this article, World War II's major security wall surrounded atomic work. It is reliably reported that anyone who even asked for this pre-war atomic article at any U. S. public library was followed home by the F. B. I., and Laurence himself tells of having a security officer remove a copy of the story from his briefcase. I suppose there was the question of whether Laurence could legally read his own article.

Peaceful power from the atom receded from public discussion for several years, but immediately after the bomb had been announced in August of 1945 there was a flood of conjecture. "Atomic energy," President Truman said, "cannot now be produced on a basis to compete with coal, oil, and falling water. There must be a long period of intensive research." England's Sir John Anderson said, "It might produce something that will revolutionize all industrial life, but it will take many years of research before an effective process is worked out."

An associate of mine—not a scientist—recalls that on the evening of the day the Hiroshima announcement was made, filled with the overconfidence of a youthful warrior home on leave after winning a war in Europe, he bet a friend that "the war will be over within thirty days and within ten years we'll be making all of our electricity with atomic energy." He admits to collecting on the first bet, and to having arranged the terms of the second so that it was neatly forgotten before he had to pay off.

"My optimism about atomic power was based entirely on complete ignorance of what I was talking about," my associate now says, adding, "I imagine you people in the business were somewhat more pessimistic."

We were, indeed. After the first burst of enthusiasm, the dawning realization of the technical, economic, and political complexity of the atomic-power problem came home, and very little was said publicly by scientists and engineers on this subject for some years immediately after the war. At a World Power Conference in 1950 there was some trouble scheduling an American paper on the subject of atomic power. Later the same year, an American scientist, writing in *Nucleonics* magazine, described the prospects for industrial nuclear power as "dismal." In the same magazine a year later, another author (whose name, according to the byline, was C. G. Suits) did not use such words as "dismal" but did say, "The economics of atomic power are not attractive at present, nor are they likely to be for a long time in the future." However, the final sentences in this article still look good on re-reading now. They say: "The future solution of difficult technical problems will be accomplished partly with knowledge now at hand, and partly with new knowledge which we are certain to gain from continuing scientific research. On a long-term project it would be a great mistake to underestimate the importance of the latter to the final solution."

In the same article I estimated the future capital cost of a 1000-megawatt nuclear power plant in the \$50 to \$100 million range. The going price for plants of this size is approaching this estimate.

In the next few years—the early 1950's—there began to appear more and more demands for the government to make it possible for private industry to get into atomic power development. Meanwhile, my scrapbook shows that in England some of the chief proponents of atomic power agreed that it would *never* be cheaper, but that more energy was needed regardless of the cost. And in this country, in a 1952 review of progress on nuclear power, one of my associates wrote: "Articles on nuclear power, and this one may be no exception, frequently merit the Churchillian paraphrase that seldom has so much been said about so little by so many."

For a number of years, of course, there was great concern about the availability and cost of uranium as a fuel. The optimistic prospector soon provided an answer for this stumbling block, and then —in the mid-1950's—confident engineers began to say such things as, "now that there is evidence that uranium will be abundant and cheap . . . and that it will not be necessary for the next several decades to go to extreme lengths to conserve this fuel, our studies of the economics of power plants show that the key to commercial atomic power is a low-cost reactor. It is our belief that this objective can be achieved by inventive design."

In the cycles between optimism and pessimism, between hope and frustration—cycles which confused the general public at least as much as the businessman and the investor—there were bound to be certain times when the prophecies were remarkably accurate. Selecting these is, of course, much easier now than it was at the time.

In 1953, Columbia University's John R. Dunning said, "The principal question that will face builders of power plants ten years from now will be 'Is this to be a nuclear or non-nuclear plant?' and the answer will be determined by the availability of conventional fuel." He could not have been more right.

But the pendulum kept swinging to and fro. Study teams from

electric utilities concluded that "no reactor could be constructed in the very near future which would be economic on the basis of power generation alone." The word "dismal" disappears from the editorials in the trade journals of the power industry, but the word "despair" is still there as they talk about the economic possibilities of nuclear power.

On the other hand, loud and clear voices began to be raised on behalf of the liberalization of governmental regulations to permit contributions by "the energy, brains, resources and techniques of private enterprise," and this was a source of encouragement. The original Atomic Energy Act had created a government monopoly in the atomic field, and left no real opportunity for the development of a private enterprise industry. So long as this remained the case, there was little incentive for the investment of private capital in an atomic power business. The growing realization that American private industry was willing and anxious to contribute its unique resources to the growth of this industry provided the motivation for a basic change in the law, which took place in the form of the Atomic Energy Act of 1954. This Act provided for the private ownership of production facilities and set forth the broad framework for the development of atomic resources in the tradition of American free enterprise.

By late 1955, concrete action–involving concrete, in fact–began to emerge, and General Electric and the Commonwealth Edison Company of Chicago signed a contract for the construction of a truly major atomic power plant. This venturesome decision–to build a massive production facility entirely with private funds–was *a* turning point in the history of atomic power, not the *only* one, by any means, but an important one.

The courage to build the Dresden plant looks the more courageous when viewed in the light of the McKinney Report [1] which was issued to the Joint Committee on Atomic Energy at about that time–1956. The report listed the prospective cost of atomic power attainable by the then future year 1960 in the range 8.1 to 9.7 mils per kilowatt hour, based on 1955 dollars. Note this fact: *our current costs are between five and ten years ahead of the average estimated power costs of this significant report.*

[1] Report of the Panel on the Impact of Peaceful Uses of Atomic Energy–January 1956.

So, today, the goal of competitive atomic power *has* been reached, well ahead of optimistic schedules. Why? How has it been achieved? The answer makes a most suitable subject for discussion during Engineers' Week, since what has happened is one of the outstanding triumphs in the history of engineering.

The engineers, aided and abetted by physicists, metallurgists, and chemists, had to not only think big; they also had to build big, because in this league size alone is a basic economic virtue. Other things being equal, which they never are, a big reactor will always be more economical than a small reactor. It would have been less of a risk to design and build an 18 megawatt plant at Dresden, but, it took about 180 megawatts to qualify in power production costs. This kind of engineering requires not only knowledge and experience—but also courage of a high order. Future large plants, and many are now in progress, are greatly benefiting by Dresden experience, which is providing an essential foundation for the further growth of the power reactor concept.

The recent announcement of plans for two very large reactor power plants by Niagara Mohawk in New York State and Jersey Central Power and Light in New Jersey will add a further new dimension to the growing edifice of atomic power production. I do not know how many reactors are in use, design, and construction world-wide, but for the company with which I am associated, the number—as of yesterday—was thirty-four.

Now for a few vital technical details that account for and give substance to the outstanding contributions which American enterprise has made to the realization of the dream of economic atomic power.

First. A significant factor in reactor power cost is determined by fuel life, and to get atomic power off dead center, it was necessary for the manufacturers to guarantee fuel life well beyond any actual experience. For example, at the time Dresden was planned, fuel life guarantees made to the customer were five times the best available actual life test data at that time. Although it is too early to make a final accounting, accumulating experience is good, and fuel life will probably exceed its guarantees. A major contribution to the optimism concerning fuel life is the remarkable stability of uranium dioxide. In early reactors the fuel was usually uranium in the form of metal or alloy, perhaps reflecting long experience with fuel at Hanford and Savannah River reactors. With high burn-up however, the metal can distort badly, or corrode, leading to intolerable mechanical problems. The discovery that uranium dioxide is practically free from this difficulty was an authentic breakthrough.

Two. One of the major design and construction costs in any reactor plant is in the metallurgical materials associated with the nuclear steam supply system. In this important area, great progress has been made and is continuing in minimizing the size and complexity of the system, and thus its requirement for premium materials. Dresden experience has provided a whole textbook of reactor metallurgy to replace the partly intuitive design that was once necessary, and we now know how to make economical use of the relatively costly metals and alloys required in reactors.

Three. Another major component of reactor plant costs is the containment vessel, which is designed to provide a super-safe housing for the atomic reaction. Initially, this housing was invariably a huge steel sphere, but we are now replacing this brute force approach by a "pressure suppressor" in which a large water reservoir replaces the tons of steel and does a better job much more economically.

Four. At one time, early in the development period, the source of uranium to supply the fuel for the atomic power industry was a real question, but prospectors—world-wide—have come through with abundant supplies for the immediate future. Whether the "immediate future" refers to a period three decades ahead, or six decades ahead, is a detail on which moderately warm arguments can be generated in some quarters. However, for the time being the problem is to build atomic power plants fast enough to keep the many presently available sources of uranium in economic production.

Five. A power reactor is simply a boiler, heated by nuclear fuel. Under the intensive study by reactor engineers, all of the basic factors of boiler design and geometry have been re-investigated, with many significant improvements in traditional performance.

These are the *principal* points on which the present economic atomic power achievement rests.

The choice of a water reactor and the decision to build Dresden are such massive highlights in the picture of atomic power that they deserve some additional comment.

There are some who say that we should have selected another coolant and moderator system, in preference to water, and if we had, *that* alternative system would now be economic. They may be right. But the ordinary water-cooled-and-moderated reactor has brought us to the point of economic atomic power far in advance of predictions made at the time the choice was made, as we have noted in our review of the McKinney Report.

So, how lucky can we be? The water reactor was a *right* choice, and it is equally clear now that *some* alternative systems selected at that time will *never* generate economic power, because the system selected was a *wrong* choice.

With respect to the Dresden decision—as this key factor in the emergence of atomic power recedes deeper into history—it is even more clear that here is an example of industrial courage of the first order. Because of the early state of the nuclear boiler art—and the lack of adequate life-test experience on many major materials, processes, and components, especially related to fuel—the engineering risks, and inevitably the related financial risks, were of major proportions. It will take some additional time to write the final chapters on the Dresden success story, but the main outlines now appear clear, and the magnitude of this giant stride forward in atomic power now seems well established.

Now, I have spent some time on a review of the emergence of atomic power for several reasons. The attainment of a major new energy source is a milestone in the progress of civilization.

For bulk electric power generation, there have been only two energy sources available—falling water and the fossil fuels—in the entire history of mankind. Now a third source—nuclear energy—has been added, and the period of time required to accomplish this modern miracle has been substantially less than a human generation. It is important to view an event of this magnitude with perspective and understanding so that, hopefully, we may look ahead and better perceive the shape of things to come.

Power from uranium *fission* is only a first step, although a giant step, on the road to the utilization of energy from the atom. Next come atomic breeder reactors—the "second generation" atomic power—and later, if we are lucky, atomic *fusion*.

The breeder reactors are now in the first phases of operation. The strategic role of breeder plants will be to greatly extend the usefulness of our uranium fuel supplies, which are only partially used up in present reactors.

Thermonuclear *fusion* as a power source is presently the physicist's dream, much as *fission* was twenty years ago. If this dream comes true, we will add a new and perhaps final energy source that will satisfy the requirements of mankind for the indefinite future. This source is basically nuclear, too, but the starting point is water, from which heavy hydrogen is extracted, rather than uranium. In a fusion reactor of the future we will, so to speak, "burn" water at hundreds of millions of degrees, forming new heavy atoms and thus releasing huge quantities of energy.

Needless to say, there remain "a few details to be worked out." However, when the New York World's Fair opens just two months from now, you will be able to witness an actual demonstration of the controlled nuclear fusion process in experimental form. How long will it take to "train" this process for electric power generation? It looks like a long time, but we should remember what we learned during the development of boiling-water reactors: *We will get from here to there—not only with the knowledge we have today—but also with a lot of additional discoveries we will make en route.* So I venture to predict that it will not take as long as we may think—now—to achieve fusion power.

Now that we may view—in perspective—the many important decisions upon which the atomic power development is based, it is appropriate to ask, "How good were these decisions?" From this point of vantage, I think the major decisions would be hard to improve upon.

The decision of Congress to establish the Atomic Energy Commission, the passing of the Atomic Energy Act and the decision to revise the act in 1954 to permit private ownership, the decision of the Commission to sponsor the development of demonstration power plants, the decision of the electric utility industry to actively participate in the building of atomic power generation, the decision of major manufacturers to concentrate on water reactors for "first generation" atomic power, the decision to build Dresden—these decisions, and many more, marked pivotal points in the developing history of atomic power. By and large these decisions were *wise*, *timely*, and in every case *courageous;* none of these decisions was made by one man; on the contrary, they all came from the free play of ideas in a democratic society. I am proud to be a member of the free democratic society that can accomplish modern wonders—like atomic power—in our lifetime.

Man-Made Diamonds—
A Progress Report

American Philosophical Society Spring Meeting, April 23, 1964, Philadelphia, Pennsylvania.

This is, first of all, a story about carbon, the most aggressively gregarious element of the periodic table. Carbon, compounded, is so versatile in nature that the major branches of chemistry are determined entirely by the presence or absence of this element. Carbon, by itself in crystalline form, offers another distinctive dichotomy: it can be slippery, messy graphite worth a few pennies a pound; and, at the other extreme, it can be magnificent diamond, nature's hardest and most glamorous substance, sometimes valued at millions of dollars an ounce.

It was once thought that to convert a plentiful, cheap, and even worthless material into something rare and valuable was a peculiar preoccupation of alchemists. If you will forgive the play on words, I will say that this is now a worthy *occupation* for *all chemists*, all physicists, and all metallurgists—at least in industrial laboratories.

There have, of course, been some spectacular successes in this effort. Witness, as examples, the conversion of nearly worthless natural silicon found in sand to the highly valuable semiconductor grade silicon on which a large segment of the semiconductor industry is based; or the conversion of cheap hydrocarbons—residues from coal, oil, and gas—to monomers and then to polymers of great utility and economic value. The planning, execution, and fruition of a successful venture in modern alchemy are one of the great satisfactions of today's scientists, technologists, and engineers.

It is not surprising that the history of science is replete with attempts to convert base carbon to noble diamond, and that the story is interlaced with claims and disclaimers, mystery and jealousy, suspense and intrigue.

The task itself is conceptually simple. Graphite is composed of carbon atoms, tightly bound in planes, with comparatively weak atomic forces holding the planes together; the flaky character of graphite is the macroscopic evidence of this atomic structure. Diamond is composed of precisely the same carbon atoms squeezed together to achieve substantially uniform interatom distances throughout the lattice. The diamond lattice is a neat packing configuration which gives each atom a tight hold on each of the four atoms arrayed around it. In other words, to turn graphite into diamond, all one must do is press it into the more compact atomic arrangement of diamond. What could be simpler?

But, as countless investigators over the years have learned to their agonizing dismay, Mother Nature did not intend her own achievement—probably performed at mysterious, unexplored depths far beneath the surface of the earth—to be easily duplicated by man. The late Professor Percy Bridgman of Harvard, a Nobel laureate for his work on high-pressure phenomena, put it succinctly. "Graphite," he said, "is nature's strongest spring."

Or, in the less grammatical phrase of one of our laboratory associates, it might be said that "it is easy to squeeze carbon atoms together, but very difficult to keep them squz."

Professor Bridgman spent many years trying to make diamonds during the 1920's, '30's, and '40's and found serious natural roadblocks at every step on this road. The most serious roadblocks were: (1) an understanding of the diamond-making process, and (2) the requirement of high pressure *and* high temperature simultaneously, both held for an appreciable interval of time. Bridgman produced pressures which were much greater than required, but the "spring" relaxed to graphite when the pressure was reduced. On the basis of what he learned, Bridgman concluded that all previous claims to success in diamond-making were based on wishful thinking and not scientific proof.

In 1951 my associates [1] and I decided to launch a new, allout assault on this problem. We resolved to delve into the process whereby diamond might be made, using not only new techniques for attaining fantastically high pressures, but also for attaining the simultaneous high temperatures that would be required to "latch" this spring: to form diamond from graphite.

I would now like to discuss three aspects of the diamond problem in turn: the pressure, the temperature, the chemistry—and then add a comment about a fourth factor: time.

Using the strongest available materials, unique prestressing techniques, and geometrically complex designs which permit the microflow of pistons and cylinders at extremes of pressure, the project to which I have referred has developed equipment capable of with-

[1] A group led by A. J. Nerad and including Drs. F. P. Bundy, H. T. Hall, H. M. Strong, R. H. Wentorf, Jr., and H. P. Bovenkerk.

standing pressures of up to 3,000,000 pounds per square inch. This is 200,000 times atmospheric pressure, or, as the scientists put it, 200 kilobars. It is the pressure that would be found some 330 miles beneath the earth's surface. Or, if you prefer, it represents the pressure of a column of mercury 100 miles high, a rather substantial barometer reading. These astounding pressures, however, are not a record by any means. Bridgman attained much higher pressures by far simpler means. The new progress, in apparatus, came from a chamber design that permitted the simultaneous attainment of high pressure *and* high temperature.

In terms of temperature, we have consistently referred to "5000 degrees." In the early days of diamond-making we meant Fahrenheit, and 5000°F is still the temperature which can be held for "a long time"—hours if need be. More recently, temperatures of 5000°K—about 9000°F—have been achieved for periods of a few hundredths of a second in our superpressure chambers. This is nearly the temperature of the sun's surface.

Chemistry was at least as important to the initial achievement of diamond-making as the temperature-pressure combination. It has been found that certain metals, which are molten in the process environment, act as catalysts and greatly enhance the rate of the required change in the lattice arrangement. The net result of catalysis is a higher yield at a higher rate, at significantly reduced temperatures and pressures. Thanks to pressure, temperature, and chemistry, the graphite "stayed squz" and emerged as diamond.

As I have indicated, it is possible in superpressure chambers to hold the chamber conditions for long periods of time and thus to achieve a steady state for reactions under study. A temperature limit of about 3000°K is set for such steady state use primarily by the available material properties, particularly the melting points of ceramic bodies. Under steady state conditions, say 1.5 million pounds per square inch and 1500°K, the diamond stable region of the carbon phase diagram is attained, but the diamond transition was not originally observed except in the presence of the catalyst. Francis Bundy showed, in static apparatus, by extending the temperature to 5000°K for transient excursions, that the *direct* conversion of graphite to diamond—that is, without catalyst action—takes place.

This is illustrated strikingly by the photograph of a bar of graphite which has been squeezed to diamond [2] in its central section by these extreme transient conditions. Still more recently, Bundy and Robert H. Wentorf, Jr., have detected direct conversions at the upper end of the steady state temperature-pressure regime. And, the answer to the question of *time* is one of nature's bounties; the time is "short"—a matter of minutes or seconds, once reaction conditions have been attained.

The announcement that diamonds had been made at the General Electric Research Laboratory appeared in February, 1955. To be honest, it should be noted that the diamonds made up to that time were small. Although you could see them with the naked eye, it was better to use a microscope, and it was important not to sneeze, for that might have done away with the entire world supply at that time.

The initial reaction to this announcement was, nevertheless, spectacular. For example, on the stock market, the total value of General Electric stock increased that day by more than $300 million. Although this was an important discovery, the market reaction was in this case, as in other cases that might be mentioned, more emotional than analytical.

To justify a $300 million increase in market price on the basis of a diamond business, even with very favorable profit ratios, would require an annual sales volume greater than $3 billion, which was the approximate level of General Electric gross sales at that time. The total world-wide sales of the diamond industry at that time were at about one-tenth of this level. After a few days, during which such calculations were undoubtedly performed, the price of General Electric stock "recovered" to nearly its former level.

The announcement released in 1955 made it clear that Man-Made diamonds were small and were not of gem quality. One might jump to the conclusion that only diamond gems are really valuable. To put this question in proper perspective one must bear in mind that the bulk of the world's diamond production from all sources, world-wide, is in industrial grades with a total market value of about $60 million. The very much smaller physical volume of gem diamond sells at a very much larger unit price to achieve a world-wide sales level of about $300 million per annum.

[2] Density of graphite 2.25—diamond 3.52.

A few points should be made concerning the *value* of diamond, the *price* of diamond, and some factors which determine both. Diamond has only one unique property, hardness, and this property is fundamental to its industrial use *and* its usefulness as a gem. But, as the mathematician might say, hardness is a necessary but not a sufficient attribute of a crystal to be used as a gem.

As a gem, diamond has other important properties such as a high refractive index and must have other additional qualities, such as size, color, and optical clarity. But, without its exceptional hardness, diamond would not be pre-eminent as a gem. A higher refractive index is available in other crystals, especially titania, and large clear crystals of many other minerals are readily available.

Perhaps you have had the experience of wearing a ruby or sapphire gem in a ring for a period of years. If you have, a careful examination of the stone will show definite evidence of wear. On a true hardness scale, however, diamond is about ten times harder than ruby and sapphire and it will accommodate the appreciable wear requirements of a gem stone. The same unique property, hardness, is the essence of diamond's industrial usefulness.

A suitable diamond saw can be used to cut concrete, which it does, much as a steel saw cuts wood. Even more important, diamond wheels are used to cut the hardest and strongest metals and alloys—like Carboloy cemented carbides—and the vital industrial usefulness of these carbides depends upon the availability of diamond for cutting these materials during manufacture.

Fortunately, diamonds do not have to be beautiful in order to be hard. The bulk of the diamonds dug from the earth are small, discolored, and not worth polishing, but their hardness is the same as diamond of gem quality. These poor cousins of the million-dollars-an-ounce gem stones are worth only a fraction of gem prices. This fraction, however, is still worth about $6,000 a pound on the basis of the lowest current price of $2.65 per carat.

Soon after the discovery of the laboratory diamond process, we had to face the question: "Can Man-Made diamonds be produced that are good enough and inexpensive enough to compete with natural diamond bort?" The answer came in a remarkably short time. A team of scientists, engineers, and manufacturing experts joined forces to make these Man-Made diamonds a competitive industrial product, and this objective was achieved in less than three years.

With this final step, America acquired for the first time an independent source of industrial diamond, a material which is very important to the industrial economy, and which is absolutely vital in key defense industries. The assurance of a steady supply of industrial diamond has encouraged broader use of this exceptional material for new applications. And, surprisingly enough, the new man-made product has turned out to be better than natural diamond for many applications.

Man-Made diamonds are grown under controlled conditions, and remarkable control can be exercised over the properties of the tiny crystals. For some purposes it is desirable to produce frangible crystals, so that fresh cutting surfaces will be exposed during use; such crystals can be grown. For other purposes, for example, metal-bonded circular saws, the ideal crystal would be an octahedron of the correct size; such crystals can be grown. This is an improvement on natural diamond because tiny crystals are not recovered from natural sources, and hence fine mesh size natural diamond is generally produced by crushing, which does not readily yield the desired crystal shape. Thus, nature's hardest substance is now subject to quality control, which has significantly enhanced its industrial usefulness.

A "diamond mine," in Detroit, part of the General Electric Company's Metallurgical Products Department, is now one of the largest single sources of industrial diamond in the world. The product is made reliably and at a price that is directly competitive with natural diamonds. The total quantity of diamonds from this mine to date is not properly measured in carats, but in tons.

Most industrial diamonds, including Man-Made diamonds now on the market, are very small: up to about half a millimeter in diameter and weighing only about one-thousandth of a carat or less. However, diamonds in these small sizes in the form of abrasive grit fill a large portion of industrial needs. Meanwhile, considerably larger Man-Made diamonds, one or two millimeters long and of good quality, can be made.

Research continues toward the development of larger industrial stones, up to and including the carat-size diamonds required for oil-drill bits and wire-drawing dies. By growing the crystals in a multistep process, adding a layer at a time, it has been possible in the laboratory to make diamonds weighing more than two carats. It must be noted, however, that these crystals are quite imperfect. They are not very strong because of inclusions, particularly between the layers, and because of internal strains in the crystal structure. They are definitely not the clear, perfect stones suitable for polishing.

The scientific achievement of Man-Made diamond, and the commercial success of the product, have served to stimulate broad interest in high-pressure research at laboratories all over the world. Approximately 200 laboratories are now equipped for superpressure research, and much new knowledge of nature's physical extremes is coming from this work.

Many of the exploratory efforts in superpressure incorporate so-called "dynamic" or explosive shock techniques in which very high temperatures and pressures up to a million atmospheres are obtained for very short periods of time, usually a few millionths of a second. Diamonds have been made by these processes. Alternatively, very high "static" pressures, such as 500 kilobars, together with high temperatures, may be attained at a sacrifice of reaction volume, and some interesting researches are being conducted by this method. Many new forms of matter are being explored at superpressures, and it is clear that the final chapters to the story of superpressure have not yet been written. One must expect exciting news from this source of scientific exploration for many years in the future.

The "belt" apparatus developed at the General Electric Research Laboratory—because of its unique combination of 200 kilobar pressures and 3000 to 5000°K temperatures—plus workable pressure-chamber volumes—has thus far provided the only economic diamond-making method as well as the research technique with greatest versatility.

Superpressure research around the world is now concerned with such subjects as the determination of the elastic constants of various materials, the properties of semiconductors at high pressure, nuclear magnetic resonance in solids and liquids under pressure, the determination of fixed points on the high-pressure scale, geochemical and geophysical studies, a variety of thermodynamic approaches to the study of activated processes under high pressure, and scores of other seemingly esoteric but inherently valuable research objectives.

In more down-to-earth language, it might be said that, obviously, it is possible to put many materials other than carbon into these chambers and subject them to high pressure and high temperature for long periods of time. Such possibilities provide a virtually infinite challenge for the research scientist.

One of the first results of this kind of exploration was a new form of boron nitride. Boron nitride, in its common form, has a structure very similar to graphite, with boron and nitrogen replacing carbon in the lattice. It is a slippery material so much like graphite in mechanical properties that it is often called "white graphite." This fact, plus the proximity of boron and nitrogen to carbon on the periodic table—they are the next-door neighbors to carbon, one on each side—prompted Wentorf at our laboratory to try superpressure techniques on boron nitride. The result was spectacular: a completely new material never found in nature.

"Borazon," as it was named, is in the same range of hardness as diamond; indeed, it is the first material other than diamond that has ever been able to scratch diamond. Because borazon is more oxidation-resistant than diamond, we believe it will eventually have important industrial applications.

In addition to borazon, more than twenty new forms of matter have been created through superpressure research in the one program with which I am most familiar. Principally, these are chemical compounds, although in some cases they are single elements converted into new crystalline arrangements. At present there is considerable scientific excitement in the laboratory concerning evidence which points to the possibility of a completely new crystal form of carbon. However, this work is still incomplete and is not yet ready for publication. In the case of both germanium and silicon, we and other workers in the field have identified some new high-density forms substantially different from the crystal structure which helps give these materials their unique value as semiconductors.

But I said at the beginning that this was to be mainly a story about carbon, that many-faceted element which is so dominant in science and life. What is the future for *carbon* as we continue to heat it and squeeze it and catalyze it?

We are gradually learning how to make larger and larger diamond crystals—how to control the nucleation of crystal growth and achieve bigger single crystals with fewer occlusions and imperfections. We are learning how to keep unwanted atoms out of the carbon structure, and it seems reasonable to hope that stronger, more perfect crystals will result. Finally, we like to look at the phase diagram for carbon and dream of the time when we can attain the presently unattainable areas of temperature and pressure. Perhaps carbon will assume a new form harder than diamond.

In any event crystallographers can conceive of atomic arrangements which might put carbon atoms even closer together, and more firmly bound, than they are in diamond. The result might be anything.

Seven Areas of Promise in Science and Technology

An address at the Engineering Services Meeting, Cherry Hill, New Jersey, September 16, 1964.

The research laboratory with which I am associated is engaged in a broad spectrum of scientific work, covering major areas of the physical sciences, new and promising fields of mathematics, and, to an increasing degree, the life sciences as well. It will come as no surprise to you that we are constantly reviewing our research programs with the aim of identifying and appraising *scientific and technological resources that promise to provide a significant component of future industrial growth opportunity.*

Now, the research areas that I will describe happen to number seven, and while this number is partly fortuitous, it is also convenient, for I wish to draw some parallels to the "Seven Wonders" of the ancient world. You will recall that these "Seven Wonders" provided some sort of symbols of human accomplishment to the ancients. With the exception of the Pyramids of Egypt, none of these artifacts has survived to this day, yet there is little doubt that, in the historical environment in which they existed, the Hanging Gardens of Babylon, the Colossus of Rhodes, the Pyramids and the ancient temples, etc. were impressive enough to qualify as authentic spectaculars.

I mention the "Seven Wonders" because, if each of the seven areas of science to be described yields its "promise" fully as a result of future research, they too will qualify as Seven Modern Wonders—authentic spectacular developments in terms of their impact on technology and the opportunities created for the growth of new business enterprise. Although we cannot hope for a ten-strike in each of these research areas, they should be considered over a sufficient period of time to allow whatever progress nature permits to develop and mature fully.

In forecasting scientific and technological progress, the period of time being forecast is crucial. For the present purpose I have in mind a "New York World's Fair Unit" of time, since these expositions provide indices that conveniently bracket some important developments from which I wish to derive perspective, and since the time interval—twenty-five years—permits or even ensures spectacular progress.

To digress a moment, I believe there is substantial evidence to support the view that scientific and technical people are frequently too optimistic in their short-range forecasts, and too pessimistic—or unimaginative—in long-range forecasting. Without discussing this observation in detail, one of the factors that is certainly involved in underestimating the future, is the human propensity to build the image of the future as an extrapolation of today.

Extrapolating today is a good start in forecasting future progress, for it provides for the least speculative component of our future scientific and technological environment. However, one or two decades is a very long time in modern science and technology, and over any such period of time there is another important component of progress that is only slightly less speculative: this component consists of the essentially unpredictable new basic discoveries that are certain to be made. Although the outlines of these discoveries may be very dim at present, their occurrence, given a decade or two, is a practical certainty. For balance, therefore, one must allow for these scientific dividends in long-range forecasts, for they will account for a substantial portion of future scientific and technological capability.

After considering several current prospects for future scientific progress, for perspective, I will turn back to the start of our last "World's Fair Unit" of time—1939—and compare the forecast which might have been made at that time with the reality we face today, at the end of the period. It is especially interesting to contemplate the nature and impact of the *unpredictable* discoveries of that period, for some of them have become major areas of new American business enterprise.

FUSION

Now, to turn to the "Seven Potential Wonders of the Modern World," let us first consider *thermonuclear fusion.* You are familiar with the fact that the present atomic power business is based on the process of atom splitting, or *fission,* and that research is now directed toward a second nuclear process—atom *fusion*—as a possible future generating process. Undoubtedly, you also know that to accomplish fusion in a useful manner it is necessary to achieve and contain extremely high gas temperatures.

Since these temperatures are higher than can be tolerated by any material structure, a magnetic field configuration, sometimes called a "magnetic bottle," is conceived to be the solution of the containment problem. This is a key technical problem, and if we can achieve and contain the high-temperature environment, we shall be able, so to speak, to "burn" heavy hydrogen or similar atomic species to produce energy. The cost of fuel for such an atomic furnace would be almost negligible, and the supply of this fuel, in the form of heavy water from the oceans, would be almost inexhaustible, and there is the prospect of greatly reduced by-product radioactivity.

It would be fair to say that there are a number of additional serious technical problems to be solved before fusion becomes a practical energy source, but some of the objectives are attractive enough to provide the motivation for a great deal of serious scientific thought. General Electric, for example, has had a fusion research program for the past seven years.

Now the detailed discussion of fusion research would take us far afield, and I intend to make only one point, and it is this. In the fusion demonstration at the New York World's Fair now in progress, we are able to heat deuterium gas to a temperature of about 60 million degrees for a few microseconds, and to observe the fusion of atoms, and thus to demonstrate the fusion process. This is, in a sense, the first practical application of nuclear fusion. At the New York World's Fair of 1939, it would *not* have been possible to demonstrate the technical feasibility of the atomic fission process—the unit process had just been discovered, and the chain reaction was not achieved until 1942. Yet, in the intervening period, fission power plants have become a major new industrial enterprise.

Since there are many scientific similarities and basic development and engineering analogs between *fission* power, and future *fusion* power, I find it difficult to avoid the conclusion that thermonuclear generation may become an economic power source before the next twenty-five years have passed.

FUEL CELLS

The *fuel cell* is a dream, which dates from about 1830, that we now hope may "come true." It is the dream of a battery that runs on gas. Of course, common "dry" cells also operate on somewhat similar electrochemical principles, except that in the case of a dry cell the "fuel" is actually the metal container of the battery, and as the fuel is used up, the battery is thus worn out. In the fuel cell proper, a gaseous fuel like hydrogen or ammonia—or a hydrocarbon gas or vapor—is continuously fed into the device, which thus continuously generates electric energy.

Recent fuel cell progress includes (a) the ion exchange type cell, which uses gaseous hydrogen fuel. These cells have very high conversion efficiency and are quite compact and simple in structure, but they require platinum as an electrode catalyst, and the fuel that they use is somewhat special. More recently, we announced (b) a low-temperature cell that operates on hydrocarbon fuels. This cell is a distinct step forward, but the problems of catalyst and structural cost are still present. A third step forward is (c) the high-temperature cell, which by virtue of its temperature operates on hydrocarbon fuels without catalysts. Recently this cell has shown fine progress, and we have also learned that important basic patent claims on this device have been allowed.

Although several technically feasible fuel cell systems are thus available, costs are still too high for broad-scale use. In the case of the low-temperature cells, the catalyst costs must be reduced. We must find substitutes for platinum, and some prospective materials are available. In the case of the high-temperature cell, we avoid the use of a catalyst, but because of the high temperature the use of such cells will be limited to applications where fairly large units can be employed, and where a warm-up time is feasible.

One can identify three stages of development of fuel cells: *Stage I. Military and Space Use.* We are now at the point of having fuel cells with performance that can qualify for military, space, and other specialty use. If we learn enough about fuel cells from these strategic applications, we may enter: *Stage II. Replace Batteries for Industrial Use.* If we learn enough about specialty fuel cells and can reduce catalyst costs, and increase current densities, we may make an entry into a large industrial market for portable power. Any industrial device that now uses storage batteries would be a prospective application for such fuel cells. If we learn enough about the industrial application, we may then enter: *Stage III. First Supplement, Then Supplant, Many Gas Engine Uses.* If we are lucky, and nature agrees, down at the end of the fuel cell road is the possibility of supplementing and finally supplanting many internal combustion engine applications by fuel cells. The gasoline engine is a very cheap power source, but it is very inefficient and it pollutes the atmosphere. The fuel cell is very efficient, does not pollute the atmosphere, but it is presently too expensive for many common applications.

A basic fact of nature is that the fuel cell is not a heat engine. It directly converts the electrochemical energy of combustion to electricity, without a heat cycle, and at very high efficiency. This is an important fact that will in the long run prevail, and I am convinced that we will eventually learn how to make cheap and efficient fuel cells. We may even conclude finally that the internal combustion engine is impractical.

SUPERCONDUCTIVITY

The discovery of *superconductivity* was made in 1911, and the phenomenon has been a fascinating laboratory curiosity ever since. Even at this date, more than fifty years after this discovery was made, the facts of superconduction are almost unbelievable. A superconductor is an example of a rare phenomenon—perfection in nature. Under superconduction conditions, which mean very low temperatures, many common metals and alloys exhibit perfect electrical conduction. If a direct current is started in a superconducting coil, the current will flow forever without attenuation.

The possibility of using superconduction, practically, in electrical equipment has been in the minds of scientists ever since the discovery was made, but there have been some major roadblocks in the way. For example, it has been expensive to attain and maintain the required low temperatures. If we could use liquid nitrogen, this problem would not be too bad, but the fact is that all known superconductors require at least liquid hydrogen temperatures, and most of them need to be within a few degrees of absolute zero, which requires liquid helium. Equally important, the superconduction effect was easily destroyed by a small magnetic field.

As a result of progressive improvements in cryogenic equipment, the attainment of extremely low temperatures is now a commonplace capability of most laboratories, and as a result of a discovery made in the Bell Telephone Laboratories a few years ago, there is now available a class of metallic materials that will permit very high magnetic fields in the superconducting state. Because of these two developments, superconduction is promising to make its way into practical specialty laboratory applications, principally for the production of very high magnetic fields for experimental and test purposes.

One additional major roadblock remains in the way of the use of superconduction in electrical power machinery; namely, use with alternating current. When alternating current is used, there is an effect, akin to hysteresis, that greatly reduces the conduction performance. Intensive work is under way to remove this final roadblock to the industrial power use of superconduction. If this final roadblock can be removed, we face the possibility of a real technological revolution in many types of a-c equipment, particularly transformers, which could be made in greatly reduced size by the use of superconductors.

INFORMATION SCIENCE

The great developments in the field of *information science*, including the development of computer machinery, new mathematical concepts for employing computers, and new technological advances in the process application of computer control, are now familiar to everyone.

For historical perspective it should be remembered that the so-called "computer" was conceived by Sir Charles Babbage in 1835. There was little development of computer machinery until World War II, when there was great need for such capability for the control of military equipment, particularly for the control of gunfire. The completely unrelated discovery of the transistor phenomenon in 1948 and its subsequent application to computers about ten years later marked the great upsurge in the commercial application of computers.

The only point I wish to make in connection with this subject is that, in addition to the very great future application of present computer technology, which is anticipated by nearly everyone, there are vast additional developments clearly evident in the broad field of information science. We are not at the end of this road by any means, but on the contrary, we are just at the genesis of furpolymers in bulk usually exhibit tensile strengths considerably below those of metals, polymers in the form of fibers—where structure is optimized and defects are minimized—exhibit strengths comparable to those of metals. As to the temperature limitations of polymers—which a disturbing number of product designers apparently still consider to be useful only near room temperature—I can remind you that at a recent American Chemical Society meeting there was described a commercially available plastic that is flexible at temperatures near absolute zero and that has a tensile strength of 5,000 pounds per square inch even at temperatures above 500°C. Some experimental polymers recently made have significant strengths even at a dull red heat.

In summary, from a business standpoint it is safe to say that polymers have long since made the transition from novelty decorative materials to structural materials of real consequence. We believe that very great additional progress is on the near research horizon, and we believe both the production and use of new plastics will be a significant factor in industrial growth during the years ahead. As examples of polymer progress, let me mention three announcements which have been made recently.

By irradiating thin polymer films—using fission fragments—it is possible to make what are perhaps the world's smallest man-made holes. These can be enlarged by etching so that they are uniform, and some very unique filters can thus be created. These filters have holes about six microns in diameter, which permit the flow of ordinary human blood but filter out cancer cells. The potential of such filters for early diagnosis of cancer is being studied by medical researchers at the Sloan-Kettering Institute.

Another new kind of polymer just announced publicly for the first time is unique because it does what polymers are not usually supposed to do: conduct electricity. Of course, we continue to look for better and better electrical *insulators* based on new synthetic polymers, but there may also be some very interesting applications for organic materials that are remarkably good conductors.

Yet another "new polymer" announcement involves some new permeable membranes that permit the flow of different gases on a selective basis. It has been known that thin films of certain plastics—notably silicone rubber—have different rates of gas permeability. For such applications, the thinner these films are the better. The catch has been to make extremely thin films without any holes, because the presence of only a few tiny holes can make the rate of "leakage" so much more rapid than the rate of permeation that the desired selectivity almost completely disappears.

Walter L. Robb, a member of our laboratory staff, has invented a way to make films out of special silicone rubber that are only about one mil thick and are so perfect that the chances of holes occurring are only about one in a billion per square foot. Some imaginative engineers are happily dreaming up simple systems for supplying submarines with air drawn from the water around them, for the purification of air in space capsules or moon stations, and for providing cheap, reliable oxygen to patients in hospitals or at home.

As a dramatic demonstration of the ability of these membranes to separate air from water—as sort of an artificial fish gill—some of our people have put together an experiment in which a pet hamster stays healthy in his own little "submarine" breathing air whose oxygen is replaced—by permeation—from the air with which the water (like all water, including the oceans to a depth of several hundred feet) is saturated. Meanwhile, the membrane gets rid of the excess carbon dioxide, which goes in the opposite direction. And some refinements of this demonstration would also make it possible for our pet hamster to get a fresh water supply through selective membranes even if his tiny "submarine" were submerged in sea water.

As I have mentioned, these are three new polymer discoveries announced in a period of a month. All of them may not be sources of industrial growth, although I would not want to bet against any one of them in the long run. They are indicative of the kind of progress that will continue to make polymer research one of the exciting areas of modern technology.

LIFE SCIENCES

Perhaps the most significant fact about the *life sciences* is that many physical scientists are turning to this area as an outstanding opportunity and challenge. After particulate matter—and gases, crystalline solids, and polymers—living substances probably do present the greatest structural complexity and challenge of all. But it is not only complex structure, but also life processes, functions, and organization that intrigue the investigator. Overlaying the scientific fascination of research in the life sciences is the well-founded belief that knowledge and understanding of life processes cannot help having useful consequences and influences on man and his environment.

One somewhat predictable area of very great potential significance is the field of photosynthesis as the basic energy storage process of living organisms. As a result of intensive research, great progress has been made in recent years on the elucidation of this complex series of processes, and although it is too early to state with assurance, one may hope that this basic life mechanism will someday be accomplished by physical/chemical synthesis.

This completes the discussion of "Seven Research Areas" and I wish now to turn to: *future discovery as a component of progress.* I have indicated that future basic discoveries of science will provide fresh starting points for new progress, and that some of these discoveries will be essentially unpredictable.

We should, of course, like to predict which discoveries will be made, and we can supply a long list of "discoveries wanted," and the mere fact that vitally important areas of need and opportunity have been identified will serve to bias scientific effort in these desired directions. One could, for example, be completely receptive to the discovery of a reversible process adaptable to bulk energy storage. The discovery of materials that would permit a two-times increase in the coefficient of performance of semiconductor refrigerating junctions would command instant attention. An economical superpressure product substantially harder than diamond would cut, among other things, a wide industrial swath. These items are, as indicated, merely random selections from a very long list. Some of these discoveries will be made during the next twenty-five years, but in addition, some completely unanticipated and unpredictable breakthroughs will be achieved, and some of them will—if history is a valid guide—become the basis of business enterprise.

Is history a valid guide? To answer this question it is most interesting to turn back to that New York World's Fair year of 1939, and to recreate the forecast of the future that might have been made at that time.

The outline of *that* future, which has since become the past, might have looked like this:

(a) MICROWAVE PHYSICS. A professor at the University of Michigan had recently succeeded in generating about one milliwatt of microwave power, and it was thought it might be useful for chemical analysis. The idea of seeing an airplane through great distances with microwave reflections had not been born, and microwave communications and radar were developed much later.

(b) FLUORESCENT LIGHTING. This new lighting development was introduced in this country at the World's Fair of 1939. It promised to become an important new light source, and it has.

(c) POLYMERS. Polymer science was primitive, and polymers of the period were principally useful for decorative and novelty purposes. However, nylon had been announced late in 1938, and this subsequently initiated a great upsurge in polymer progress.

(d) JET ENGINES. Jet engines existed in the minds of very few people, principally Whittle in England and Dr. Sanford Moss in General Electric. The intervening period since then has seen jet aircraft develop as a major new medium of transportation.

(e) COMPUTERS. Although the idea of the computer had been conceived much earlier, there was almost no embodiment of the concept in 1939. Dr. Bush at MIT was experimenting with "integrating machines" as analog computers, but the discovery of the transistor phenomena in 1948 was the breakthrough that later made the modern computer feasible. Few people were interested in IBM stock in 1939.

(f) SOLID-STATE ELECTRONICS. The basic theories of solid-state electronic conduction phenomena were well developed in 1939, but it was an academic exercise that had no demonstrable experimental support. The development of germanium purification and diodes in 1943, and the transistor in 1948, started this field on its long and still growing path of scientific and industrial progress.

(g) TELEVISION. Television in 1939 was taken seriously by relatively few people except for some of those men who were in the forefront of the rapidly growing electronic industry. The problems of high-frequency power generation, entertainment programming, and the mass production of small compact sets that might function reliably in a living room seemed to present very substantial problems.

What were the discoveries—unpredictable in 1939—that have since been made? There were many, but these major events come to mind:

(a) ANTIBIOTICS. Even though Fleming's penicillin molds had been reported—and ignored—some years before, I think it must be recognized that the fantastic number and diversity of antibiotic drug discoveries during and following World War II were completely unpredictable in 1939.

(b) ATOMIC FISSION AND ATOMIC ENERGY. This discovery was so revolutionary that, at the time, a great deal of analysis and interpretation by the world's outstanding physicists was required to understand what had been discovered.

(c) SPACE SCIENCE, TECHNOLOGY, AND EXPLORATION. Although rockets were known to the ancients, it is safe to say that *man in space*, and all of the attributes, connotations, and significance of this concept had not yet been born in the year 1939.

Our retrospection reveals that a twenty-five-year period is sufficient to permit enormous additions to human scientific and technological capability. It seems that almost anything technical can be accomplished in a couple of decades, and parenthetically, almost nothing in a couple of years. Man's whole way of life—his framework of security, the size of his travel world, his political environment, his control of disease, and his outlook on the universe—may change drastically as a consequence of scientific and technological progress.

For our present purpose it is important to note that some extremely difficult technical objectives can be accomplished, given enough time. As unlikely as it seemed in 1939, it *is* possible to see aircraft hundreds of miles away, by radio echo; it *is* possible to fly at the speed of sound, using a blast of hot air for propulsion; and hardly a home in America is without a television set—and many have a second highly compact portable set. Every one of the developments described in 1939 perspective has become an important business.

I believe that history is a valid guide, but not a blueprint, and that when the next New York World's Fair rolls around some of today's most challenging industrial research objectives will be accomplished facts, and some *completely new dimensions of technological capability* will have been added by new discovery.

Selectivity—
A Modern Research Necessity

An address at the American Management Association, Chicago, Illinois, September 15, 1964.

For my readers, there is little need to belabor the obvious point that trying to decide what research programs a scientific laboratory should undertake has always been difficult and is getting more so all the time.

Specifically, the increasing difficulty of this task arises from an American phenomenon that I will call the "Selectivity Gap." This is the widening area between the expenditures for industry-sponsored research, development, and engineering, which has been growing recently at the rate of 10 to 12 per cent per year, and the expenditures for research, development, and engineering for the country as a whole, which has been growing at a 20 to 30 per cent rate per year. An *average* growth rate of industrial technical work of 10 per cent per year still leaves a little statistical room for a few sporty laboratories to grow at a very much higher rate, but not for many. For industrial laboratories that plod along with an expenditure growth rate of 10 per cent or so a year, the continuing inflation in salaries of nearly this magnitude imposes some neat problems. For example, how does one achieve a net increase in effort and output to balance increased cost? Without some increase in output, how does one accommodate the increasing area of technical opportunity and need? In other words, how does one manage the "Selectivity Gap"?

Laboratory operations *are* of course, becoming more productive. We employ equipment and facilities that are constantly improved, and computers play a large role in some areas of research. However, I for one do not believe that research productivity is growing as rapidly as costs, and I observe few laboratory managers who voluntarily reduce research staff in the belief that productivity is rising.

The principal management strategy that *is* available to combat research cost inflation and the "Selectivity Gap" is–selectivity. We must be more thoughtful and deliberate in allocation of the available effort. Since selectivity has long since been the main professional occupation and preoccupation of research directors, developing some new *vision* here is something like developing extrasensory perception in your spare time. It presents a problem.

"National research, development, and engineering" includes

many things, and the potentially endless arguments as to whether it includes "real research" in significant amounts, or whether it is merely a statistical hodgepodge of DOD odds and ends, does not provide much comfort, and does not detract from the fact that—whatever it is—it is growing at the rate of 20 to 30 per cent per year, and it has now grown to a $20 billion annual expenditure level. Any technical activity as large as that must be a matter of some gross consequence. One can, I think, come to some conclusions without further argument.

First of all, it is difficult to escape the conclusion that, as a result of this lush growth in technical work in many fields, there is much more research acreage out on the expanding frontier waiting to be plowed.

Second, of course, there is no absolute guarantee now, or has there been in the past, that any particular plot of this new-found acreage will be fertile.

Third, the problem of "cross fertilization" has new dimensions. Many people are roaming the scientific frontier, and progress in unrelated fields may be important to you, and may provide new solutions to your old problems. In that event, *you* should know first.

Fourth, we must, of course, be still more productive in the research we elect to do. We must be more efficient in our use of scientist, assistant, and supporting service capabilities, and we must have the most modern tools, facilities, and environment. But these are the classical objectives of research resource management, and most laboratories have studied these relationships *ad nauseam*, in an effort to achieve optimum operations and perhaps to avoid an absolutely unavoidable conclusion, which is that the most important research is done by good minds in deep thought. Provide a favorable environment for *that*, and productive research will follow. Lack *that*, and no amount of research equipment and assistance will compensate.

For these and many other reasons, selectivity *is* a modern research necessity. A conscious effort at selection does not ensure that successful choices will be made—this is still a speculative endeavor—but it is much better than flying blind.

Actually, were it not for the annoying proviso that the selection of research areas should be done in a way that gives the sponsor of the research at least his money's worth—and preferably more—the picking and choosing would really be quite easy. At least a dozen simple selection techniques are readily available. Many of them are quite frequently used. I would like to take a moment to discuss them—not necessarily in the order of their relative merit, and I may indicate some reservations concerning merit.

One of the easiest approaches is to undertake research in areas that are currently *fashionable*. At any given time there are always two or three programs that *any* self-respecting scientific organization *must* be participating in to avoid losing its license. Research enthusiasm is contagious, and it sometimes grows to epidemic proportions as some new concept or discovery or idea presents exciting prospects. Not uncommonly, an exciting new idea will be further embellished, expanded, and elaborated as the word traverses research circles, and especially as it appears in popular print. Sometimes, things get out of hand, and I will mention some examples later.

Alternatively, research projects can be selected because they will attract *government support*. Indeed, if a laboratory is living by government R & D, it is important to select projects that will attract government support. But some selectivity is desirable here, too, and not all projects are equally attractive in terms of the contractor's offering of applicable experience and special technical capabilities, or the government agency's analysis and appraisal of need and significance of the work. Both of these situations may greatly affect the continuity of the project and the mutual profit opportunity. Using the availability of government support as the sole method of selecting industrial research projects can give managers—to say nothing of scientists themselves—a distinct sense of insecurity.

Our friends in marketing and product planning are not unwilling to suggest directions for research, and when they do, it is well to listen carefully. For example, we are reminded about once a week that it would be welcome and even profitable if we take time out from more important things and invent a new kind of storage battery that would be at least two orders of magnitude better than anything now available. Only the fact that no one seems to have even the vaguest idea as to how to start to achieve this worthy objective has kept us from launching an allout project attack. I shall have more to say later about *need,* represented by perceived markets, and *opportunity,* represented by technological ability-to-do.

Another useful device for the research planner is to find something that has been *art* for centuries and then see if this art cannot be *turned into a science* with resultant opportunities for new technology. The metallurgists did this with considerable success a decade or two ago, and their friends the ceramists are following the pattern now. The best strategy is to find a very important old art that no one has thought of approaching scientifically. Unfortunately, we seem to be running out of arts that have not already become scientific. Even the old art of managing has been transformed into a modern science, and our hosts at this meeting have played a key role in this magical transformation.

It is quite possible to select research programs based on the *ease of hiring scientists* to do the work. Just as certain scientific areas go in and out of fashion, so certain scientific disciplines also have their ups and downs. One of the disadvantages of this tactic is that it brings to you many skills for doing things that you may not want to do. On second thought, I will withdraw this suggestion. There must be better ways of selecting research projects.

Research programs can be selected because they are *publishable,* which makes them popular with some scientists, or because they are *publicizable,* which is ranked high in importance by certain corporate types. Almost any area of solid-state physics, regardless of its applicability to the needs of the sponsor, will provide opportunity for writing papers that a very large number of journals will be willing to review. If publicity is the aim, I can only suggest limiting all technical work to the areas of spaceflight, curing the common cold, or studying human mating habits.

Closely related to the technique of choosing research programs because they are fashionable is that of choosing them for *insurance.* Frequently, the research director's most careful analysis may prove to him that a particular project has very little if any chance of success, but since many other laboratories—with obviously less knowledgeable directors—are spending a great deal of money in this same area, there is no alternative but to keep on with it—*just in case.*

Another approach to the insurance-type project is *group insurance,* which works out this way. In addition to the projects that you can staff and support adequately, there will be a considerable list of things that you cannot prove to be unimportant—conclusively. If the question should be asked in high places, you may have to respond "Yes, of course, we're working on that." So, it is customary here to take this list of twenty or thirty marginal projects and to assign one-third, or one-fourth, or even one-fifth of a scientist to each. In this way you can do a lot of research, in many important fields, at very little expense; you are thus carrying research group insurance. And I might add that it is a great waste of money.

There is another research program stratagem that has a peculiar appeal and fascination in some fields of science and technology, and that is to do something on a *very large scale.* This approach will not succeed unless the *something* will be the "world's largest" something, and numerous successful examples could be cited. If your something has really adequate size, weight, scope, and especially cost, these factors alone will almost ensure support. Needless to say, only the Federal Government will have the money, and the need to enhance national scientific prestige and image will impart momentum and overwhelm all opposition. But this stratagem succeeds only if you think *really big;* any compromise is fatal. Your choice of "something" is not too important—it can be an astronomical telescope, or a radio telescope, or an accelerator, or simply a hole in the earth—the possibilities are endless.

Finally, you might undertake a research program because some very able scientist on your staff wants to do it. This is of course a last resort, but if you cannot think of anything better, it is worth considering.

I think at this point I should remove my tongue from my cheek and try to grapple with some of the practical realities of research program selection. My discussion will be illustrated by examples of good and bad practice, and, in common with most Research Directors, I can find examples of both in the operations under my responsibility. It may help to identify steps in the selection process:

1. *Delineate the general area of effort* in science and technology appropriate to the objectives of your sponsor. Relevance of the work to present interests and future plans of the sponsor is essential.

2. *Choose the specific areas* of relevant science and technology where research may be expected to yield progress in the desired directions.

At this point the plans of your sponsor need to be examined in detail, particularly with respect to timing. If you are planning an advance in present technology it may be accomplished in a year or two. If you plan relatively basic work in new fields, opportunities for resultant new technology are several years off, and opportunities for new products or services lie several years beyond that. It is important to determine if you want to be at *that* point—at *that* time.

3. *Establish research projects and appropriate level-of-effort.* Project selection and level-of-effort are not independent quantities. Some research work is not worth doing with a "sub-critical" effort. Like an atomic reactor, it will not produce heat if it is too small. Other work can be done effectively at any desired level the sponsor can afford. An important objective and—hopefully—a consequence of adequate effort is research leadership. Modern research is competitive, and industrial research is a key element in a competitive industrial system where leadership can be translated into lead time of consequence to the sponsor. Hence, being there first is more than a matter of prestige; it may be a vital component of the success of business enterprise originating in research.

All of the steps described require adequate consideration at both the management and research scientist level. To accomplish the desired selection and planning in an effective manner, it is essential that the managers of such work have a significant background and experience in research work, and that the principal scientists concerned become knowledgeable in the operation and objectives of their sponsor's business.

Perhaps the most important question that industrial laboratories should face in their selective analysis is the possibility that a proposed research venture might be *completely successful.* Since in modern technical organizations it is nearly always possible to achieve an objective, at a price, it is thus increasingly necessary to contemplate the consequences of research success. One of the most expensive ways to do research is to arrive at some difficult technical goal, and *then* find that you do not really want to be there.

The key decisions that enter into any step-wise process of selection concern *science-oriented* questions and *sponsor-oriented* questions—*relevance.*

Let us consider some examples that bring these factors into perspective, and illustrate the relationships that bear on selection decisions. A currently interesting case is that of the *laser.*

The laser concept and embodiment has risen like a bright star on the science horizon, and has attracted tremendous scientific and popular attention. From a scientific standpoint coherent light *is* needs and wants is in error, or is very difficult to accomplish. An example is the electric toothbrush, which in its earliest concept required the most careful appraisal and consideration, and it is safe to say that some skeptics were heard at this point. "Who needs an electric toothbrush?" is a fair question that almost answers itself. It could be argued with some conviction that, of all the labor-saving machinery that may be conceived to ease human burdens, perhaps the labor saved during tooth brushing may be the least urgently needed. Yet the electric toothbrush has become a remarkably successful product, and confirmed users would not consider turning back to what is now regarded as a primitive method of dental hygiene.

On still other occasions, consternation is created in the intended orderly process of proceeding through *need* and *ability* to product by the unexpected appearance of a new ability-to-do that was not on the want list—a solution in search of a problem. This causes a great deal of trouble, and only our most agile industrial minds are able to cope with the resulting confusion. In such cases the confusion may extend to the terms *unexpected* and *unwelcome*, which may be used interchangeably as though they meant the same thing. A full discussion of the complexities of this situation would occupy much time.

In the meantime, it is well to remember that "matching" needs and ability-to-do can be done from *both* ends. When it is done successfully, there must be some understanding of *needs* on the part of the researcher, and *ability-to-do* on the part of the marketer. Hopefully, in this event, effective two-way communications will exist with resultant understanding of the problems and opportunities at both ends of the creative enterprise.

A few moments ago, I outlined a three-step selectivity process. Now, as an example of how this process can be applied I would like to evoke experience in an actual case in the company with which I am associated. The case concerns the broad field of energy conversion, with special reference to the generation of electrical energy. It will be evident that in this particular example the first step—the identification of the general area of effort in science and technology that will benefit the supporter of the research—presented no serious problems. The generation and utilization of electricity has been a major interest since the founding of the company

—and continues to be. It is also obvious that there is a long-range trend toward the need for new portable power sources. Present portable power sources are fundamentally more complicated than any engineer would prefer to have them, and they create a problem of atmospheric pollution, which is rapidly becoming a very serious matter. The inevitable growth of space exploration will require new kinds of power plants. The world's population explosion requires that we use our existing fuels with ever greater efficiency, and that we explore alternative generating methods and alternative and additional energy sources.

So, in this case, it was not hard to decide that General Electric should be interested in the scientific study of unique methods of energy conversion—new ways to generate electricity. Virtually all of the electric power we now use comes to us through the courtesy of Michael Faraday and his invention of electromagnetic induction. But well over twenty other ways to turn some other kind of energy into electrical energy have been discovered, and it is toward this score of unique generating techniques that we properly focused the second step of our selectivity process.

What are the more specific areas of science where research may provide progress in the desired direction? Here, of course, one of the key jobs is determining whether there is a *tangible scientific opportunity*. For at least half of the twenty-plus techniques, the theoretical limits of efficiency can be very positively established. For some of the others, there is a history of unproductive effort sufficient for quick disqualification. In several cases, the basic science required is in areas of botany and biology and other life sciences, where we do not feel it is appropriate for our laboratory to seek leadership. (Although, let me state parenthetically that interest in and effort devoted to the life sciences is increasing very rapidly in many parts of our laboratory, as in many other so-called "physical science" laboratories throughout the world.)

To make a long story short, we identified that on the basis of our abilities and interests in such disciplines as polymer chemistry, surface chemistry, catalysis, vacuum technology, cathodic emission, solid-state physics in general and semiconductor studies in particular, plasma physics, high-temperature metallurgy, ceramic science, and photoconductive processes, we were in a position to consider projects related to generating electricity by fuel cells, thermionic converters, thermoelectric junctions, magnetohydrodynamics (MHD), and photovoltaic cells, as solar energy converters. I will omit, in this discussion, the rather special energy conversion case of thermonuclear fusion.

Having identified what we believe are the key technical problems of several types of fuel cells—both high-temperature and low-temperature, using both exotic and conventional fuels—we have established several full-scale fuel cell projects. We remain optimistic about the long-range potential of the fuel cell as a portable power source and, in the even more distant future, perhaps as a central-station technique.

In thermionic converters we undertook projects aimed at improving both gas-filled and vacuum devices. As the work progressed, it became evident that the potential efficiency of gas-filled thermionic converters was substantially higher than that of those using a high vacuum. The latter project was dropped and effort concentrated on the former, which has both spaceflight and ground-based power plant potential. Thermionic converters are applicable to use with nuclear energy sources, and here our company's interest in nuclear reactor technology has been a factor in deciding that continuing research on thermionic converters is justified.

The case of MHD is somewhat different. We entered upon a rather substantial MHD project, during which we learned a great deal about the behavior of plasmas, especially the conductivity of seeded ionized gases. We explored a variety of electrode materials and configurations. We built some small-scale MHD generators. Then, a few years ago, we decided that the trend of the work in the direction of building bigger and bigger MHD generators was premature. The timing was wrong. We were, and still are, facing some very fundamental problems relating to the conductivity of plasmas and the high-temperature strength of electrode materials, and we felt the need for significant research progress in these areas before embarking on large-scale power generating ventures. This is somewhat contrary to currently fashionable work in MHD, but we think we will make more rapid progress by a fundamental approach to the formidable basic problems of plasma and materials.

Meanwhile, work on MHD for space power applications continues in our company's Space Sciences Laboratory, where some excellent progress is being made. Here the operating requirements of an MHD generator are somewhat special, and we have an opportunity to apply the method now, while we continue to learn more about MHD fundamentals.

The story on thermoelectricity is somewhat similar. A brief burst of progress in improving semiconductor junctions for both thermoelectric cooling and thermoelectric power generation occurred shortly after the discovery of the transistor focused tremendous attention on semiconductor materials. But then a stone wall was reached—not too short of the goal, in fact, but far enough short to make thermoelectricity unattractive for either refrigeration or power generation at the present. This stone wall appears to have no discernible cracks in it at this time; they may develop, but if they do, it will be as the result of basic studies of semiconductor crystals—which we are doing—not through a specific project on thermoelectricity, which we feel is no longer justified.

The final possibility from among the score of unique energy conversion methods—solar energy—got no further at our laboratory than a "project analysis," a study by that group of senior scientists I mentioned that reduced to specific terms what would have to be discovered for solar energy to become practical. *Economic* facts of life for proposed areas of usefulness made it clear that present energy storage techniques would have to be improved by one order of magnitude, and photovoltaic energy conversion cells would have to be improved by two orders of magnitude before solar energy might even begin to be attractive. *Scientific* facts of life made it clear that there is no reason to believe that either of these improvements is on the horizon, based on existing knowledge or reasonable expectations for future progress. Again, the timing seems wrong. Solar energy is an area that must be re-evaluated at appropriate intervals, but where substantial judgment, insight, and courage must be employed in resisting the temptation to establish a major effort too early.

In brief, the research director's continual problems in the area of selectivity, that modern research necessity, are:

(1) How to get the best kind of continuing objective review of the current and future science of relevance to the sponsor;

(2) How to make the necessary changes in emphasis and get these changes implemented without so much delay that the value of the research is lost, yet not so quickly and arbitrarily that morale

and confidence are lost. The key to success in this effort is the participation of the research staff in the appraisals, and the judgments, and the decisions that provide the guidelines in the process of selection. No research organization can "do everything." It may not be humanly possible for any research organization to do well all that it undertakes. But authentic progress in science is so important in the modern world, that we must not only put forth our best research efforts, but we must also make our most thoughtful selections of the area of scientific opportunity.

Index